HANDBOOK OF
POLYCYCLIC
AROMATIC
HYDROCARBONS

HANDBOOK OF
POLYCYCLIC AROMATIC HYDROCARBONS

VOLUME 2

Emission Sources and Recent
Progress in Analytical Chemistry

Edited by

ALF BJØRSETH
Norsk Hydro-Petroleum
Research Center
Bergen, Norway

THOMAS RAMDAHL
Central Institute for
Industrial Research
Oslo, Norway

MARCEL DEKKER, INC.   New York and Basel

Library of Congress Cataloging-in-Publication Data
(Revised for volume 2)

Main entry under title:

Handbook of polycyclic aromatic hydrocarbons.

Includes bibliographical references and index.
Partial contents: v. 2. Emission sources and recent
progress in analytical chemistry / edited by Alf
Bjørseth, Thomas Ramdahl.
1. Hydrocarbons--Handbooks, manuals, etc.
2. Polycyclic compounds--Handbooks, manuals, etc.
I. Bjørseth, Alf. II. Ramdahl, Thomas, [date]
QD341.H9H317 1983   547'.61   83-1903
ISBN 0-8247-1845-3 (v. 1)
ISBN 0-8247-7442-6 (v. 2)

MARCEL DEKKER, INC.
270 Madison Avenue, New York, New York 10016

Current printing (last digit):
10 9 8 7 6 5 4 3 2 1

PRINTED IN THE UNITED STATES OF AMERICA

Preface

The historical evolution of knowledge about chemical carcinogens and environmentally induced carcinogenesis is strongly linked to characterization of polycyclic aromatic hydrocarbons (PAH). The environmental importance of this class of compounds is as high today as ever.

Environmental and biochemical studies of PAH form a large and comprehensive area of research. Very few people have the ability and the opportunity to engage actively in all aspects of such research. We have seen the need for an extensive and systematic treatise describing our present knowledge of PAH in areas relevant to researchers and educators, as well as industrial hygienists and representatives of environmental agencies.

Three years ago the first volume of this handbook was published, reviewing the state of the art for important topics such as sampling of PAH, analytical methods, and occurrence in important environmental matrices. However, the research in this subject is developing so fast that we found it necessary to update some important areas, and to add new information on selected topics. The present volume of the handbook therefore concerns sources of PAH, their emission factors, and relative importance. Further, it deals with exposure, uptake, metabolism, and detection of PAH in the human body. Finally, the handbook contains an update of information in selected research areas.

In recent years, many studies have been devoted to derivatives of PAH, such as nitrogen-, sulfur- and oxygen-containing compunds. This is reflected also in some chapters in this volume. The term polycyclic aromatic compounds (PAC) has been introduced to cover all these derivatives, including PAH, and this will also be used throughout the book where appropriate.

It is our hope that the readers will find this book a good contribution to their knowledge about PAH and a useful supplement to the information given in Vol. 1.

We would like to express our sincere thanks to all the contributors to this handbook, for their cooperation, their hard work, and their patience while their manuscripts were being prepared for press. We are also indebted to our colleagues at the Central Institute for Industrial Research, in Oslo, Norway, for their valuable advice and encouragement.

Alf Bjørseth
Thomas Ramdahl

Contents

Preface   iii

Contributors   vii

Contents of Volume 1   ix

 1 Sources and Emissions of PAH   1
    Alf Bjørseth and Thomas Ramdahl

 2 PAH Emissions from Coal-Fired Plants   21
    Kristofer Warman

 3 PAH Emissions from Combustion of Biomass   61
    Thomas Ramdahl

 4 PAH Emissions from Automobiles   87
    Ulf R. Stenberg

 5 Recent Progress in the Determination of PAH by High
    Performance Liquid Chromatography   113
    Stephen A. Wise

 6 Recent Advances in the Analysis of Polycyclic Aromatic
    Compounds by Gas Chromatography   193
    Keith D. Bartle

 7 Determination of Occupational Exposure to PAH by
    Analysis of Body Fluids   237
    Georg Becher and Alf Bjørseth

 8 Analysis of 6-Nitrobenzo[a]pyrene in Mammalian Cells
    and Microsomes by High-Pressure Liquid Chromatography   253
    Samuel Tong and James K. Selkirk

 9 Nitrogen-Containing Polycyclic Aromatic Compounds in
    Coal-Derived Materials   265
    Douglas W. Later

10 Atmospheric Reactions of PAH   351
    Karel A. Van Cauwenberghe

11 Reference Materials for the Analysis of Polycyclic Aromatic
    Compounds   385
    Walter Karcher

Index   407

Contributors

KEITH D. BARTLE Department of Physical Chemistry, University of Leeds, Leeds, United Kingdom

GEORG BECHER Toxicological Department, National Institute of Public Health, Oslo, Norway

ALF BJØRSETH Petroleum Research Center, Norsk Hydro, Bergen, Norway

WALTER KARCHER Materials Department, Joint Research Centre of the Commission of the European Communities, Petten, The Netherlands

DOUGLAS W. LATER[1] Biology and Chemistry Department, Battelle Pacific Northwest Laboratory, Richland, Washington

THOMAS RAMDAHL[2] Environmental Chemistry Department, Central Institute for Industrial Research, Oslo, Norway

JAMES K. SELKIRK The University of Tennessee-Oak Ridge Graduate School of Biomedical Sciences, and Biology Division, Oak Ridge National Laboratory, Oak Ridge, Tennessee

ULF R. STENBERG[3] Department of Analytical Chemistry, University of Stockholm, Stockholm, Sweden

SAMUEL TONG The University of Tennessee-Oak Ridge Graduate School of Biomedical Sciences, and Biology Division, Oak Ridge National Laboratory, Oak Ridge, Tennessee

KAREL A. VAN CAUWENBERGHE Chemistry Department, University of Antwerp, Antwerp, Belgium

KRISTOFER WARMAN[4] Fuel Technology Department, Studsvik Energyteknik AB, Nyköping

STEPHEN A. WISE Organic Analytical Research Division, National Bureau of Standards, Gaithersburg, Maryland

---

Current affiliations:
[1]Lee Scientific, Inc., Provo, Utah
[2]Statewide Air Pollution Research Center, University of California, Riverside, California
[3]KabiVitrum AB, Stockholm, Sweden
[4]Studsvik Environmental Consultants AB, Nyköping, Sweden

Contents of Volume 1

1 Physical and Chemical Properties of Polycyclic Aromatic
Hydrocarbons
Maximilian Zander

2 Sampling, Extraction, and Analysis of Polycyclic Aromatic
Hydrocarbons from Internal Combustion Engines
Frank Sen-Chun Lee and Dennis Schuetzle

3 Extraction of Polycyclic Aromatic Hydrocarbons for
Quantitative Analysis
Wayne H. Griest and John E. Caton

4 Profile Analysis of Polycyclic Aromatic Hydrocarbons in Air
Gernot Grimmer

5 High-Performance Liquid Chromatography for the Determination
of Polycyclic Aromatic Hydrocarbons
Stephen A. Wise

6 Analysis of Polycyclic Aromatic Hydrocarbons by Gas
Chromatography
Bjørn Sortland Olufsen and Alf Bjørseth

7 Mass Spectrometric Analysis of Polycyclic Aromatic
Hydrocarbons
Björn Josefsson

8 Optical Spectrometric Techniques for Determination of
Polycyclic Aromatic Hydrocarbons
E. L. Wehry

9 Analysis of Polycyclic Aromatic Hydrocarbons by Thin-Layer
Chromatography
Joan M. Daisey

10 Determination of Polycyclic Aromatic Hydrocarbons in
Sediments and Marine Organisms
Bruce P. Dunn

11 Polycyclic Aromatic Hydrocarbons in Foods
Thomas Fazio and John W. Howard

12 Long-Range Transport of Polycyclic Aromatic Hydrocarbons
    Alf Bjørseth and Bjørn Sortland Olufsen

13 Determination of Polycyclic Aromatic Hydrocarbons in
    Coal-Derived Materials
    Curt M. White

14 Analysis of Metabolites of Polycyclic Aromatic Hydrocarbons
    by GC and GC/MS
    Jürgen Jacob

15 Polycyclic Aromatic Hydrocarbons in River and Lake Water,
    Biota, and Sediments
    Joachim Borneff and Helga Kunte

16 Polycyclic Aromatic Hydrocarbons in Work Atmospheres
    Richard B. Gammage

Appendix: List of Dicyclic and Polycyclic Aromatic Hydrocarbons,
           Their Structure, Molecular Weight, Melting and
           Boiling Points

HANDBOOK OF
POLYCYCLIC
AROMATIC
HYDROCARBONS

1
Sources and Emissions of PAH

ALF BJØRSETH / Petroleum Research Center, Norsk Hydro, Bergen, Norway

THOMAS RAMDAHL* / Environmental Chemistry Department, Central Institute for Industrial Research, Oslo, Norway

I.   Introduction   1

II.   Formation of PAH   2

III. Sources of PAH   4

IV.   Emission Factors   4

    A. General   4
    B. Emission factors for specific sources   6

V.   Total PAH Emissions   13

VI.   Conclusions   17

    References   17

I. INTRODUCTION

Since the British surgeon Sir Percival Pott in 1775 reported [1] that chimney sweeps in Britain often developed cancer of the scrotum, there has been an awareness of the harmful effects of soot, tar, and pitch. However, more than 150 years passed until carcinogenic constituents in pitch [2] were identified as polycyclic aromatic hydrocarbons (PAH). By 1976 more than 30 PAH compounds and several hundred derivatives of PAH were reported to have carcinogenic effects [3,4], making PAH the largest single class of chemical carcinogens known today.

Potential hazards from the occurrence of PAH in the environment have been noted in the drinking water standards set forth by the World Health Organization's Committee on the Prevention of Cancer [5], as well as by several national agencies concerned with PAH in food, working atmospheres, and effluents from industries and mobile sources [6,7].

It was also fairly recently that chemical characterization of sources to PAH emissions was initiated. Reliable analysis of PAH emissions requires instrumentation which has been available only for the last 30-40 years, and even today the quantitative information about PAH emissions from anthropogenic as well as natural sources is limited. In this chapter we review, briefly the formation mechanisms and sources of PAH, estimate emission factors for known antropogenic sources of PAH, and establish their relative importance.

---

*Current affiliation: Statewide Air Pollution Research Center, University of California, Riverside, California

II. FORMATION OF PAH

PAH can be formed by thermal decomposition of any organic material containing carbon and hydrogen. Formation is based on two major mechanisms:

1. Pyrolysis or incomplete combustion
2. Carbonization processes

A pioneering contribution to their understanding has been made by Badger and coworkers [8,9]. The chemical reactions in flames proceed by free-radical paths, and a synthetic route based on this concept has also been postulated for the formation of PAH. Based on the results of a series of pyrolysis experiments, Badger suggested a stepwise synthesis of PAH from C_2 species during hydrocarbon pyrolysis. These pyrolysis studies were conducted by passing the hydrocarbon vapor in nitrogen through a silica tube at 700°C. Badger's data have been supported by Boubel and Ripperton [10], who found that benzo[a]pyrene (BaP) is also produced during combustion at high percentages of excess air, although the amount of BaP is larger at lower percentages of excess air.

Badger and Spotswood [11] also pyrolyzed toluene, ethylbenzene, propylbenzene, and butylbenzene and obtained high yields of BaP with butylbenzene, a potential intermediate in Badger's reaction scheme. Obviously, it is not necessary to completely break down the starting material to a two-carbon radical in order to form BaP. Any component of the combustion reaction that can contribute intermediate pyrolysis products of the structure required for BaP synthesis would also be expected to lead to increased yields of BaP.

Other and more recent studies of the PAH formation models also tend to confirm most of the mechanisms proposed by Badger. Crittenden and Long [12] determined the chemical species formed in rich oxyacetylene and oxyethylene flames. Compounds identified suggest that the C_2 species react to form C_4, C_6, and C_8 species, and that reactions involving styrene and phenylacetylene are probably important in the formation of PAH. Also, a $C_{10}H_{10}$ species was detected in the gases of both flames which corresponds to the C_4-substituted benzene postulated by Badger.

In spite of the tremendous number of different PAH that might be formed during the primary reactions, only a limited number of PAH enter the environment. Many of the PAH formed by primary reactions will have short half-lives under pyrolysis conditions, and will stabilize in subsequent reactions. At high temperatures the thermodynamically most stable compounds will be formed in corresponding quantitative ratios. These are mainly the unsubstituted parent PAH. Irrespective of the type of material to be burned, surprisingly similar ratios of PAH are formed at a defined temperature. For example, thermal decomposition of pit coal, cellulose, tobacco, and of polyethylene and polyvinyl chloride carried out at 1000°C yields very similar PAH profiles [13]. Consequently, PAH profiles seem to depend more on the combustion conditions than on the type of organic material burned.

The absolute amount of PAH formed under defined pyrolysis conditions depends, however, on the reaction temperature as well as the material. The amount of BaP formed by pyrolysis of various substances at 840°C in a nitrogen stream is shown in Table 1. The table demonstrates the relative unimportance of the oxygen-containing carbohydrates in the generation of BaP compared to the C_{32} paraffin or β-sitosterol, a common plant sterol. Furthermore, the absolute amount of PAH formed during incomplete combustion is dependent on temperature, as shown in Table 2. Under comparable conditions, 1 g of tobacco yields 44 ng of BaP at 400°C and 183,500 ng of BaP at 1000°C [15].

Table 1. Levels of Benzo[a]pyrene Produced on Pyrolysis (840°C, N_2)

Substance pyrolyzed	μg BaP/g pyrolyzed
Glucose	47.5
Fructose	98.4
Cellulose	288.8
Stearic acid	1200
Dotriacontane	3130
β-Sitosterol	3750

Source: Ref. 14. Reprinted with permission of Raven Press, New York.

At temperatures below approximately 700°C, the pyrolysis products contain, in addition to the parent PAH, larger amounts of alkyl-substituted PAH, mainly methyl derivatives. A typical example is tobacco smoke, which yields soot quite abundant in alkyl-substituted PAH [16]. Figure 1 shows schematically the relative abundance of PAH as a function of the number of alkyl carbons at different formation temperatures as given by Blumer [17]. As indicated by the figure, the number of alkyl carbons present as side chains on PAH correlates closely with the temperature at which the compounds are formed.

Table 2. Formation of Benzo[a]pyrene and Benzo[e]pyrene from Blend Tobacco (100 g) as a Function of Temperature (1000 ml N_2/min)

Temperature (°C)	BaP (μg)	BeP (μg)
400	4.4	3.2
500	12.9	8.4
600	32.0	19.0
700	56.0	29.4
700[a]	88.6	45.2
800	270.0	155.0
900	1,820.0	824.0
900[a]	4,725.0	2,015.0
1000[a]	18,350.0	6,710.0

[a]At 1500 ml/min to avoid a back reaction of the condensate.
Source: Ref. 13.

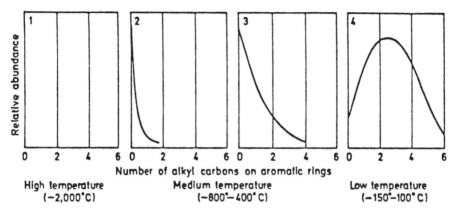

Figure 1. The relative abundance of polycyclic aromatic hydrocarbons as a function of the number of alkyl carbons at different formation temperatures. From Polycyclic aromatic compounds in Nature, Max Blumer [17]. (Copyright © 1976 by Scientific American, Inc. All rights reserved.)

PAH is not formed solely by high temperature and open flames. Various processes of carbonization that occur (e.g., during the generation of mineral oil and coal) lead to the formation of PAH from decaying biological material at low temperatures (below 200°C) and high pressure over a period of millions of years. The transformation resembles pyrolysis, but the reactions are exceedingly slow because of the modest temperatures involved.

III. SOURCES OF PAH

Although it is recognized that there are natural sources of PAH [17,18] (i.e., volcanic activity and biosynthesis), the anthropogenic sources are predominant and by far the most important to air pollution [13]. This chapter therefore deals mainly with anthropogenic sources.

Table 3 lists the main sources, divided into two categories: stationary and mobile sources. The stationary sources include industrial sources, power and heat generation, residential heating, incineration, and open fires. The second category is that of the mobile sources, which include gasoline-engine automobiles, diesel-engine automobiles, trucks, airplanes, and sea traffic. In all the processes listed, organic material is burned or strongly heated, and will in most cases result in emissions of PAH.

IV. EMISSION FACTORS

A. General

The amount of PAH released from any process is largely dependent on raw materials and the combustion technology. Therefore, a given process with a fixed set of process parameters may produce a specific and invariable amount of PAH. The amount can be correlated to the process parameters, and thus a relative emission factor for the process can be established. Emissions from industrial sources are determined primarily by the production level of the plants emitting PAH. The emission factor can be expressed as weight of PAH per ton of product. In the case of mobile sources, the consumption of fuel could be chosen as the emission-determining unit, giving an emission factor expressed as micrograms of PAH per kilogram of fuel consumed.

Table 3. Sources of PAH

Stationary sources

1. Residential heating

     Furnaces, fireplaces, and
        stoves (wood and coal)

     Gas burners

2. Industry

     Coke production

     Carbon black production

     Petroleum catalytic cracking

     Asphalt production

     Aluminum smelting

     Iron and steel sintering

     Ferroalloy industry

3. Power and heat generation

     Coal- and oil-fired power
        plants

     Wood- and peat-fired
        power plants

     Industrial and commercial
        boilers

4. Incineration and open fires

     Municipal and industrial
        incinerators

     Refuse burning

     Forest fires

     Structural fires

     Agricultural burning

Mobile sources

5. Gasoline-engine automobiles

     Diesel-engine automobiles

     Rubber tire wear

     Airplanes

     Sea traffic

Although numerous publications deal with PAH, very few studies link the
emission to any process parameter. These studies are therefore not suited
for emission factor development, as such. Another problem is the inconsis-
tency of reported PAH compounds. For different reasons BaP has been
measured more frequently than any other PAH. In these cases, BaP can
help in judging how a given source contributes to the total PAH emission by
comparing it to values of BaP emission from other sources. However, BaP
is only a minor component, usually less than 5% of the total mount of PAH.
In studies where more PAH than BaP are measured, the number of PAH analyzed may vary, depending on the analytical technique employed. Hangebrauck
et al. [19] analyzed 10 different PAH in emission samples by column chromatography and ultraviolet-visible spectrophotometry, whereas modern glass
capillary gas chromatography can separate and quantify more than 40 PAH
in comparable samples [20]. The term "total PAH emissions" therefore depends strongly on the number of compounds determined. Because measurements of the same source often give large differences, the emission factors

presented consist of a minimum, a maximum, and an intermediate value when possible. *The emission of bicyclic aromatic compounds is not included in the calculation of PAH emission factors.*

Emission factors for PAH are available in only a few reports. However, the U.S. Environmental Protection Agency has published a comprehensive report [21]. Most of the emission factors in this report were developed from data reported by Hangebrauck et al. in their relatively comprehensive 1967 study [19]. A literature search conducted by EPA to update this information resulted in very few additional data. The emission factors discussed below have been updated wherever new results were available.

B. Emission Factors for Specific Sources

1. Residential Heating

Wood- and coal-burning stoves and wood-burning fireplaces are frequently used for residential heating. The combustion of wood and coal is often incomplete, due to slow, low-temperature burning with insufficient access to air at the burning surface. This results in the formation of large amounts of PAH, which may be released directly to the atmosphere.

Two Scandinavian studies of wood stoves yielded emission values between 1 and 36 mg of PAH (22-26 individual PAH compounds determined) per kilogram of dry fuel [22, 23] while an American study [24] measured 270 mg/kg (26 compounds). These differences may reflect variations in design, sampling methodologies, test cycles, and so on, and illustrate the difficulty of calculating a single emission factor for a given source. In Table 4 the emission factor is given, with an intermediate value of 40 mg PAH/kg dry wood.

Residential coal firing gives an emission factor of 60 mg/kg [19,25]. However, this factor can very over five orders of magnitude with respect to the type of coal used [26]. Emissions from residential heating with oil vary with the size of the burner. A 100,00 Btu/hr (30 kW) burner has been found to emit 150 μg/kg oil [19], whereas a smaller, 25,000 Btu/hr (7.5 kW) unit has been found to emit 10 mg/kg oil [27]. Further measurements are needed to establish good emission factors for residential oil heating.

2. Aluminum Production

Production of primary aluminum is based on two different technologies, using Söderberg or prebaked anodes. PAH may be emitted from the carbon electrodes containing tar and pitch as binder [28]. The emissions from the prebaked electrodes are usually 1-10% of that of the Söderberg electrodes. To the best of our knowledge, only one plant have been studied to determine the total PAH emission from the production of aluminum [29]. The PAH emmission to air was determined to approximately 35 metric tons of PAH at a plant with an annual production of 83,500 tons of aluminum using the Söderberg process. The PAH in this study represent the sum of some 25 compounds.

Assuming that the Söderberg and the prebaked process each cover 50% of the worldwide production, and that the prebaked process emits 5% of that of the Söderberg process, an emission factor of 235 g/ton aluminum produced is established for aluminum production, as shown in Table 5. It should be stressed, however, that the data base is limited to one technology and one aluminum plant. The emission factors calculated here therefore reflect considerable uncertainty.

Table 4. Emission Factors for Residential Heating

| Source | Benzo[a]pyrene | | PAH | | |
	Range (μg/kg)	Typical (μg/kg)	Range (mg/kg)	Typical (mg/kg)	References
Wood stoves	1-10,000	500	1-370	40	22-24
Fireplaces		700		29	24
Coal furnaces		1500	1-1200	60	19,25
Oil (30 kW)	2.0-4.4	2.2	0.006-0.75	0.15	19
Oil (7.5 kW)			0.9-21.6	10	27

Table 5. Emission Factors for Industrial Sources

Source	Benzo[a]pyrene		PAH		References
	Range (mg/ton)	Typical (mg/ton)	Range (mg/ton)	Typical (g/ton)	
Aluminum smelting	10,000–38,000	15,000	$2 \times 10^5 – 6.5 \times 10^5$	235	29
Anode baking		66		50	29
Coke production				15	30
Petroleum catalytic cracking					
No control	0.03–0.71	0.41	1.4–3200	0.03	19,21,31
CO waste heat boiler	0.007–0.16	0.04	0.32–2.4	0.001	19,21,31
Carbon black production			220–490	0.3	21,32,33
Asphalt production					
Air blowing	0.16–750	1	2.8–4100	0.05	19,21
Shingle saturators	0.08–0.4	0.001	1–50	0.005	19,21
Iron and steel sintering	0.6–1100	17		0.34	21,34
Ferrous foundries				7.7	33
Ferroalloy industry				10	35
Iron works				60[a]	35

[a]Based on Söderberg process.

3. Coke Production

It has long been known that coke ovens are major sources of PAH emission.
The leaks from the oven doors and battery-top lids are found to be the major
sources [30], giving an emission factor of 15 g/ton coal charged. Again the
remark should be made that the data base is limited.

4. Petroleum Catalytic Cracking

The catalytic cracking process is used to upgrade heavy petroleum fractions
by breaking up long-chain hydrocarbons to produce high-octane gasoline.
An emission factor weighted by the cracker population in the United States
is 25 mg/m^3 fresh feed without any emission control [19,21,31]. Assuming a
density of 900 kg/m^3 fresh feed, this yields an emission factor of 28 mg/ton.
When the regenerator exhaust gas is recirculated and burned in waste heat
boilers, the emission factor is 1 mg/ton [19,21,31].

5. Other Industrial Sources

Another important industrial source is the production of carbon black, which
is widely used as pigment (i.e., in rubber tires). Literature data indicate
an emission factor of 300 mg/metric ton of product [21,32,33].

The production and modification of asphalt involve various processes that
may give PAH emissions. For instance, air blowing has an emission factor
of 50 mg/metric ton and shingle saturators 5 mg/ton [19,21].

Another source of significant air emissions is iron and steel sintering.
An emission factor of 17 mg BaP/ton has been reported [21,34]. Assuming
that BaP represents 5% of the total PAH, this source has an emission factor
for PAH of 340 mg/ton.

The ferroalloy industry also has PAH emissions, due to contact between
hot metal and tarry products in electrodes and shutters. Only data for
PAH emissions to water have been found for this source [35]. However,
assuming equal emission to water and air (50% efficiency in the wet scrubbers),
this source has an emission factor of 10 g PAH/ton alloy.

In a study in a Norwegian iron works, emissions to water were determined
[35]. PAH emission is due to the use of Söderberg electrodes. Making the
same assumptions as above with respect to the scrubber efficiency, the iron
works emit 60 g PAH/ton iron produced. It must be stressed that this emission factor is valid only for this particular process. In ferrous foundries
the shakeout process of casting metal emits PAH, resulting in an emission
factor of 7.7 g/ton casting produced [33].

6. Power and Heat Generation

A survey of the emission factors for power generation is given in Table 6.
Each of the sources listed is discussed below.

a. Coal and oil. Coal-fired power plants may have very different construc-
tions, such as vertically, front-wall, tangentially, cyclone-fired plants, and
plants using a moving grate. In an emission study a weighted emission factor of seven types of coal-fired power plants was found to be 19 μg/kg coal
charged [19,21,36]. Coal-fired industrial boilers are usually smaller units.
However, they generally give a higher emission relative to energy produced.
A weighted emission factor for six types of boilers was found to be 41 μg/kg
fuel [19,21,36].

Table 6. Emission Factors for Power and Heat Generation

Source	Benzo[a]pyrene Range (μg/kg)	Benzo[a]pyrene Typical (μg/kg)	PAH Range (μg/kg)	PAH Typical (μg/kg)	References
Pulverized coal-fired power plants	1.1-2.7	1.6	0.5-32	19	19,21,36,37
Coal-fired fluidized bed	150[a]		36,000[a]		37
Coal-fired industrial boilers		0.93		41	19,21,36
Oil-fired industrial boilers		1.1	5.3-100	23	19,21
Oil-fired commercial boilers		40		820	19,21
Gas-fired intermediate boilers		10	490-1,100	1,000	19,21
2 MW hot water boiler					
Wood	10-36	22	1,180-3,390	2,000	40
Peat				15,000	40

[a]The plant was not operated properly.

Extensive studies of coal-fired power plants have been carried out in Scandinavia. Three coal-fired power plants (25-600 MW) have a PAH emission of less than 0.5 µg/kg coal charged. Thus, when properly operated, these power plants appear to contribute little to overall PAH emission [37]. Two other Swedish coal-fired power plants (350-400 MW) have been measured to emit 30 µg PAH/kg coal [38]. Improperly operated, the emissions may be higher. For a more detailed discussion, see Chap. 2.

Different technologies have also been studied. A small fluidized-bed coal-fired power plant (2 MW) emitted 36 mg/kg coal [37]. It was claimed that the plant was not operated properly at the time of sampling, and this clearly demonstrates the importance of optimalization of operation. In terms of the total emission per year, a few heavy emission periods may overwhelm the otherwise low and steady emission.

In a recent study at a well-tuned 500-MW oil-fired power plant, no detectable amount of PAH could be found [39]. The detection limit was 0.05 µg of each component per sample. Calculated BaP emission is less than 0.01 g/hr. Other oil-fired power plants in Sweden have been estimated to emit 10 µg PAH/kg oil [38]. Oil-fired intermediate boilers (6-8 MW) have an emission factor of 23 µg/kg [19,21].

b. Gas. Some fairly old data for fired intermediate boilers (1.2 MW) heated by premix burners indicate an emission factor of 1 mg/kg gas burned [19, 21].

c. Biomass. Biomass represents an alternative fuel, and considerable interest has been paid to the environmental aspect of use of such fuels. These fuels are treated further in Chap. 3, where such topics as whether wood and peat are a potential fuel for power generation are discussed. In a study of a small 2 MW hot water boiler, the combustion of wood and peat gave emission factors of 2 and 15 mg PAH/kg fuel, respectively [40]. Other studies have shown much lower emissions for biomass combustion. As for the other sources, the combustion technology is of main importance. In some cases [41,42] no PAH could be detected in the emissions.

7. Incineration and Open Fires

A survey of the emission factors established for incinerators and open fires is given in Table 7.

a. Incinerators. There are many different types of incinerators with various types of combustion systems. A geometric mean for large incinerators with emission control systems is 17 µg/kg refuse charged [19,21,43]. Smaller commercial incinerators burning 2-5 tons/day have an emission factor of 6.8 mg/kg [19,21].

b. Open burning. Open burning includes sources such as burning of automobile tires, coal refuse, forest fires, and agricultural burning. Open burning of automobile tires emits 240 mg/kg and municipal refuse, 1.4 mg/kg [19, 21,44]. Burning of coal refuse banks results in large emissions of PAH [3], but no emission factor can be established. Forest fires are potential sources of PAH emission. An emission factor of 20 mg/kg was calculated from laboratory studies [45]. Large quantities of PAH are suspected to be emitted from agricultural burning. Although no data for PAH emission from this source have been found, a rough estimate of 20 mg/kg, similar to that of forest fires, is suggested. Furthermore, structural fires will, by burning of organic material, emit PAH, but no emission factor for this source could be established.

Table 7. Emission Factors for Incineration and Open Fires

Source	Benzo[a]pyrene		PAH		References
	Range (μg/kg)	Typical (μg/kg)	Range (mg/kg)	Typical (mg/kg)	
Incinerators (U.S.A., geometric mean)		0.13		0.017	19,21,43
Incinerators (2-5 tons/day)		200		6.8	19,21
Open burning					
Automobile tires				240	19,21,44
Municipal refuse				1.4	19,21,44
Forest fires	17-140	100	3.5-31.5	20	45
Agricultural burning				20	

8. Mobile Sources

A survey of the emission factors developed for mobile sources is given in Table 8.

a. Gasoline-fueled automobiles. Surprisingly few studies are available for the determination of emission factors for total PAH. Grimmer et al. [46] characterized the PAH (59 compounds) emitted from two different vehicles and developed an emission factor of 10 mg/kg for noncatalytic cars. Catalytic converters for the exhaust gases seem to reduce this figure significantly. An American study reports the emission factor for BaP to be 4.6 µg/kg without catalyst and only 0.36 µg BaP/kg with catalyst, a reduction of 92% [47]. Approximately 50% of American cars now have catalytic converters. Other factors such as air/fuel ratio, age of engine, fuel aromaticity, driving mode, oil consumption, and cold start, also have an important impact on the emission [46,48].

Recent results indicate that engine design developments and alternative fuels such as alcohol and liquefied petroleum gas (LPG) reduce fuel consumption and PAH emissions considerably [49]. It is therefore to be expected that the emission factor given above may be adjusted as new studies are published.

b. Diesel engines. Diesel emissions have been studied carefully for the last few years [50]. However, few publications are suitable for the purpose of establishing relative emission factors. BaP emissions from diesel engines can vary by two orders of magnitude depending on engine design alone [50]. BaP emission from the diesel engine having the lowest emissions are the same order of magnitude as those from gasoline engines with catalytic converters [50]. An emission factor of 10 mg/kg is estimated, based on available data. PAH emissions from heavy-duty diesel trucks have been estimated at 5 mg PAH/kg [51].

c. Other mobile sources. Motorcycles, snowmobiles, lawn mowers, and chain saws all have two-stroke engines with low fuel efficiency. An emission factor for motorcycles of 3.6 mg BaP/kg has been reported [3]. No estimates are available on other two-stroke engine emissions.

Degradation of automobile tires releases carbon black particles to the atmosphere. Carbon blacks are used in tires and contain PAH. No direct emission factor could be established, but the NAS study [3] estimated the annual BaP emissions to be 11 tons/year in the United States [3]. The total U.S. mileage is approximately 2×10^{12} km/year, which gives a BaP emission of 5 µg/km.

Shabad [52] reports that a jet airplane emits 2-10 mg BaP/min. No other studies were available, but a recent report from Sweden shows a decreasing amount of fallout PAH at increasing distance from the runway [53]. No data are available with respect to sea traffic.

V. TOTAL PAH EMISSIONS

Another important parameter for the total emission estimate is the total production or consumption of the various sources. It is the product of the relative emission factor and the production or consumption capacity that finally show the emission severity of the source. Based on statistical data and the relative emission factors given previously, total PAH emissions are calculated for the United States, Sweden, and Norway (Table 9). These data are discussed next.

Table 8. Emission Factors for Mobile Sources

Source	Benzo[a]pyrene		PAH		
	Range	Typical (μg/kg)	Range (mg/kg)	Typical (mg/kg)	References
Automobiles					
Gasoline, no catalyst		50	5-50	10	46
Gasoline, no catalyst		4.6			47
Gasoline, catalyst		0.36			47
Diesel	0.2-31.7 μg/mile		5-100	10	
Diesel					50
Trucks, diesel				5	51
Two-stroke engines		3600			3
Airplane, jet	2-10 mg/min				52

Table 9. Estimated PAH Emission[a]

Source	Norway Metric tons/year	Norway %	Sweden Metric tons/year	Sweden %	U.S.A. Metric tons/year	U.S.A. %
Residential combustion						
Wood, coal	48	16	96	38	700	12
Oil, gas	14.5	5	36	14	15	0.3
Industrial production						
Coke manufactoring	5.1	2	18	7	630	11
Carbon black	0.1	<0.1	<0.1	<0.1	3	<0.1
Asphalt production			0.3	<0.1	4	<0.1
Aluminum production	160	54	35	14	1000	17
Iron and steel works	34	12	1	0.4		
Ferroalloy industry	3.5	1				
Petroleum cracking			<0.1	<0.1		
Power generation						
Coal and oil-fired power plants	0.1	<0.1	<0.1	<0.1	1	<0.1
Peat, wood, straw			6.5	3		
Industrial boilers	1.2	0.4	6.5	3	400	7

Table 9. Continued

Source	Norway		Sweden		U.S.A.	
	Metric tons/year	%	Metric tons/year	%	Metric tons/year	%
Incineration						
Municipal incineration	0.3	<0.1	2.2	0.9	50	0.8
Open burning	0.4	<0.1			100	2
Forest fires	7	2	1.3	0.5	1000	17
Agricultural burning	6	2				
Mobile sources						
Gasoline automobiles	13	4	33	13	2100[b]	35
Diesel automobiles	7	2	14	6	70	1
Air traffic	0.1	<0.1	<0.1	<0.1		
Total	295		250		6000	

[a]Numbers are encumbered with great uncertainties. To be used as an indicator of order of magnitude.
[b]Approximately 50% of U.S. cars have catalytic converters. This is not corrected for in this number.

In the United States vehicular traffic seems to be the major source, emitting approximately 2100 tons of PAH per year, 35% of the total emission. This is twice as much as any other source. The aluminum industry is also a major source in the United States. Based on the emission factor from Table 4, this industry emits 1000 tons of PAH annually. However, this figure is very uncertain and must be considered only as a very rough estimate, as the emission factor is developed on the basis of a single aluminum reduction plant.

Coke manufacturing is estimated to emit 630 tons of PAH per year. Industrial sources are responsible for 26% of the total emission. Residential heating (wood, coal, oil, and gas) is another important source category, making up 12% of the total emission. Residential combustion of wood has gained greater popularity in the United States recently, due to rising oil prices [54,55], giving rise to large local emissions. Of particular importance is the potential use of relative emission factors in areas near major sources to determine possible environmental impact as a function of geographical and meteorological factors.

Similar data for Sweden and Norway are included in Table 9. As revealed in this table, the relative importance of different sources varies depending on the industrial profile of each country. For example, in Norway, which is a large producer of aluminum, PAH emissions from this industry represent a major source. It must be stressed that the figures in Table 9 are very uncertain, caused by the uncertainties in the emission factors. They should therefore be considered only as estimates.

VI. CONCLUSIONS

Emission factors require a certain degree of caution and a certain degree of consideration, for the following reasons. Many of the emission factors are obtained from a relatively small number of measurements, in some cases even only one. Thus it is not always certain that they are representative for the exact average of the prevailing cases.

Emission factors are meant to be representative of average values and thus do not correspond to the true values for each case. Therefore, an emission factor will depend not only on management and the state of maintenance, but also on the special measures that have been taken to limit the extent of the emission.

The lack of standardization of analytical procedures and the number of PAH reported make the term "total PAH" very uncertain. To reduce the uncertainties, more measurements are needed for many sources, such as gasoline and diesel engines, wood and coal residential heating, aluminum smelting, fluidized-bed coal-fired power plants, air and sea traffic, steel industry, petroleum cracking, coke production, and municipal incinerators.

REFERENCES

1. P. Pott, *Chirurgical Observations Relative to the Cataract, the Polypus of the Nose, the Cancer of the Scrotum, the Different Kinds of Ruptures, and the Mortification of the Toes and Feet,* L. Hawes, W. Clarke, and R. Collins, London, 1775.
2. J. Cook, C. L., Hewett, and I. Hieger, J. Chem. Soc., 395 (1933).
3. *Particulate Polycyclic Organic Matter,* National Academy of Sciences, Washington, D.C., 1972.

4. A. Dipple, in *Chemical Carcinogens*, ACS Monograph 173, C. E. Searle
    (Ed.), American Chemical Society, Washington, D.C., 1976.
5. World Health Organization, *International Standards for Drinking Water*,
    3rd ed., Geneva, 1971.
6. E. A. Walker, Pure Appl. Chem. *49*:1673 (1977).
7. U.S. Department of Labor, OSHA, Fed. Regist. *41*:467441 (1976).
8. G. M. Badger, Natl. Cancer Inst. Monogr. *9*:1 (1962).
9. G. M. Badger, *The Chemical Basis of Carcinogenic Activity*, Charles
    C. Thomas, Springfield, Ill. 1962.
10. R. W. Boubel, and L. A. Ripperton, J. Air Pollut. Control Assoc. *13*:
    553 (1963).
11. G. M. Badger and T. M. Spotswood, J. Chem. Soc., 4420 (1960).
12. B. D. Crittenden and R. Long, in *Carcinogenesis—A Comprehensive
    Survey*, Vol. 1, R. I. Freudenthal and P. W. Jones (Eds.), Raven Press,
    New York 1976, p. 209.
13. G. Grimmer, *Environmental Carcinogens: Polycyclic Aromatic Hydro-
    carbons*, CRC Press, Boca Raton, Fla. 1983.
14. I. Schmeltz and D. Hoffmann, in *Carcinogenesis—A Comprehensive
    Survey*, Vol. 1, R. I. Freudenthal and P. W. Jones (Eds.), Raven Press,
    New York, 1976, p. 225.
15. G. Grimmer, A. Glaser, and G. Wilhelm, Beitr. Tabaksforsch. *3*:415
    (1966).
16. R. F. Severson, W. S. Schlotzhauer, R. F. Arrendale, M. E. Snook,
    and H. C. Higman, Beitr. Tabaksforsch. *9*:23 (1977).
17. M. Blumer, Sci. Am. *234*:34 (1976).
18. J. B. Andelman, and M. J. Suess, Bull.WHO. *43*:479 (1970).
19. R. P. Hangebrauck, D. J. von Lehmden, and J. E. Meeker, *Sources
    of Polynuclear Hydrocarbons in the Atmosphere*, U.S. Department of
    Health, Education and Welfare, Public Health Service, AP-33, PB174-
    706, Washington, D.C., 1967.
20. A. Bjørseth, and G. Eklund, Anal. Chim. Acta *105*:119 (1979).
21. Energy and Environmental Analysis Inc., *Preliminary Assessment of the
    Sources, Control and Population Exposure to Airborne Polycyclic Matter
    (POM) As Indicated by Benzo(a)pyrene (BaP)*, Report to EPA under
    contract 68-02-2836, 1978.
22. L. Rudling, B. Ahling and G. Löfroth, *Statens Naturvårdsverk*,
    Report pm 1331, Solna, Sweden, 1980 (in Swedish).
23. T. Ramdahl, I. Alfheim, S. Rustad, and T. Olsen, Chemosphere *11*:
    601 (1982).
24. D. G. DeAngelis, D. S. Ruffin, J. A. Peters, and R. B. Reznik,
    Source Assessment: Residential Combustion of Wood, EPA-600/2-80-
    042b, Research Triangle Park, N.C., 1980.
25. D. G. DeAngelis, and R. B. Reznik, *Source Assessment: Residential
    Combustion of Coal*, EPA-600/2-79-019a, Research Triangle Park, N.C.,
    1979.
26. H. Beine, Staub—Reinhalt. Luft *30*:334 (1970).
27. A. Herlan, Combust. Flame *31*:297 (1978).
28. A. Bjørseth, O. Bjørseth, and P. E. Fjeldstad, Scand. J. Work Environ.
    Health *4*:212 (1978).
29. I. Alfheim, and L. Wikström, Toxicol. Environ. Chem. *8*:55 (1984);
    and A. Bjørseth and L. Wikström, unpublished results, 1979.
30. A. R. Trenholm, and L. L. Beck, *Assessment of Hazardous Organic
    Emissions from Slot-Type Coke Oven Batteries*, Internal EPA Report,
    Durham, N.C. 1978.

31. The Oil and Gas Journal, *Worldwide Directory: Refining and Gas Processing 1977-1978*, 1977.
32. A. E. Vandegrift, L. J. Shannon, *Handbook of Emissions, Effluents, and Control Practices for Stationary Particulate Pollutant Sources*, Report of NAPCA Contract CPA 22-69-104, 1970.
33. F. Vena, Environment Canada, personal communication, 1981.
34. R. St. Louis, Pennsylvania Department of Environmental Resources Emission Test Results, Harrisburg, Pa., 1977.
35. L. Berglind, and E. Gjessing, Report A3-25, Norwegian Institute for Water Research, Oslo, 1980, (in Norwegian).
36. N. H. Supernant, *Preliminary Emissions Assessment of Conventional Stationary Combustion Systems*, Vol. 2, EPA-600/2-76-046b, 1976.
37. J. Bergström, *Emissions from Combustion of Coal and Oil in Powerplants*, Studsvik Technical Report EK-81/103, Studsvik, Sweden, 1981 (in Swedish).
38. K. Erikson, *PAH, NO_x and Metal Emissions, etc., from Coal and Oil Combustion*, ÅF-Energikonsult 1980-06-04, 1980 (in Swedish).
39. *Report to Södertälje (Sweden) Community*, Swedish Steam Users Association, 1980 (in Swedish).
40. T. Alsberg, and U. Stenberg, Chemosphere 8:487 (1979).
41. SSVL (Stiftelsen Skogindustriernas Vatten- och Luftvårdsforskning), *Emissions of polycyclic organic compounds from bark combustion*, SSVL 99:1 (1980) (in Swedish).
42. G. Eklund, *Determination of Heavy Metals and Organic Compounds in Emissions from Växjö Woodfired Power Plant*, Studsvik Arbetsrapport E2-80/159, 1980 (in Swedish).
43. I. W. Davies, R. M. Harrison, R. Perry, O. Fatanyaka, and R. A. Wellings, Environ. Sci. Technol. *10*:451 (1976).
44. EPA, *Compilation of Air Pollution Emission Factors*, 3rd ed., AP-42, 1977.
45. C. K. McMahon, and S. N. Tsoukalas, in *Carcinogenesis, Vol. 3: Polynuclear Aromatic Hydrocarbons*, P. W. Jones and R. I. Freudenthal (Eds.), Raven Press, New York, 1978, p. 61.
46. G. Grimmer, H. Böhnke, and A. Glaser, Erdoel Kohle *30*:411 (1977).
47. G. P. Gross, *Third Annual Report on Gasoline Composition and Vehicle Exhaust Gas Polynuclear Aromatic Content*, CRC APRAC Project CAPE-6-68, EPA Contract 68-0400-25, APTD 1560, PB 218-873, Detroit, Mich., 1972.
48. C. R. Begeman, and J. C. Burgan, *Polynuclear Hydrocarbon Emission from Automotive Engines*, SAE Paper 700469, Society of Automotive Engineers, Detroit, Mich., 1970.
49. K. E. Egebäck, and B. M. Bertilsson, *Chemical and Biological Characterization of Exhaust Emissions from Vehicles Fueled with Gasoline, Alcohol, LPG and Diesel*, National Swedish Environment Protection Board, Report SNV pm 1635, Solna, Sweden, 1983.
50. National Research Council, *Impacts of Diesel-powered Light-Duty Vehicles; Health Effects of Exposure to Diesel Exhaust*, National Academy Press, Washington, D.C., 1981.
51. J. Santodonato, D. Basu, and P. Howard, *Health Effects Associated with Diesel Exhaust Emission; Literature Review and Evaluation*, EPA-600/1-78-063, Research Triangle Park, N.C., 1978.
52. L. M. Shabad, and G. A. Smirnov, Atmos. Environ. *6*:153 (1972).

53. A. Colmsjö, *Report to the Health Authority of Stockholm (Sweden) Community*, Stockholm, 1980.
54. M. M. Waldrop, Science 211:914 (1981).
55. Wood Heating Alliance, *Proceedings Documents for Wood Heating Seminars 1980/1981*, Washington, D.C., 1981.

2
PAH Emissions from Coal-Fired Plants

KRISTOFER WARMAN* / Fuel Technology Department, Studsvik Energyteknik
AB, Nyköping, Sweden

I.   Introduction   1

II. Sampling and Analysis   22

    A. Sampling equipment   22
    B. Sample collection   24
    C. Sample analysis   25
    D. Total PAH versus vaporous and particle-associated PAH   27

III. PAH Emission Data from Coal Combustion   27

    A. Comparability of results   27

IV. Conclusions   55

    References   56

I. INTRODUCTION

In most cases, the firing of coal in energy-producing plants falls far short
achieving complete combustion. As a consequence, compounds other than
carbon dioxide and water are created. This incomplete combustion is generally considered to be the main industrial source of PAH emission [1,2].
PAH emission is dependent on the type and condition of the specific process,
control equipment used, feeding characteristics, and operating conditions.

About 85% of the coal consumed by power generation facilities is burned
in boilers fired with pulverized coal. In approximately 50% of these boilers
tangential firing is used, 32% use front firing, and only 15% use opposed firing
[3]. However, it is expected that opposed firing will become more common in
the future. Other widespread methods of coal burning include moving grate
furnaces, cyclone furnaces, and fluidized-bed combustors (FBC).

Large plants (500-1500 MW) are dominated by pulverized-coal firing,
whereas utilities with capacities below 500 MW often utilize the moving-grate
type with surface feeding, stoker, or spreader stoker. Fluidized-bed combustion is a technique that has been known for more than 50 years but has
undergone intensive development during the last 10 years. The FBC oper-
ates at lower combustion temperatures than pulverized-coal and even moving-
grate combustors, which can result in lower emissions of nitrogen oxides.
Limestone or dolomite can be used in the bed to reduce the presence of
sulfur oxides in the combustion effluents. FBC combustors are rarely built

---

*Current affiliation: Studsvik Environmental Consultants AB, Nykoping,
Sweden

with a capacity exceeding 50 MW. For the purpose of generating electric power, the energy-conversion efficiency can be increased still further by pressurizing the fluidized-bed combustor to approximately 10 bar. The additional power is then generated by expanding the hot, pressurized combustion gases through a gas turbine.

The environmental debate connected with coal combustion has focused on the emission of compounds that cause acidification. The basic goal in pulverized-coal combustion has been to achieve complete combustion and limit the emission of nitrogen oxices by reduction of the temperature levels throughout the flame. Results from FBC development units indicate that FBC can meet new NO_x and SO_x emission standards. The presence of carbon monoxide and char in the bed during the combustion acts to surpress the emission of NO_x by reduction of NO to N_2.

Unfortunately, combustion at a decreased temperature and a decreased total excess-air ratio leads to a dramatic increase in the organic emission level [4]. Attempts have been made to retain the benefits of low-temperature and low-excess-air combustion while maintaining the completeness of combustion. It seems at present that the most practical approach makes use of multistage combustion.

PAH emissions from coal-fired plants can depend on a considerable number of factors. Among these are the size of the plant, the load factor, the type of coal, the type of burner, and the flue-gas cleaning equipment. The load factor [5,6] and boiler type [6-8] are undoubtedly the most crucial factors in determining emissions. One study [2] suggests that stoker-fired installations generally produced higher-weight-per-volume emissions. As for the other variables listed above, very careful controlled experiments need to be performed to determine their importance.

The goal of this chapter is to summarize fully emissions data from the measurement of PAH when burning coal. The short descriptions of plants and their operating parameters will provide the reader an opportunity to judge if the type of facility, fuel composition, operating conditions, and emission of PAH are related.

To eliminate nonsystematic errors, a large investigation has been chosen as the subject for discussion. This investigation, the Swedish Coal-Health-Environment Project, was performed between 1980 and 1982 by the same team of scientists. The project was performed under contract from the Swedish State Power Board.

All frequently used types of plants are represented in the overview below: pulverized-coal-fired plants (A-C), plants with a moving grate (D-F), plants with a fixed grate (G), and fluidized-bed combustors (H-K). In the final part of this chapter (Table 33) the results of a U.S. study performed by the Electric Power Research Institute (EPRI) [3] are cited.

None of the plants discussed in this chapter fired brown coal. Thus the results presented here consider only effluents arising from the firing of pit coal.

II. SAMPLING AND ANALYSIS

A. Sampling Equipment

To be acceptable, the sampling equipment should satisfy some fundamental demands:

The organic compounds have to be collected quantitatively.
It must be possible to recover the sampled material without changing the
    character and composition of the sample.

A good review of the sampling methods for PAH collection at the stationary source is presented by Jones [9]. The earlier methods employed for PAH sampling were based on collection of PAH in impinger bottles. Method 5, of the U.S. Environmental Protection Agency (EPA) became most prevalent among procedures using impingers. Method 5 was published for the first time in the *Federal Register* in 1971 [10,11]. This method was, however, not worked out for the sampling of organic matter. In spite of this many early PAH collections were made using method 5.

Work on the use of high surface adsorbents for collecting vaporous samples was reported at the beginning of the 1970s [12-14]. At the same time a new concept in flue gas sampling of organic emissions was presented. This new idea combined conventional filtration with collection of the organic vapors on the resins [15-17].

Attempts were made to standardize the sampling equipment. As a result of the EPA project the SASS (source assessment sampling system) train was developed [18,19]. Unfortunately, none of the sampling systems are widely accepted as standard equipment. At present most of the stationary investigations are carried out with the SASS train. For portable equipment, the Battelle train and a modification of it are used in most cases [20-26]. The choice of PAH sampling procedure thus depends on local circumstances and plant operating parameters.

Adams et al. [27] have presented a comparison of the properties of the various resins. They also discuss the resins suitability for sampling organic vapors. Two adsorbents are especially favored: Amberlit XAD-2 (Rohm & Haas; a styrene-divinylbenzene polymer) and Tenax GC (a polymer of 2,6-diphenyl-*p*-phenylene oxide). None of the adsorbents can be accepted without reservation, but XAD-2 seems more suitable for collecting vaporous PAH. The resins have not been criticized for their collecting efficiency—the retention volume of the resins is quite sufficient—but for the difficulties encountered in obtaining acceptable blanks. Investigations using Fourier transform infrared spectroscopy (FT/IR) and gas chromatography/mass spectrometry (GC/MS) [28] show that better blanks can be achieved by multisolvent sequential extraction for XAD-2 than for Tenax GC.

There is considerable problem with the poor long-term stability of the high surface resins. The cleaned absorption bed maintains its blank for only a couple of weeks, even when stored in a cool, dark place and in a nitrogen atmosphere. Tenax GC decomposes spontaneously into 2,6-diphenylquinone and 2,6-diphenylquinol, while the XAD-2 decomposition products are benzoic acid, ethylbenzaldehyde, dimethylbenzoic acid, diethyl phthalate, and diphenoxybenzene [28,29]. All these degradation products can cause difficulty when gas chromatographic analysis is performed.

The particles collected when sampling the flue gases are usually extracted by Soxhlet extraction or treated by vacuum sublimation with the purpose of separating out the particle-associated PAH. It is a matter of fundamental importance that the procedure be completed.

Sonnichsen et al. [30] discuss the difficulties connnected with the extraction procedures when particulate matter is present. According to the authors, there is growing evidence that conventional extraction of PAH from fly ash is not completely efficient, and this remains a potential source of the poor recovery levels.

When choosing the sampling equipment for collecting the PAH from coal-fired plants, one should take into account the risk of secondary reactions between the collected PAH and other components in the flue gases. Since PAH are known to react intensely with electrophilic agents, they can presumably react with the nitrogen oxides and nitric acid that are present in flue gas. The products of these reactions may be nitrated, nitrozated, and oxidized PAH [31,57]. Such secondary reactions are caused by the sampling equipment's design and must be considered to be sampling artifacts.

B. Sample Collection

In this section we present a brief overview of the sampling methods employed to generate the results reported in Tables 1 to 32.

1. Method A

This sampling equipment was utilized for collecting the samples in the Coal-Health-Environment Project. The equipment described was employed to collect the samples in most of the investigations cited in this chapter. The exceptions are plants A, B (Sept. 11-13, 1979), J, and K.

The sampling system for semivolatile and high-boiling organic compounds was a modified version of the SASS train. The equipment [25] was developed to enable isokinetic sampling with an all-glass system. The sampling train is shown schematically in Fig. 1.

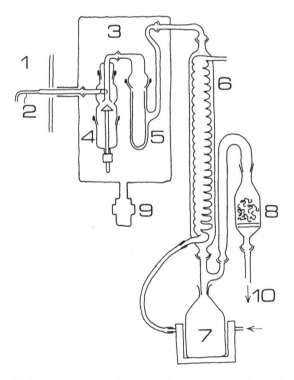

Figure 1. Studsvik sampling system. 1 stack; 2 glass probe; 3 heated oven 160°C; 4 glass cyclone; 5 quartz fiber thimble; 6 cooler; 7 condensate collector; 8 XAD-2 adsorbent bed; 9 heater; 10 to the pump and volume meter.

The interchangeable probe extended into an oven module which was kept at 160°C during sampling. The oven contained a variable-range cyclone and a high-purity tissue quartz thimble filter. At a flow rate of 4 m^3/hr the cut size of the cyclone was adjusted to 1.5 μm. After leaving the filter the flue gas was drawn through a high-efficiency cooler, a condensate collector, and an XAD-2 sorbent trap. The coolant was maintained at a temperature below −10°C so that the temperature of the gas entering the XAD-2 adsorbent bed did not exceed 5-10°C. At the conclusion of the sampling run the probe and the cooler were washed with acetone. The residue was then treated as a part of the sample. The flue gas volumes were between 10 and 20 m^3 in each sample.

All glassware used for sampling was carefully rinsed and then heated to 450°C overnight. The quartz fiber filter was purified by heating to 530°C. Amberite XAD-2 was purified by successive washing with water, methanol, and acetone. The resin was then packed in the sampling ampulla and Soxhlet-extracted with acetone and dichloromethane for 48 hr with intermediate changes of the solvent. The blank extracts were preconcentrated and examined using gas chromatography.

2. Method B

This sampling unit was utilized to measure effluents from plants A, B, (Sept. 11-13, 1979), and J. The sampling train was a combination of the source sampling system as described by Jones et al. [15] and the EPA Method 5 train. The sampling train consisted of an interchangeable steel probe, a glass fiber filter, a glass cooler, a condensate collector, and XAD-2 absorption bed, a dry gas meter device, and a pump. The XAD-2 resin was Soxhlet-extracted before use with water (24 hr), methanol (24 hr), and dichloromethane (24 hr). All sampling equipment was carefully rinsed with ethanol before use.

3. Method C

Murthy et al. [32] utilized the source assessment sampling system (SASS) to collect samples for organic analysis. In three tests, the stainless steel condenser module normally supplied with the SASS train was replaced with a glass module of similar dimensions. The reason for replacing it was corrosion of the stainless module. More details on the sampling equipment and analytical procedures used in the investigation are reported in Ref. 32.

C. Sample Analysis

There were some significant differences between the methods utilized for sample preparation and analysis.

1. Method A

This analytical procedure was applied the samples in investigations involving plants C, F, G, I, and J [26]. The condensates were extracted in a separatory funnel with dichloromethane at two different pH values: 1-2 and 11-12. The XAD-2 adsorbent beds and the particulate fractions were Soxhlet-extracted with dichloromethane for 24 hr. The extracts were concentrated in Kuderna-Danish apparatus, dried, and then the internal standards were added. The volume of the extract was adjusted using cyclohexane. This extract was called the main extract.

In some cases the extracts from the condensate and XAD-2 adsorbent were combined. The separation of the main extract was performed according to the scheme in Fig. 2. The aim of this procedure was to separate the main extract into three fractions: acidic, basic, and neutral.

The fractions were analyzed by gas chromatography. All substances were identified by gas chromatography/mass spectroscopy (GC/MS). The mass sprectra were compared with those of a standard mass spectra library. The identification of PAH was confirmed by comparing the GC retention times with those of the prepared PAH standards. The peaks in the chromatograms were quantified by comparing their peak areas with the peak area of the nearest internal standard. The results were not adjusted with relative response factors.

2. *Method B*

This method was used for extraction separation and analysis of samples from plants B (Nov. 5 and 6, 1981) D, E, and H [33]. The analytical procedure for treatment of the condensates and XAD-2 adsorbent was identical with that of Method A. The particulate fractions were extracted first with pentane, then with benzene, and finally with ether. After the last extraction the water-free pyridine was added to the ether extract, ether was evaporated, and the pyridine solution was silyated. All the extracts were analyzed by gas chromatography after addition of the internal standards.

All substances were identified by GC/MS and a computer-based program for retention index calculations. The peaks in the chromatograms were quantified by comparing their peak areas with the peak areas of the internal standards.

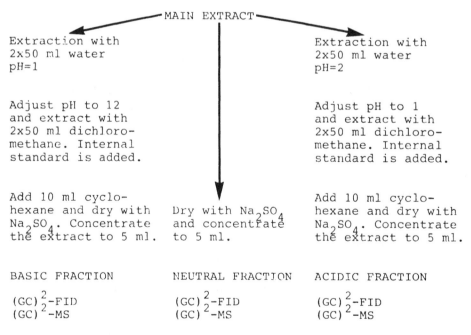

Figure 2. Treatment of extracts.

3. Method C

This method was utilized for the cleaning and analysis of the samples collected
from plants A and B (Sept. 11-13, 1979). [23]. The particle fractions and
XAD-2 absorption beds were Soxhlet-extracted with cyclohexane 16 hr after
the internal standards were added. Condensates were extracted twice with
cyclohexane. Cyclohexane extracts were liquid/liquid extracted with a mixture of dimethylformamide-water (9:1) and partitioned in accordance with
the method described by Bjørseth [34]. The qualitative and quantitative
analyses were carried out using gas chromatography.

D. Total PAH Versus Vaporous and Particle-Associated PAH

Most of the sampling techniques used to collect the PAH measure a total of
both particulate and vaporous PAH. The reason for this is that it is difficult to determine the relative amounts of the PAH occurring as vapors and
on particles.

Since the stack gas temperature for a typical coal-fired power plant is
not constant along the gas duct, the relative amounts of the vaporous PAH
and the particle-associated PAH are different at different sampling points
[35]. As adsorption of the PAH on particles depends on both the temperature and the surface-area-to-volume ratio of the particle, the PAH will tend
to condense onto the smaller particles. For the same reasons, in stacks with
lower temperatures, higher particulate loadings, and finer particles, more
of the PAH will occur in particulate form [36].

Natush and Tomkins [37] have shown in a theoretical analysis that most
PAHs will exist as vapors at temperatures below 160°C, which is the prevalent stack temperature range for a typical coal-fired plant. When the stack
gas exits the stack, the temperature of the gas decreases rapidly and most
of the PAH will condense or adsorb onto particles.

The sampling system that divides the collected material between particle-
associated PAH and vaporous PAH works at the temperature chosen by the
operator. As this temperature is predetermined and represents only some
part of the flue gas duct, there is some doubt if the discussion of distribution of the PAH between particulate matter and vapor is significant. The
division between particle-associated PAH and vaporous PAH in the results
from an investigation represents the situation at only one point in the duct
of the coal-fired plant. These figures are also a product of the sampling
system and its way of filtering the particles. Finally, the distribution
between particles and vapor should be different for flue gas that leaves
the stack from that being collected in the sampling system. To report total
organic stack emissions seems, in the absence of a universally applicable
standard sampling procedure, to be a very questionable procedure.

III. PAH EMISSION DATA FROM COAL COMBUSTION

A. Comparability of Results

The absence of standarized methods causes difficulty with any comparison.
Furthermore, the comparability of the results obtained for the various combinations of the sampling and analytical techniques employed is questionable,
even for the same PAH species. Therefore, the most meaningful discussion
can be carried on when results from the same testing series are considered.

Given this standard, the emission data generated in larger tests, when many PAH sources are examined by the same group of scientists, are particularly interesting. Even if methodological errors are made, it is possible to draw conclusions about differences between plants, firing conditions, and so on.

1. Pulverized-Coal-Fired Plants

a. Plant A (see Tables 1 to 3). This plant was a pulverized-coal-oil-fired power plant capable of producing 350 MW_{th} when operating at full load. The boiler was equipped with eight coal oil burners located at three levels at the front of the boiler. The flue gas was cleaned in an electrostatic precipitator.

b. Plant B (see Tables 4 to 6). This plant was a pulverized-coal-fired power plant capable of producing 270 MW_e (600 MW_{th}) when operating at full load. The boiler was equipped with a high-efficiency electrostatic precipitator. The unit fired a mixture of low-sulfur coals.

c. Plant C (see Tables 7 to 9). This plant was a pulverized-coal-fired power plant, of the Benxon type, capable of producing 720 MW_e when operating at full load. The flue gas cleaning system consisted of an electrostatic precipitator, and the desulfurization unit of a wet-lime scrubber. About 20% of the total gas flow was treated in the desulfurization unit. The SO_2 removal efficiency was better than 80%. The untreated flue gas was bypassed.

Table 1. Composition of Flue Gas During Tests, Plant A

Compound	Mean value	Minimal value	Maximal value
CO (%)	<0.02[a]		
	<0.02[b]		
CO_2 (%)	13.1[a]	12.5[a]	14.0[a]
	13.3[b]	12.5[b]	14.0[b]
NO_x (ppm)	290[a]	230[a]	340[a]
	310[b]	220[b]	360[b]

[a]Sept. 4, 1979 [23].
[b]Sept. 6, 1979 [23].

Table 2. Composition of Coal Fired During Tests, Plant A[a,b]

Parameter	
Total moisture (wt %)	7.8
Ash (wt %)	15.8
Heating value (MJ/kg)	26.1
S (wt %)	1.8

[a]Sept. 4, 1979 [23].
[b]Sept. 6, 1979 [23].

Table 3. PAH Concentration in Flue Gas from Plant A ($\mu g/m^3$)

Compound	Concentration	
	a	b
Phenanthrene	0.55[a]	0.19[b]
Anthracene	0.04	
Methylphenanthrene/anthracene, isomer 1	0.07	0.04
Methylphenanthrene/anthracene, isomer 2	0.10	0.05
2-Methylanthracene	0.02	0.04
4,5-Methylenephenanthrene	0.01	0.02
Methylphenanthrene/anthracene, isomer 3	0.06	0.04
1-Methylphenanthrene	0.07	0.04
Fluoranthene	1.24	0.35
Pyrene	0.44	0.18
4-Methylpyrene	<0.02	<0.02
1-Methylpyrene	<0.02	<0.02
Benz[a]anthracene	0.15	<0.02
Chrysene/triphenylene	0.49	0.25
Benzo[b]fluoranthene	0.23	0.07
Benzo[k]fluoranthene	0.12	0.03
Benzo[e]pyrene	0.13	0.03
Benzo[a]pyrene	0.09	0.04

[a]Sept. 4, 1979 [23].
[b]Sept. 6, 1979 [23].

Table 4. Composition of Flue Gas During Tests, Plant B

Compound	Mean value	Minimal value	Maximal value
O_2 (%)	6.3[a]		
	6.7[b]		
CO (%)	0.030[c]		
	<0.02[d,e]		
	−1.5[a]	−3.6[a]	−0.2[a]
	−0.9[b]	−2.7[b]	−1.6[b]
CO_2	12.2[c]	11.7[c,d]	12.8[c]
	12.0[d]		12.6[d]
	12.1[e]	11.8[e]	12.4[e]
	11.9[a]	11.4[a]	12.5[a]
	12.4[b]	12.0[b]	12.9[b]
SO_2 (ppm)	631[a] 647[b]	606[a,b]	653[a]
			667[b]
NO_x (ppm)	480[c,d]	450[c,d]	500[c]
			520[d]
	560[e]	490[e]	610[d]
	468[a]	452[a]	494[a]
	537[b]	495[b]	584[b]

[a] Nov. 5, 1981 [38].
[b] Nov. 6, 1981 [38].
[c] Sept. 11, 1979 [23].
[d] Sept. 12, 1979 [23].
[e] Sept. 13, 1979 [23].

Table 5. Composition of Coal Fired During Tests, Plant B[a-c]

Parameter	
Total moisture (wt %)	8.6
Ash (wt %)	15.9
Heating value (MJ/kg)	25.9
S (wt %)	0.4

[a]Sept. 11, 1979 [23].
[b]Sept. 12, 1979 [23].
[c]Sept. 13, 1979 [23].

Table 6. PAH Concentration in Flue Gas from Plant B ($\mu g/m^3$)[a]

Compound	Concentration
Phenanthrene	1.42[b]
	1.11[c]
	<0.02[d]
Anthracene	0.14[b,c]
	<0.02[d]
	<0.01[e,f]
Anthracenone/phenanthrenone	0.28[e]
	<0.01[f]
Methylphenanthrene/anthracene, isomer 1	0.11[b]
	0.07[c]
	<0.02[d]
Methylphenanthrene/anthracene, isomer 2	0.13[b]
	0.09[c]
	<0.02[d]
2-Methylanthracene	0.02[b]
	0.03[c]
	<0.02[d]

Table 6. Continued

Compound	Concentration
4,5-Methylenephenanthrene	0.02^b
	0.01^c
	$<0.02^d$
Methylphenanthrene/anthracene, isomer 3	0.09^b
	0.06^c
	$<0.02^d$
1-Methylphenanthrene	0.09^b
	0.06^c
	$<0.02^d$
Fluoranthene	0.60^b
	0.53^c
	$<0.02^d$
	$<0.01^{e,f}$
Pyrene	0.29^b
	0.24^c
	$<0.02^d$
	$<0.01^{e,f}$
4-Methylpyrene	$<0.02^{b-d}$
1-Methylpyrene	$<0.02^{b-d}$
Benz[a]anthracene	0.06^b
	0.05^c
	$<0.02^d$
	$<0.1^{e,f}$

Table 6. Continued

Compound	Concentration
Chrysene/triphenylene	0.23^b
	0.31^c
	$<0.02^d$
	$<0.01^{e,f}$
Benzo[b]fluoranthene	0.03^b
	0.04^c
	$<0.02^d$
Benzo[k]fluoranthene	$0.02^{b,c}$
	$<0.02^d$
Benzo[e]pyrene	$0.01^{b,c}$
	$<0.02^d$
	$<0.01^{e,f}$
Benzo[a]pyrene	0.02^b
	0.03^c
	$<0.02^d$
	$<0.01^{e,f}$
Indeno [1,2,3-cd]pyrene	$<0.01^{e,f}$
Perylene	$<0.01^{e,f}$
Benzo[ghi]perylene	$<0.01^{e,f}$

[a] Conversion factor µg/MJ: 0.35.
[b] Sept. 11, 1979 [23].
[c] Sept. 12, 1979 [23].
[d] Sept. 13, 1979 [23].
[e] Nov. 5, 1981 [39].
[f] Nov. 6, 1981 [40].

Table 7. Composition of Flue Gas During Tests, Plant C

Compound	Mean value	
	a	b
O_2 (%)	7.9	5.2
CO (ppm)	0.0	0.0
CO_2 (%)	11.2	12.6
SO_2 (ppm)	450	62
NO_x (ppm)	350	370

[a] Dec. 2, 1981. The SO_2 scrubber was out of service during the tests [41].
[b] Dec. 3, 1981. The flue gas was sampled after the SO_2 scrubber [41].

Table 8. Composition of Coal Fired During Tests, Plant C

Parameter	a	b
Total moisture (wt %)	2.1	2.2
Ash (wt %)	15.1	15.5
Volatiles (wt %)	27.4	27.2
Fixed carbon (wt %)	55.5	55.1
Heating value (MJ/kg	27.1	27.0
C (wt %)	70.0	69.8
H (wt %)	4.1	4.1
N (wt %)	1.4	1.6
O (difference) (wt %)	8.5	8.0
S (wt %)	0.92	0.93
Cl (wt %)	0.05	0.06

[a] Dec. 2, 1981 [41].
[b] Dec. 3, 1981 [41].

Table 9. PAH Concentration in Flue Gas from Plant
C ($\mu g/m^3$)[a,b]

Compound	Concentration
Phenanthrene	<0.1
Anthracene	<0.1
Fluoranthene	<0.1
Pyrene	<0.1
Benzo[ghi]fluoranthene	<0.1
Chrysene	<0.1
Benz[a]anthracene	<0.1
Benzo[e]pyrene	<0.1
Benzo[a]pyrene	<0.1
Indeno[1,2,3-cd]pyrene	<0.1
Perylene	<0.1
Benzo[ghi]perylene	<0.1

[a]Dec. 2, 1981 [41]. Conversion factor $\mu g/m^3$ to $\mu g/MJ$:
0.41.
[b]Dec. 3, 1981 [41]. Conversion factor $\mu g/m^3$ to $\mu g/MJ$:
0.44.

2. *Plants with Moving Grate*

a. Plant D (see Tables 10 to 12). This plant was a coal-fired industrial
steam boiler (40 tons of steam per hour at 6 MPa and 440°C). The low-
sulfur (0.9%) coal was fired on a moving grate. The grate area was 23 m^2.
During the tests the load was 78% (25MW$_{th}$). The flue gas cleaning equipment consisted of a multiple-cyclone unit and a secondary cyclone circuit.

b. Plant E (see Tables 13 to 15). This plant was a coal-fired steam boiler
of the La Mont type. The steam was heat-exchanged and used for district
heating. The low-sulfur ocal (0.7%) was fired on a moving grate similar to
that of plant D. The boiler capacity was 35 tons of steam per hour at 2.7
MPa and 425°C. The samples were collected in a flue gas stream that was
cleaned in a baghouse filter. The baghouse filter was installed at the plant
for test purposes.

c. Plant F (see Tables 16 to 18). This plant was a spreader stoker, coal
and oil fired, with a water tube boiler. The boiler capacity was 130 tons
of steam per hour at 11.5 MPa and 540°C. The flue gas was cleaned with
a two-step multiple-cyclone unit, with 440 little cyclones in step 1 and two
big cyclones in step 2.

Table 10. Composition of Flue Gas During Tests, Plant D[a]

Compound	Mean value
O_2 (%)	9.8
CO (ppm)	15
CO_2 (%)	9.3
SO_2 (ppm)	394
NO_x (ppm)	49

[a]March 25, 1981 [42].

Table 11. Composition of Coal Fired During Tests, Plant D[a]

Parameter	
Total moisture (wt %)	8.8
Ash (wt %)	14.7
Volatiles (wt %)	27.5
Fixed carbon (wt %)	49.0
Heating value (MJ/kg)	25.7
C (wt %)	64.4
H (wt %)	4.8
N (wt %)	1.0
S (wt %)	0.85
Cl (wt %)	0.01

[a]March 25, 1981 [42].

Table 12. PAH Concentration in Flue Gas from Plant D ($\mu g/m^3$)[a]

Compound	Concentration
Phenanthrene	<0.1
Anthracene	<0.1
4,5-Methylenephenanthrene	<0.1
Fluoranthene	<0.1
Pyrene	<0.1
Benzo[ghi]fluoranthene	<0.1
Chrysene	<0.1
Benz[a]anthracene	<0.1
Benzo[e]pyrene	<0.1
Benzo[a]pyrene	<0.1
Indeno [1,2,3-cd]pyrene	<0.1
Perylene	<0.1
Benzo[ghi]perylene	<0.1

[a]March. 25, 1981 [43,44]. Conversion factor $\mu g/m^3$ to $\mu g/MJ$: 0.51.

Table 13. Composition of Flue Gas During Tests, Plant E

Compound	Mean value	Standard deviation	Minimal value	Maximal value
O_2 (%)	1.11[a]	0.45[a]	10.2[a]	12.1[a]
	10.1[b]	0.47[b]	9.1[b]	11.3[b]
CO (ppm)	11.0[b]	2.0[b]	0[b]	<10[a]
				20[b]
CO_2 (%)	8.9[a]	0.42[a]	8.0[a]	9.7[a]
	9.8[b]	0.40[b]	8.8[b]	10.7[b]
SO_2 (ppm)	345[a]	24[a]	307[a]	403[a]
	377[b]	22[b]	322[b]	431[b]

[a]Nov. 18, 1980 [45].
[b]Nov. 19, 1980 [45].

Table 14. Composition of Coal Fired During Tests, Plant E

Parameter	a	b
Total moisture (wt %)	11.8	12.4
Ash (wt %)	11.8	12.4
Volatiles (wt %)	26.5	26.2
Fixed carbon (wt %)	50.7	49.0
Heating value (MJ/kg)	26.5	26.0
C (wt %)	65.4	66.9
H (wt %)	4.9	
N (wt %)	1.0	1.0
S (wt %)	0.65	0.65
Cl (wt %)	0.13	0.14

[a]Nov. 18, 1980 [45].
[b]Nov. 19, 1980 [45].

Table 15. PAH Concentration in Flue Gas from Plant E ($\mu g/m^3$) [a-c]

Compound	Concentration
Phenanthrene	<0.1
Anthracene	d
Fluoranthene	<0.01
Pyrene	<0.07
Benzo[ghi]fluoranthene	<0.01
Chrysene	<0.01
Benz[a]anthracene	<0.01
Benzo[e]pyrene	<0.01
Benzo[a]pyrene	<0.01
Indeno[1,2,3-cd]pyrene	<0.01
Perylene	<0.01
Benzo[ghi]perylene	<0.01

[a]Conversion factor $\mu g/m^3$ to $\mu g/MJ$: 0.50.
[b]Nov. 18, 1980 [46,47].
[c]Nov. 19, 1980 [46,47].
[d]Compound can be detected but not quantified.

Table 16. Composition of Flue Gas During Tests, Plant F[a]

Compound	Mean value	Standard deviation
O_2 (%)	9.7	0.3
CO (ppm)	40	10
CO_2 (%)	9.8	0.2
SO_2 (ppm)	571	22
No_x (ppm)	218	9

[a]Dec. 17, 1981 [48].

Table 17. Composition of Coal Fired During Tests, Plant F[a]

Parameter	
Total moisture (wt %)	2.1
Ash (wt %)	12.7
Volatiles (wt %)	32.1
Fixed carbon (wt %)	53.1
Heating value (MJ/kg)	28.7
C (wt %)	69.4
H (wt %)	4.6
N (wt %)	1.4
O (difference (wt %)	10.6
S (wt %)	1.30
Cl (wt %)	0.04

[a]Dec. 17, 1981 [48].

Table 18. PAH Concentration in Flue Gas from Plant F $(\mu g/m^3)^a$

Compound	Concentration
Phenanthrene	0.29
Anthracene	<0.1
Fluoranthene	<0.1
Pyrene	<0.1
Benzo[ghi]fluoranthene	<0.1
Chrysene	<0.1
Benz[a]anthracene	<0.1
Benzo[e]pyrene	<0.1
Benzo[a]pyrene	<0.1
Indeno[1,2,3-cd]pyrene	<0.1
Perylene	<0.1
Benzo[ghi]perylene	<0.1

[a]Dec. 17, 1981 [48]. Conversion factor $\mu g/m^3$ to $\mu g/$ MJ: 0.43.

3. Plant with Fixed Grate

a. Plant G (see Tables 19 to 21). This plant was a coal-fired hot water boiler with a capacity of 2.1 MW_{th}. The grate was immovable and deslagging of the furnace was done manually twice a day. No deslagging was carried out during the test. The boiler was equipped with multiple cyclones with full flow circulation.

4. Fluidized-Bed Combustors

a. Plant H (see Tables 22 to 24). This plant was a small atmospheric fluid-bed combustor arranged mainly for demonstration but also producing hot water for district heating. The boiler was water cooled. The coal-loading attachment (a stoker) was located at the bed level. The freeboard height of the bed combustor was 4 m. The capacity of the boiler was 4.7 MW_{th} when operating at full load, and the samples were collected at 87% of full capacity. The FBC was equipped with a compact cell filter containing four casettes. During the tests limestone was not used to reduce effluents from the combustor.

b. Plant I (see Tables 25 to 27). This plant was a circulating atmospheric fluid-bed combustor of the Pyroflow type. Its capacity was 90 tons of steam per hour at 8.4 MPa and 500°C. The FBC was equipped with an electrostatic precipitator. No limestone was used during the test.

Table 19. Composition of Flue Gas During Tests, Plant G[a]

Compound	Mean value	Standard deviation	Minimal value	Maximal value
O_2 (%)	4.6	0.9	1.8	6.2
CO (ppm)	280	395	0	>5000
CO_2	14.5	0.9	12.7	17.3
SO_2 (ppm)	344	26	295	450
NO_x (ppm)	141	22	110	240

[a]Feb. 24, 1982 [49].

Table 20. Composition of Coal Fired During Tests, Plant G[a]

Parameter	
Total moisture (wt %)	5.1
Ash (wt %)	6.5
Volatiles (wt %)	32.1
Fixed carbon (wt %)	56.3
Heating value (MJ/kg)	30.0
C (wt %)	73.5
H (wt %)	5.0
N (wt %)	1.3
O (difference) (wt %)	13.3
S (wt %)	0.47
Cl (wt %)	0.12

[a]Feb. 24, 1982 [49].

Table 21. PAH Concentration in Flue Gas from Plant G (μg/m^3)[a]

Compound	Concentration
Phenanthrene	15.3
Anthracene	<0.1
4,5-Methylenephenanthrene	<0.1
Methylanthracene	<0.1
Fluoranthene	22.1
Pyrene	0.91
Benz[c]acenaphthylene	<0.1
Phenylnaphthylene	1.7
Methylpyrene	<0.1
Benzofluorene	<0.1
Ethyl-4,5-methylphenanthrene	2.4
Benzo[ghi]fluoranthene	1.4
Chrysene	<0.1
Benz[a]anthracene	1.4
Benzo[b,k,j]fluoranthene	0.88
Benzo[e]pyrene	0.81
Benzo[a]pyrene	0.79
Perylene	<0.1
Binaphthyl	<0.1
Benzo[ghi]perylene	<0.1
Indeno[1,2,3-cd]pyrene	<0.1
Dibenz[a,h]anthracene	<0.1
Coronene	<0.1

[a]Feb. 24, 1982 [50]. Conversion factor μg/m^3 to μg/MJ: 0.31.

Table 22. Composition of Flue Gas During Tests, Plant H[a]

Compound	Mean value
O_2 (%)	6.7
CO (%)	0.2
CO_2 (%)	12.9
SO_2 (ppm)	335
NO_x (ppm)	320

[a]Dec. 16, 1980 [51].

Table 23. Composition of Coal Fired During Tests, Plant H[a]

Parameter	
Total moisture (wt %)	13.1
Ash (wt %)	13.8
Volatiles (wt %)	27.0
Heating value (MJ/kg)	22.7
C (wt %)	59.2
H (wt %)	3.7
N (wt %)	0.92

[a]Dec. 16, 1980 [51].

Table 24. PAH Concentration in Flue Gas from Plant H ($\mu g/m^3$)[a]

Compound	Concentration
Phenanthrene	657
Anthracene	109
Methylphenanthrene and/or anthracene	2.2
4,5-Methylenephenanthrene	14
Fluoranthene	227
Benz[e]acenaphthylene	21.7
Pyrene	205
Dimethylphenanthrene and/or anthracene	8.9
Ethyl-4,5-methylenephenanthrene	8.5
Benzylnaphtalene	1.3
Benzofluoren and/or methylfluoranthene and/or methylpyrene	27.6
Trimethylphenanthrene and/or anthracene	1.9
Benzo[c]phenanthrene	14.4
Dibenzo[a,c]fluoren	20
Benzo[ghi]fluoranthene	23
Dihydrobenz[a]anthracene	4.1
Benz[a]anthracene	32.2
Chrysene and/or triphenylene	18.4
Benzo[b]fluoranthene and/or benzo[j,k]fluoranthene	26.0
Benzo[a]fluoranthene	7.5
Benzo[e]pyrene	12.5
Benzo[a]pyrene	12.6
Perylene	2.1
Indeno[1,2,3-cd]pyrene	5.0
Dibenz[ac,ah]anthracene	1.0
Benzo[ghi]perylene	3.9

[a]Dec. 16, 1980 [51,52]. Conversion factor $\mu g/m^3$ to $\mu g/MJ$: 0.35.

Table 25. Composition of Flue Gas During Tests, Plant I

	Mean value	
Compound	a	b
O_2 (%)	9.4	8.4
CO (ppm)	43	26
CO_2 (%)	10.1	10.7
SO_2 (ppm)	326	366
NO_x (ppm)	293	240

[a] April 22, 1982 [53].
[b] April 23, 1982 [53].

Table 26. Composition of Coal Fired During Tests, Plant I

Parameter	a	b
Total moisture (wt %)	8.1	8.1
Ash (wt %)	14.7	14.3
Volatiles (wt %)	27.3	27.7
Fixed carbon (wt %)	49.9	49.8
Heating value (MJ/kg)	26.4	26.6
C (wt %)	64.6	64.7
H (wt %)	4.8	4.8
N (wt %)	1.9	1.9
O (difference (wt %)	13.5	13.7
S (wt %)	0.60	0.61
Cl (wt %)	0.004	0.004

[a] April 22, 1982 [53].
[b] April 23, 1982 [53].

Table 27. PAH Concentration in Flue Gas from Plant I ($\mu g/m^3$)

Compound	Concentration	
	a	b
Phenanthrene	33.2	0.28
Anthracene	1.8	<0.1
Methylphenanthrene	1.41	<0.1
Fluoranthene	11.0	<0.3
Pyrene	5.8	<0.1
Benzo[ghi]fluoranthene	<0.1	<0.1
Chrysene	<0.1	<0.1
Benz[a]anthracene	<0.1	<0.1
Benzo[e]pyrene	<0.1	0.1
Benzo[a]pyrene	0.1	<0.1
Indeno[1,2,3-cd]pyrene	<0.1	<0.1
Perylene	<0.1	<0.1
Benzo[ghi]perylene	<0.1	<0.1

[a]April 22, 1982 [53]. Conversion factor $\mu g/m^3$ to $\mu g/MJ$: 0.39.
[b]April 23, 1982 [53]. Conversion factor $\mu g/m^3$ to $\mu g/MJ$: 0.38.

Table 28. Composition of Flue Gas During Tests, Plant J[a]

Compound	Mean value
O_2 (%)	5.29
CO (%)	0.06
CO_2 (%)	13.1
SO_2 (ppm)	369
NO_x (ppm)	250

[a]Sept. 8, 1982 [63].

Table 29. Composition of Coal Fired During Tests, Plant J[a]

Parameter	
Total moisture (wt %)	5.8
Ash (wt %)	7.2
Volatiles (wt %)	31.3
Fixed carbon (wt %)	55.7
Heating value (MJ/kg)	30.7
C (wt %)	77
H (wt %)	4.6
N (wt %)	1.18
O (difference) (wt %)	9.0
S (wt %)	0.59

[a]Sept. 8, 1982 [63].

c. Plant J (see Tables 28 to 30). This plant was an atmospheric fluid-bed combustor with a bed area of 10 m^2. The boiler was equipped with a superheater and an economizer. The fuel was fed to the bed through an over-bed feed duct, located below the bed level. The FBC capacity was 16 MW_{th} or 20 tons of steam/per hour at 3.2 MPa and 425°C. The flue gas was cleaned in two steps: first by a multiple cyclone unit and then by a baghouse filter. No limestone was used during the tests.

Table 30. PAH Concentration in Flue Gas from Plant J ($\mu g/m^3$)[a]

Compound	Concentration
Phenanthrene	<1.0
Fluoranthene	<1.0
Pyrene	<1.0
Benz[a]anthracene	<0.6
Chrysene	<0.6
Benzo[a]pyrene	<0.2

[a]Sept. 8, 1982 [63]. Conversion factor $\mu g/m^3$ to $\mu g/$MJ: 0.50.

Table 31. Composition of Flue Gas During
Tests, Plant K

Compound	Mean value
O_2 (%)	5.5
CO (ppm)	53
CO_2 (g/m^3)	24
SO_2 (ppm)	28
NO_x (ppm as NO_2)	70

Source: Ref. 54.

d. Plant K (see Tables 31 and 32). This plant was a pressurized fluid-bed combustor constructed for research by Exxon. The facility had a 0.32 m diameter reactor which operated at 890°C, 900 kPA, 1.2 m/sec superficial velocity, and 40% excess air. Coal consumption was 75 kg/hr and 11.0 kg of dolomite sorbent/per hour was fed to the reactor. For the test reported [54], a molar ratio of 1.25 Ca/S was calculated. The flue gas was cleaned using a two-stage cyclone separator. No information about the coal composition was reported.

5. *Different Coal-Fired Utilities*

In the EPRI paper [3] a comprehensive summary of chemical and biological measurements on organic emissions from fossil-fuel-fired electric power generators is given. The principal results are summarized in Table 33.

Table 32. PAH Concentration in Flue Gas from Plant K
(μg/m^3)

Compound	Concentration
Anthracene/phenanthrene	0.053
Methylanthracenes	0.005
Fluoranthene	0.026
Pyrene	0.009
Methylpyrene/fluoranthene	0.001
Benzo[c]phenanthrene	0.0002
Chrysen/benz[a]anthracene	0.0038
Benzofluoranthenes	0.001
Benzo[a]pyrene	0.0005

Source: Ref. 54.

Table 33. Total Organic Stack Emissions from Coal-Fired Conventional Power Plants as 10^{-5} ng/J

Compound		Type of firing					
	Not mentioned	Vertically fired[a]	Front-walled[a]	Tangentially fired[a]	Opposed downward inclined[a]	Cyclone[a]	Spreader stoker[a]
Anthanthrene				0.46			
Anthracene/phenanthrene	0.6[b,c] 8.3[b,c] 7.6[b,c]						
Benz[a]anthracene	2.40[b,d]						
Benzo[a]pyrene	1.9[e] 20.9[e] 0.42[b,d] 0.0035[b,c] 0.007[b,c] 0.09[b,c]	1.8 1.8 5.2 12.3	1.6 2.0	13.3	13.3 2.0 2.1	35.0 7.2	2.3 1.4 1.4
Benzo[c]phenanthrene	0.0035[b,c] 0.007[b,c] 0.007[b,c]						

Table 33. Continued

Compound	Type of firing						
	Not mentioned	Vertically fired[a]	Front-walled[a]	Tangentially fired[a]	Opposed downward inclined[a]	Cyclone[a]	Spreader stoker[a]
Benzo[e]pyrene	37.7[e]	3.9	5.2	8.0	39.8	64.5	5.8
	0.34[b,d]	7.5			10.4	10.4	
	0.0035[b,c]				6.8		
Ethylbenzene	2635[g,f]						
Fluoranthene	18.9[e]	19.9	15.2	37.0	19.9	10.4	5.6
	7.1	18.0	1.2		6.2	4.2	3.0
	0.09[b,d]	38.9			5.2		2.0
	0.07[b,d]	8.0					
	0.5[b,d]						
	2.1[b,d]						
Indeno[1,2,3-cd]pyrene	0.0035[b,c]						
	0.007[b,c]						
	0.007[b,c]						

Compound					
m,p-xylenes	9193[g]				
Methylanthracenes	0.08[b,c]				
	0.5[b,c]				
Methylbenzopyrenes	1.6[b,c]				
3-Methylcholanthrene	0.0035[b,c]				
	0.007[b,c]				
Methyl chrysenes	0.007[b,c]				
Methyl pyrene/fluoranthene	0.0035[b,c]				
	0.25[b,c]				
	0.45[b,c]				
o-Xylene	2028[g,f]				
Perylene	1.6[e]	2.2		6.7	3.2
	0.7[b,d]				
	0.0035[b,c]				
	0.03[b,c]				
	0.08[b,c]				
Phenanthrene			19.0	3.0	

Table 33. Continued

Compound		Type of firing					
	Not mentioned	Vertically fired[a]	Front-walled[a]	Tangentially fired[a]	Opposed downward inclined[a]	Cyclone[a]	Spreader stoker[a]
Benzofluoranthene	0.0035[b,c]						
	0.007[b,c]						
	0.009[b,c]						
Benzo[ghi]perylene	18.9[e]	7.9	1.3	14.2	104.3	34.1	
	0.0035[b,c]				18.0	3.4	
	0.007[b,c]				14.2		
	0.007[b,c]						
Benzoperylene	0.42[b,d]						
Chrysene	1.68[b,d]						
Chrysene/benz[a]anthracene	0.0035[b,c]						
	0.0035[b,c]						
	0.22[b,c]						

Compound							
Coronene	0.5	0.7	0.8	1.8	4.1	1.0	0.9
	0.0035[b,c]						
Dibenz[a,h]anthracene	0.007[b,c]						
	0.007[b,c]						
Dibenz[ai + ah]anthracene	0.0035[b,c]						
	0.007[b,c]						
	0.007[b,c]						
Dibenzo[c,g]carbazole	0.0035[b,c]						
	0.007[b,c]						
	0.007[b,c]						
7,12-Dimethylbenz[a]anthracene	0.0035[b,c]						
	0.007[b,c]						
	0.007[b,c]						
2,4-Dimethyl pentane and benzene	1520[e,f]						

Table 33. Continued

Compound	Type of firing						
	Not mentioned	Vertically fired[a]	Front-walled[a]	Tangentially fired[a]	Opposed downward inclined[a]	Cyclone[a]	Spreader stoker[a]
Pyrene	15.2[e]	18.0	18.9	13.3	12.3	170.6	5.6
		11.4	15.2		4.8	23.7	3.0
	0.4[b,c]				3.7		2.0
	1.1[b,c]						
Toluene	1230[g,f]						

[a]Ref. 6.
[b]Assume 0.35×10^{-6} m^3/MJ stack gas.
[c]Ref. 59.
[d]Ref. 60.
[e]Ref. 61.
[f]Ref. 62.
[g]Measured as methane.

Source: Reprinted with permission from EPRI, copyright Sept 29, 1983, Electric Power Research Institute, EPRI Report EA-1394, *Inventory of Organic Emissions from Fossil Fuel Combustion for Power Generation.*

IV. CONCLUSIONS

Comparisons of emission data obtained from different investigations must be made with extreme care to account for the effects on the data of the different techniques employed in sampling and analysis. The knowledge of the chemical reactions that can occur during the sampling and transportation to the laboratory is limited [57,58]. The choice of cleanup procedures can also favor some compounds or groups of compounds with specific properties. As a result, these compounds may appear more clearly in the analytical results. The available results are probably adequate for the ranking of the sources examined but are not sufficient to generate the order of magnitude of PAH emissions for other coal-fired plants.

In cases where the operating conditions are normal and controlled, the PAH emissions from coal-fired power plants are small. Where there is faulty operation or an uncontrolled combustion, indicated by the amount of carbon monoxide increasing, emission of organic compounds is also large. The carbon monoxide concentration in the flue gas constitutes a measure of the degree of oxidation reached by the combustion gas. It is an indicator of the combustion efficiency. Comparison of the concentration of CO in the flue gas with the PAH concentrations shows that large emissions of PAH occur together with high CO levels [55].

High PAH emissions often occur in connection with abnormal circumstances at a plant or in cases where it is hard to fully control the functioning of the plant. For example, the emission level of plant H differs substantially from that of the other simply because of the difficulties involved in controlling the combustion process.

It is difficult to find a characteristic PAH profile for coal-fired power plants. The most important and unquestionable observation is that the low molecular PAH are found in higher concentrations in coal combustion effluents than are the high molecular PAH. Nielsen [56] investigated fly ashes from the pulverized-coal-fired power plant. He established the fact that the concentration of pyrene, fluoranthene, and benzo[k]fluoranthene increased in the fly ash when the content of unburned material in the fly ash increased. The same fly ash showed decreasing concentrations of fluorene and phenanthrene as the content of unburned material in the fly ash was increased. No explanation of the phenomenon was given.

The results obtained do not indicate that there is a conncection between the type of fuel and the emission of organic compounds. It is known, however, that the type of fuel can indirectly influence the flue gas composition [24]. Varying fuel quality can make it difficult to control combustion conditions, thus causing higher emission levels of PAH and other organic compounds.

ACKNOWLEDGMENTS

Some of the studies described in this chapter were performed under a contract with the Coal-Health-Environment Project of the Swedish State Power Board. I want to acknowledge my appreciation for assistance from several institutes and companies in Sweden, Norway, and Denmark which have made this program possible. I am also grateful to the Department for Fuel Technology at Studsvik for supporting the writing of this review.

REFERENCES

1. G. M. Badger, Mode of formation of carcinogens in human environment. Presented at the *Symposium on the Analysis of Carcinogenic Air Pollutants,* Cincinnati, Ohio, Aug. 29-31, 1961.

2. B. T. Commins, *Atmos. Environ.* 3:565 (1969).

3   S. G. Zelenski, N. Pangaro, and J. M. Hall-Enos, *Inventory of Organic Emissions from Fossil Fuel Combustion for Power Generation,* EPRI EA 1394, Apr. 1980.

4. B. Leckner, B. Jansson, O. Lindqvist, and B.-M. Nielsen, Emissions from a 16 MW$_{th}$ FBC Boiler. Presented at The 7th International Conference on Fluidized Bed Combustion, Philadelphia, Oct. 1982.

5. S. T. Cuffe, R. W. Gerstle, A. A. Ornig, and C. H. Schwarts, J. Air Pollut. Control Assoc. *14:*353 (1964).

6. R. P. Hangebrauck, D. J. von Lehmden, and J. E. Meeker, *Sources of Polynuclear Hydrocarbons in the Atmosphere,* U.S. Department of Health, Education and Welfare, Bureau of Disease Prevention and Environmental Control, Cincinnati, Ohio, 1967, p. 3.

7. S. T. Cuffe and R. W. Gerstle, *Emissions from Coal-Fired Power Plants: A Comprehensive Summary,* U.S. Department of Health, Education and Welfare, National Air Pollution Control Administration, Durham, N.C., 1967, p. 19.

8. E. K. Diehl, F. du Braenil, and R. A. Glenn, J. Eng. Power, 276 (1967).

9. P. W. Jones, VDI-Berichte *358:*23 (1980).

10. EPA, Fed. Regist. *36*(247):24888 (Dec. 23, 1971).

11. EPA, Fed. Regist. *41*(111):23076 (June 8, 1976).

12. A. Dravnieks, B. K. Krotoszynski, J. Whitfield, A. O'Donnell, and T. Burgwald, Environ. Sci. Technol. 5:1220 (1971).

13. A. Zlatkis, H. A. Lichtenstein, and A. Tishbee, Chromatographia, *6:*60 (1973).

14. J. P. Mieure and M. W. Dietrich, J. Chromatogr. Sci. *11:*559 (1973).

15. P. W. Jones, R. D. Giammar, P. E. Strup, and T. B. Stanford, Environ. Sci. Technol. *10:*806 (1976).

16. P. W. Jones, J. E. Wilkinson and P. E. Strup, *Measurement of Polycyclic Organic Materials and Other Hazardous Organic Compounds in Stack Gases,* EPA-600/2-77-202, 1977.

17. P. E. Strup, R. D. Giammar, T. B. Stanford and P. W. Jones, in *Carcinogenesis,* Vol. 1: *Polynuclear Aromatic Hydrocarbons: Chemistry, Metabolism, and Carcinogenesis,* (R. I. Freundenthal and P. W. Jones (Eds.), Raven Press, New York, 1976.

18. W. Feairheller, P. J. Mann, D. H. Harris and D. L. Harris, *Technical Manual for Process Sampling Strategies for Organic Materials,* EPA-600/2-76-122 1977.

19. J. W. Hamersma, S. L. Reynolds, and R. F. Maddalone, *IERL-RTP Procedures Manual, Level I Environmental Assessment* EPA-600/2-76-106a, 1977.

20. E. L. Merryman, A. Levy, G. W. Felton, K. T. Liu, J. M. Allen, and H. Nack, *Method for Analyzing Emissions from Atmospheric Fluidized-Bed Combustor,* EPA-600/7-77-034, 1977.

21. R. L. Hanson, R. L. Carpenter, G. J. Newton, and S. J. Rothenberg, J. Environ. Sci. Health A*14:*223 (1979).

22. R. L. Carpenter, S. Weissman, G. J. Newton, R. L. Hansson, E. R. Peele, M. H. Mazza, J. Kovach, P. A. Green, and U. Grimm, *Characterization of Aerosols Produced by an Experimental Fluidized Bed Coal Combustor Operated with Sub-bituminous Coal*, Lovelace Biomedical and Environmental Research Inst., Albuquerque, N. Mex., 1978.

23. K. Eriksson, *PAH, NO$_x$ and Metal Emissions from Coal- and Oil-Fired Plants*, Report 816427-04, ÅF-Energikonsult, Stockholm, July 1980 (in Swedish).

24. T. Ramdahl, I. Alfheim, S. Rustad, and T. Olsen, Chemosphere *11*: 601 (1982).

25. J. Bergström, *Emissions of PAH. Development of the sampling system*, Studsvik Technical Report, E2-81/12, Studsvik, Sweden, 1981 (in Swedish).

26. J. G. T. Bergström, G. Eklund, and K. Trzcinski, in *Proceedings of the 6th International Symposium on Polynuclear Aromatic Hydrocarbons*, Battelle, Columbus, Ohio, Oct. 27-29, 1981.

27. J. Adams, K. Menzies, and P. Levins, *Selection and Evaluation of Sorbent Resins for the Collection of Organic Compounds*, EPA-600/7-77-044, 1977.

28. M. B. Nehr and P. W. Jones, Anal. Chem. *49*:512 (1977).

29. R. L. Hansson, C. R. Clark, R. L. Carpenter, and C. H. Hobbs, Environ. Sci. Technol. *15*:701 (1981).

30. T. W. Sonnichsen, M. W. McElroy, and A. Bjørseth, in *Proceedings of the 4th International Symposium on Polynuclear Aromatic Hydrocarbons*, Batelle, Columbus, Ohio, Oct. 1979.

31. J. Jäger and V. Hannūs, J. Hyg. Epidemiol. Microbiol. Immunol. *24*(1): 1 (1980).

32. K. S. Murthy, J. E. Howes, H. Nack, and R. C. Hoke, in *Proceedings of the U.S. EPA Symposium on Process Measurements for Environment Assessment*, Atlanta, Ga. Feb. 13-15, 1978.

33. K. Kveseth, *Organic Analysis of Emissions from Coal- and Oil-Fired Plants*, Report 80 08 07-2, Central Institute for Industrial Research, Oslo, 1981 (in Norwegian).

34. A. Bjørseth, Anal. Chim. Acta *94*:21 (1977).

35. R. L. Hanson, R. L. Carpenter, and G. J. Newton, *Chemical Characterization of Polynuclear Aromatic Hydrocarbons in Airborne Exhaust Effluents from an Experimental Fluidized Bed Combustor*, Inhalation Toxicology Research Institute, Lovelace Biomedical and Environmental Research Institute, Albuquerque, N. Mex., 1979.

36. P. C. Siebert, R. Coleman, E. Burns Coffey, and C. Craig, *State-of-the-Art Estimation of Atmosphere POM Emissions and Exposures*, Energy and Environmental Analysis, Inc., Arlington, Va., AIChE Symp. Ser. 196, *76* (1980).

37. D. F. S. Natush and B. A. Tomkins, in *Carcinogenesis—A Comprehensive Survey*, Vol. 3: *Polynuclear Aromatic Hydrocarbons, Second International Symposium*, P. W. Jones and R. I. Freudenthal (Eds.), Raven Press, New York, 1978, p. 145.

38. B. Nilsson, *Emission from Firing of Coal and Oil*, Studsvik Technical Report, EK-81/26, Studsvik, Sweden, 1981. (in Swedish).

39. G. Eklund and B. Strömberg, *Emission of Organic Compounds in Flue Gas from the Asnaes Plant*, Studsvik Technical Report, EK-81/45, Studsvik, Sweden, 1981 (in Swedish).

40. K. Kveseth, *Organic Analysis of Effluents from the Asnaes Plant in Denmark*, Report 80 08 07-1, Central Institute for Industrial Research, Oslo, 1981 (in Norwegian).

41. B. Nilsson, *Emissions from Coal and Oil Firing. Flue Gas Measurements at Wilhelmshaven Power Station*, Studsvik Technical Report, EB-82/22, Studsvik, Sweden, 1982.

42. G. Nyström, *Perstorp, Boiler No. 3, The Emission Measurement*, Report 868694, ÅF-Energikonsult, Stockholm, 1981 (in Swedish).

43. G. Eklund and B. Strömberg, *Emission of Organic Compounds from the Perstorp Plant*, Studsvik Technical Report, EK-81/65, Studsvik, Sweden, 1981 (in Swedish).

44. K. Kveseth, *Organic Analysis of Effluents from the Coal-Fired Power Plant*, Report 80 08 07-8, Central Institute for Industrial Research, Oslo, 1981 (in Norwegian).

45. L. Gustafsson, *Emissions from Coal Firing*, Studsvik Technical Report, EK-81/47, Studsvik, Sweden, 1981 (in Swedish).

46. G. Eklund, *Emission of Organic Compounds in Flue Gas from Norrkoeping*, Studsvik Technical Report, EK-81/50, Studsvik, Sweden, 1981 (in Swedish).

47. K. Kveseth, *Organic Analysis of Effluents from the Coal-Fired Power Plant (Moving-Grate Boiler)*, Report 80 08 07-3, Central Institute for Industrial Research, Oslo, 1981 (in Norwegian).

48. G. Nyström, *H. C. Oersted Power Plant, Boiler No. 6, Emission Measurement*, Report 883458, ÅF-Energikonsult, Stockholm, 1982 (in Swedish).

49. G. Nyström, *Nilssons Handelstraedgaard. Hot Water Boiler. Emissions Measurement*, Report 402749, ÅF-Energikonsult, Stockholm, 1982 (in Swedish).

50. G. Eklund and B. Strömberg, *Analysis of Organic Compounds in Flue Gas from Plant I*, Studsvik Technical Report, EB-82/28, Studsvik, Sweden, 1982 (in Swedish).

51. K. Trzcinski, *Emissions from Coal-Firing*, Studsvik Technical Report, EK-81/61, Studsvik, Sweden, 1981 (in Swedish).

52. K. Kveseth, *Organic Analysis of Effluents from the Coal-Fired Plant Under Disturb Operation*, Report 80 08 07-4, Central Institute For Industrial Research, Oslo, 1981 (in Norwegian).

53. C. Andersson, *Coal-Health-Environment, Emission Measurements at Coal-Fired Pyroflow Boiler in Kauttua, Finland*. Studsvik Technical Report, EB-82/130, Studsvik, Sweden, 1982 (in Swedish).

54. K. S. Murthy, J. E. Howes, H. Nack, and R. C. Hoke, Environ. Sci. Technol. *13*:1 (1979).

55. K. Warman, *Organic Micropollutants in Stack Gases. Sampling, Analysis and Relation Between Emission of Carbon Monoxide and Hydrocarbons*, Studsvik Report, 83/1, Studsvik, Sweden, 1983.

56. P. Nielsen, *Emission of Polycyclic Aromatic Hydrocarbons (PAH) from Coal-Fired, Power Plant Boilers*, Dansk Kedelforening, Copenhagen, 1982 (in Danish).

57. T. Nielsen, *A study of the Reactivity of Polycyclic Aromatic Hydrocarbons*, Nordic PAH-Project, Report 10, Central Institute for Industrial Research, Oslo, 1981.

58. E. Brosström and A. I. Lindskog, *Degradation of Polycyclic Aromatic Hydrocarbons During Sampling*, Nordic PAH-Project, Report 12, Central Institute for Industrial Research, Oslo, 1981.

59. R. Bennett, K. T. Knapp, P. W. Jones, J. E. Wilkinson, and P. E. Strup, in *Proceedings of the Third International Symposium on Polynuclear Aromatic Hydrocarbons*, P. Lieber and P. W. Jones (Eds.), Ann Arbor Science, Ann Arbor, Mich. 1979.

60. D. F. S. Natusch, Environ. Health Perspect. *22*:79 (1978).

61. M. R. Guerin, *Energy Sources of Polycyclic Aromatic Hydrocarbons*, EPA, CONF-770-130-2, 1977.

62. F. E. Littman, R. W. Griscom, and O. Klein, *Regional Air Polluting Study, Point Source Emission Inventory*, EPA, EPA-600/4-77-014, 1977.

63. I. -B. Andersson, E. Björkholm, O. Lindqvist, E. Ljungström, and B. -M. Nielsen, Emissions from a fluidized bed combustor fired with Coal. Chalmers University of Technology and University of Gothenburg, Department of Inorganic Chemistry, Sweden, Oct. 1982.

3
PAH Emissions from Combustion of Biomass

THOMAS RAMDAHL* / Environmental Chemistry Department, Central
Institute for Industrial Research, Oslo, Norway

I.   Introduction   61

II. Combustion of Biomass   62

    A. Wood chemistry   62
    B. Wood combustion   62

III. Sampling   64

IV. PAH Emissions from Different Combustion Systems   64

    A. Residential stoves and fireplaces   64
    B. Central heating boilers   65
    C. Larger equipment for biomass combustion   76

V.   PAH Emissions from Open Burning of Biomass   77

    A. Forest fires   77
    B. Agricultural burning   77

VI. Possible Source-Specific PAH Compounds from Wood
    Combustion   77

VII. Ambient Impact of PAH from Wood Combustion   82

VIII. Conclusions   83

    References   83

I. INTRODUCTION

Many of the oil-importing countries of the world now suffer from the high
cost of imported petroleum. This has increased the search for alternative
energy sources, the combustion of biomass being one of them. Additionally,
in many of the developing countries, combustion of biomass is the major
source of energy today.

The simplest form of biomass is firewood, but other solid fuels, such as
bark, straw, wood chips, different kinds of pellets, and charcoal should be
included. Biomass may be used as fuel in a number of ways [1], but only
direct combustion is considered in this chapter. The direct combustion of
biomass often involves particulate and gaseous emissions to the atmosphere.

---

*Current affiliation: Statewide Air Pollution Research Center, University
of California, Riverside, California

This issue has been addressed in several symposia [2,3] and recent reports [4-9]. Combustion of biomass is often incomplete and polycyclic aromatic hydrocarbons (PAH) are formed as by-products, as shown in a number of studies [10-16]. In this chapter we discuss the PAH emissions from various biomass combustion units. We do not discuss the relevant literature at any length, but many pertinent examples are given. Representative emission factors for PAH for some relevant combustion processes are given in Chap. 1.

II. COMBUSTION OF BIOMASS

Biomass occurs in various stages of development: living, growing vegetation; dormant live vegetation; dead and dry vegetation; and naturally decomposing vegetative matter. Wood chemistry and combustion, as well as other biomass fuels on a general scale, are discussed in the following sections. A more detailed description may be found elsewhere [17].

A. Wood Chemistry

Wood is a complex material of chemically different components, including cellulose, hemicellulose, lignin, and a group of extractables (oils, pigments, minerals, and other organic substances). Cellulose, hemicellulose, and lignin constitute up to 90-95% of the weight of oven-dry wood. The woods of various species consist of the same principal components but in different proportions; the largest difference exists between the two main groups, conifers and deciduous trees. Conifers have a high proportion of lignin (28-34%); deciduous trees on average have less lignin (18-27%). The organic fraction of the extractables (5-10%) consists of many classes of compounds, including aliphatic and aromatic hydrocarbons, alcohols, aldehydes, gums, and sugars.

Although the number of organic compounds is large, 90-95% of the dry weight of wood is composed of the three components, with cellulose predominating. Thus there is a chemical limitation in its use as a fuel that would minimize the types and quantities of emissions during the pyrolytic decomposition and combustion of wood. The bulk of the emissions generated will be from these three components.

The moisture content of green wood (freshly cut) varies from 30 to 60%. Air drying for approximately 1 year reduces the moisture content to 18-25%. Wet bark may contain 80% or more moisture, while air-dried bark approximates 5-10%.

There are several properties of wood that relate to combustion and its attendant emissions. Dry wood is very hygroscopic and the amount of moisture absorbed depends mainly on the relative humidity and temperature. The exceptions are species having a high extractives content, such as cedars and redwoods. In green wood, the cell walls are saturated with moisture, while the cell cavities may be incompletely or completely full of water. The moisture in the cell walls represents "bound" water, while the moisture in the cell cavities is called "free" water. Free water removal has little or no effect on many of the properties of the wood, whereas the removal of bound water affects its properties.

B. Wood Combustion

The degradation of wood by heat may be divided into two different processes:

1. In pyrolysis, heating is done in the absence of oxygen, with resultant charcoal, organic tars, and gaseous emission products. Variations exist in the pyrolysis products of different trees and tree components, but the greatest variation exists between hardwoods and conifers.
2. In combustion, wood is heated in the presence of oxygen, and in complete combustion, the hot volatile combustible gases are in contact with sufficient oxygen, ideally producing only carbon dioxide, water vapor, and inorganic ash.

1. Pyrolysis

Pyrolysis is decomposition of wood by heating in the absence of air. The decomposition and the amount of resultant products depend mainly on the heating temperature, duration of heating, the surrounding medium, and the wood species. When wood is exposed to a heat flux, the wood heats up by conduction and when the surface layer becomes hot enough, water vapor starts to evolve. As heating progresses, the surface layer starts to char and other gaseous volatiles evolve. On large pieces of wood, as the interior becomes hotter, the pyrolysis effects go deeper, but the outward flow of volatiles convects heat back to the surface. Thus on very large pieces of wood, charring effects are observed on the surface, which acts as a thermal insulating barrier.

The hemicellulose deomposes first, then the cellulose, and then the lignin. The extractables evolve on the basis of their volatility and reactivity at the higher temperatures. The time-temperature relationship of heating and their products has been studied extensively. The course of pyrolysis can be presented on the basis of zones of heating in relation to the temperature applied. The initial heating results in an endothermic reaction where the gaseous products are largely noncombustible. Further heating results in an exothermic reaction (about 280-300°C) with the liberation of large amounts of carbon dioxide, carbon monoxide, and a liquid distillate containing acetic acid and its homologs, methanol, and light tars. Furthermore, high-temperature heating results in the production of hydrogen and heavy tars, including PAH.

2. Combustion

Combustion is the thermal degradation of the material in air. The volatile vapors escape from the surface and mix with oxygen, yielding a flame if conditions are right. A low-energy fire (270°C or less) undergoes intermolecular dehydration, resulting in a phenomenon called glowing combustion in which char and water vapor are produced. In the presence of oxygen, the char sustains glowing combustion, with the final products of carbon dioxide and more water vapor. A high-energy fire (340°C or higher) undergoes depolymerization, resulting in the phenomenon called flaming combustion, in which carbon monoxide and hydrogen are produced. In the presence of oxygen and an ignition source, a highly exothermic gas-phase reaction takes place with extensive flaming. The complete combustion of wood requires a heat flux (temperature gradient), adequate oxygen content (air supply), and sufficient duration.

Pyrolysis and combustion go hand in hand, and the burning conditions dictate the proportion of the types of emissions released into the atmosphere. PAH formation is inevitably a result of incomplete combustion of wood and other biomass, as of other carbonaceous materials.

III. SAMPLING

The objective of a PAH sampling method is that it be able to sample PAH
components from the source emissions in a representative, nondiscrimina-
ting, and nondestructive way. The sample should be suitable for trans-
ference to compatible solvents for chemical characterization. Sampling of
source emissions has been addressed in Chap. 2. Only a brief discussion
will be given here.

One important requirement of the method is that the amount of sample
be sufficient for carrying out a chemical analysis of PAH. The sampling
method and workup techniques must therefore be on a preparative scale.
The temperature of the emissions from combustion sources is most often in
the range of 50-400°C. This means that a significant part of the compounds
is in the vapor phase. The sampling system must therefore be designed to
collect all parts of the emission and can generally be divided into three parts:

1. A sampling probe and the particle collection system
2. A sampling system for compounds in the vapor phase (b.p. 150-300°C),
    most often solid sorbents
3. Gas flowmeter, flow control, and pumping equipment

Due to high concentrations of reactive gases and high temperatures there
are possibilities for chemical reactions during sampling. The sampling probe
should therefore preferably be made of glass, which is more inert than steel.
It is important that the temperature of the filter be above the dew point of
the emissions to avoid condensation and blocking of the filter. After passing
the filter, the sample flow is gradually cooled; the condensate is collected
by condensers and impingers and organic vapors are collected preferably
by solid sorbents such as Tenax GC or XAD-2 [18,19]. More comprehensive discussions of different source sampling methods have been published
[19,20]. Additional methods of extraction and analysis of PAH from emission samples have been discussed thoroughly in the first volume of this handbook [21].

IV. PAH EMISSIONS FROM DIFFERENT COMBUSTION SYSTEMS

The PAH emissions are very dependent on the combustion conditions and
therefore of the equipment used for combustion. In the following the PAH
emission from combustion of biomass is discussed as a function of this equipment.

A. Residential Stoves and Fireplaces

The traditional method of biomass combustion is the use of firewood in a
stove. Although stoves have been used for a long time, only recently has
their design been a matter of interest. This means that the majority of
wood stoves used for residential heating are of old and simple construction,
with no provision for complete combustion. A number of studies has shown
that the PAH emissions from such stoves may be considerable.

The PAH emissions from fireplaces and from baffled and nonbaffled stoves
determined by DeAngelis et al. [13] are given in Table 1. A large number
of PAH compounds were identified. From the numbers given it is clear that
emissions from stoves are larger than those from open fireplaces. This may
be explained by the higher combustion temperature and richer air supply in

fireplaces. There does not seem to be any marked difference in emission
levels with regard to the types of wood used; the combustion conditions seem
more important. The PAH emissions factors given in this study seem very
high compared to those in a study by Hubble et al. [14].

The results of the study by Hubble et al. are given in Table 2 and the
numbers are considerably less than those given in Table 1. One reason
for this could be that Hubble et al. did not use a solid sorbent for collection of vapor-phase PAH, only cold traps. This has been shown to be inefficient for collection of vapor-phase compounds and will result in too low
a figure. In addition, the combustion conditions may have been better in
Hubble's study and as no fuel was added during the tests, the emissions
were lower.

Reports by Rudling et al. [10] and Ramdahl et al. [15] also give PAH
emission levels lower than those given by DeAngelis et al. [13]. Table 3
give the number reported by Ramdahl et al. [15]. These samples were
taken under steady burning, and no fuel was added during the test. Here
the effect of different combustion conditions are shown. Burning spruce
under starved air conditions results in considerably higher emissions than
under normal air supply conditions. For the birch samples, an important aspect of sampling is demonstrated. The sample of normal burning of birch
shows the same level of PAH as the sample taken during the starved air
conditions. This may be explained by the fact that the normal burning of
birch was measured after the burning of birch under starved air conditions.
PAH may have been deposited in the stack with the creosote and been re-
evaporated during the later run with normal burning, as the stack was not
cleaned between the samples. This demonstrates the importance of a good
sampling strategy to obtain representative samples. A PAH chromatogram
of a wood combustion emission sample is given in Fig. 1. The figure demonstrates the variety of PAH compounds; however, only the major peaks in
the chromatogram are indicated.

There have been several attempts to reduce the PAH emissions from residential wood stoves. These include controlled secondary air systems, use
of smaller split wood, better insulation of the stove, and catalytic combustors [2,3]. To date, none of these modifications have demonstrated any
dramatic decrease in PAH emission levels and there is room for considerable
improvement.

A different approach to the emission problem is to refine the fuel. Charcoal is easily made from wood. This fuel is pure carbon with a very low
volatile compound content and a low water content. A specially designed
residential stove has been developed for this fuel which enables very efficient combustion [22]. The PAH emissions from this stove were less than
those from wood stoves, as shown in Table 3 [15]. This may be further
improved by the use of catalytic combustion. The emission is so clean that
it has also been proposed as a CO_2 source in greenhouses for growth enhancement [23].

B. Central Heating Boilers

Central heating boilers are often utilized for larger heating demands than
may be covered by residential stoves. Biomass fuels for such boilers may
include whole wood and wood chips, straw, and peat. Such boilers may be
simple through-burning furnaces with a horizontal cylindrical combustion
chamber. Recently, more advanced boilers have been developed, having a

Table 1. PAH Emission from Various Types of Residential Wood Combustion Facilities (g/kg fuel)[a]

Compounds	Fireplace			Baffled stove		Nonbaffled stove		
	Seasoned oak		Green pine POM train	Seasoned oak POM train	Seasoned pine POM train	Seasoned oak POM train	Green pine	
	POM train	SASS train					POM train	SASS train
Anthracene/phenanthrene	0.0082	0.0114	0.0069	0.0745	0.1463	0.0618	0.1034	0.0104
Methylanthracenes/phenanthrenes	0.0027	0.0034	0.0083	0.0211	0.0510	0.0167	0.0513	0.0028
C_2-alkylanthracenes/phenanthrenes	0.0014[b]	0.0011	0.0014	0.0040	0.0070	0.0045	0.0094	0.0008
Cylopentaanthracenes/phenanthrenes	0.0014[b]	0.0004	0.0014	0.0032	0.0086	0.0030	0.0047	0.0002
Fluoranthene	0.0014[b]	0.0026	0.0016	0.0180	0.0316	0.0208	0.0188	0.0012
Pyrene	0.0014[b]	0.0026	0.0016	0.0156	0.0240	0.0169	0.0188	0.0013
Methylfluoranthenes/pyrenes	0.0014[b]	0.0023	0.0016	0.0128	0.0167	0.0103	0.0142	0.0016
Benzo[ghi]fluoranthene	—	0.0009	0.0014	0.0048	0.0067	0.0047	0.0047	0.0004
Cyclopenta[cd]pyrene	0.0014[b]	0.0010	0.0014	0.0048	0.0089	0.0051	0.0138	0.0005
Benzo[c]phenanthrene	—	0.0004	0.0013	0.0016	0.0023	0.0016	0.0046	0.0002
Benz[a]anthracene/chrysene	0.0014[b]	0.0020	0.0014	0.0125	0.0138	0.0076	0.0371	0.0013

Methylbenzanthracenes/benzophenanthrenes/chrysenes	0.0014[b]	0.0013	0.0016	0.0062	0.0104	0.0062	0.0048	0.0009
C$_2$-alkylbenzanthracenes/benzophenanthrenes/chrysenes	—	0.0009	0.0014	0.0055	0.0044	0.0037	0.0047	0.0005
Benzofluoranthenes	0.0014[b]	0.0022	0.0016	0.0128	0.0159	0.0112	0.0141	0.0015
Benzopyrenes/perylene	0.0014[b]	0.0017	0.0014	0.0083	0.0116	0.0084	0.0094	0.0011
Methylcholanthrene	—	—	—	0.00007	—	—	—	—
Indeno[1,2,3-cd]pyrene/Benzo[ghi]perylene	—	0.0013	0.0015	0.0045	0.0099	0.0043	0.0048	0.0011
Anthanthrene	—	—	—	—	—	—	—	—
Dibenzanthracenes/phenanthrenes	—	0.0003	0.00005	0.0007	0.0014	0.0010	0.00005	0.0002
Dibenzocarbazoles	—	—	—	—	—	—	—	—
Dibenzopyrenes	—	0.0007	0.0001	0.0011	0.0010	0.0007	0.00002	0.0005
Total[c] (= column sum)	0.0249	0.0365	0.0360	0.2121	0.3715	0.18851	0.3187	0.0265

[a]POM train used an XAD-2 resin and cooler (reduction to 21°C) to trap organic gases. SASS (source assessment sampling system) train used three cyclones, a filter for particle size fractionation, an XAD-2 trap, and a trace inorganic impinger trap. Dash indicates not detected.
[b]Compound was identified but not quantified, because of the detection limits of the analytic method.
[c]The detection limit was taken as the emission factor for compounds that were identified but not qualtified.
Source: Ref. 13.

Table 2. PAH Emission from an American Residential Horizontally Baffled Updraft Wood Stove[a,b]

Compounds	Emission factor (mg/kg)	
	0.12-m logs, 0.82-kg/hr burn rate	0.06-m logs, 7.73-kg/hr burn rate
Phenanthrene/anthracene	0.88	2.30
C_1-Phenanthrenes/anthracenes	0.42	0.18
C_2-Phenanthrenes/anthracenes	0.11	0.04
Cyclopentaphenanathrenes/ anthracenes	ND	ND
Pyrene	0.33	1.39
Fluoranthene	0.25	0.10
Benzo[a]fluorene	0.26	0.26
Unidentified PAH	0.70	1.31
C_1-Fluoranthenes/pyrenes	0.10	ND
Benzo[b]fluorene	0.04	0.10
9-Phenylanthracene	0.04	0.20
C_3-Phenanthrenes/anthracenes	0.20	0.03
Benzo[ghi]fluoranthene	ND	ND
C_2-Pyrenes	0.08	ND
Cyclopenta[cd]pyrene	ND	ND
Benzo[c]phenanthrene	ND	ND
Benz[a]anthracene/chrysene	0.44	1.35
Higher molecular weight PAH	0.25	1.26
Total	4.10	8.52

[a]ND, not detected.
[b]Samples consisted of particulate matter, condensate, and volatiles collected in a cold trap.
Source: Data from Ref. 14.

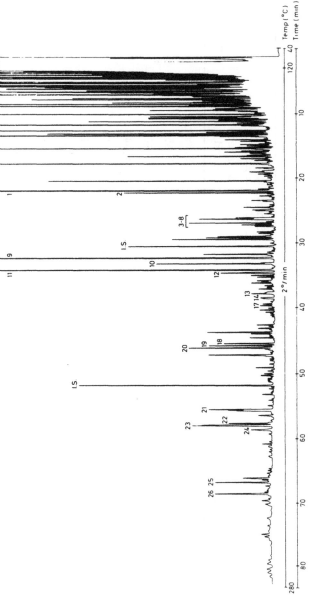

Figure 1. Chromatogram of the PAH fraction in the emission from burning birch in an airtight wood stove under a rich air supply. For peak identities, see Table 3. (Reprinted from Ref. 3 with permission from Pergamon Press.)

Table 3. PAH Emission from a Norwegian Horizontally Baffled Updraft Wood Stove and a Recently Developed Residential Charcoal Stove (μg/kg fuel)[a]

No.	Compound	Birch		Spruce		Charcoal
		Normal cond.	Starved cond.	Normal cond.	Starved cond.	
1	Phenanthrene	5,821	5,060	834	8,390	45.6
2	Anthracene	822	1,654	119	1,859	3.9
3	3-Methylphenanthrene	245	1,302	78	767	
4	2-Methylphenanthrene	627	925	116	1,046	
5	2-Methylanthracene	68	847	26	518	
6	4,5-Methylenephenanthrene	507	398	56	919	
7	4- and/or 9-Methylphenanthrene	162	792	42	986	
8	1-Methylphenanthrene	187	877	141	2,213	
9	Fluoranthene	2,843	1,334	296	3,245	16.0
10	Benz[e]acenaphthylene	688	423	74	1,465	1.2
11	Pyrene	2,690	1,488	232	3,822	20.1
12	Ethylmethylenephenanthrene	545	297	91		3.0

13	Benzo[a]fluorene	151	527	18	845	0.9
14	Benzo[b]fluorene	166	393	11	b	0.7
15	4-Methylpyrene	98	396		b	
16	2-Methylpyrene	100	838		950	
17	1-Methylpyrene		597		816	
18	Cyclopenta[cd]pyrene	381	169	40	720	
19	Benz[a]anthracene	476	326	60	781	
20	Chrysene and triphenylene	615	481	b	829	1.6
21	Benzo[b,j, and k]fluoranthenes	1,002	179	108	1,016	0.2
22	Benzo[e]pyrene	350	166	35	347	0.5
23	Benzo[a]pyrene	560	237	46	617	0.3
24	Perylene	247	25	23	38	
25	Indeno[1,2,3-cd]pyrene	415	170	33	291	0.2
26	Benzo[ghi]perylene	544	188	34	285	0.1
	Total identified PAH	20,870	20,085	2560	32,270	95.1

[a]Samples consisted of particulate matter, condensate, and volatiles trapped on XAD-2.
[b]Covered by interfering compound.
Source: Data from Ref. 15. Reprinted with permission from Pergamon Press.

secondary combustion chamber and higher combustion temperatures. In a recent report [24] three such boilers for whole wood were compared with regard to PAH emissions, and in addition a simple straw-heated boiler [16] was tested (see Table 4). PAH emissions from the boilers are given in Table 5. The different levels of PAH emission show great variation. The importance of water content in the fuel is demonstrated by the samples from boiler A. Sample 1 has a greater content of PAH than sample 2, due partly to higher water content (33% versus 12%). This was also shown by a higher CO level in the emissions (0.6% versus 0.3%). Sample 3 was taken under starved air, very poor combustion conditions (closed air supply, 1.5% CO), and the resulting PAH level is very high. The emissions from boiler A are greater than those from boilers B and C, probably a result of boiler design. Boiler A is a simple through-burning boiler, whereas boilers B and C utilize under-combustion and a secondary combustion chamber which leads to reduced organic emissions. The secondary air is supplied by a damper in boiler B and an electrical fan in boiler C. In this test the two systems seem to be quite similar in efficiency with regard to organic emissions.

The emissions from straw combustion in boiler D are quite high, although the fuel was continuously fed in small portions by a stoker. The high emissions are probably a result of poor boiler design, since boiler D is similar to boiler A. Comparing these two boilers, straw combustion results in lower emissions than those from whole wood combustion.

More refined fuel for biomass-heated boilers included wood chips and peat powder. A study by Alsberg and Stenberg [11] reports PAH emission from a 2-MW boiler fired with wood chips (the boiler was designed for oil combustion) and their results are presented in Table 6. The emissions are somewhat lower than those from wood-heated boilers with secondary combustion.

Table 4. Data for Different Biomass Fired Central Heating Boilers Tested[a]

	Boiler			
	A	B	C	D
Fuel	Wood	Wood	Wood	Straw
Effect, max. (kW)	75	40	25	55
Feeding	Manual	Manual	Manual	Stoker
Combustion chamber, (liters)	800	83	~100	800
Secondary combustion chamber	No	Yes	Yes	No
Secondary air supply	Damper	Damper	Fan	Damper
Combustion type	Through	Under	Under	Through

[a]The PAH emissions are given in Table 5.
Sources: Refs. 16 and 24.

Table 5. PAH Emissions from Different Types of Biomass-Fired Boilers ($\mu g/m^3$ flue gas)[a]

Boiler:	A			B		C		D	
Compound Sample:	1	2	3	4	5	6	7	8	9
Phenanthrene	4,980	3,200	36,700	210	95	70	150	1,200	350
Anthracene	770	370	7,500	30	15	20	70	140	90
3-Methylphenanthrene	190	170	2,200	10		4	20	5	20
2-Methylphenanthrene	490	250	2,600	55		6	30	60	30
2-Methylanthracene	50	90	1,200		1	1	10	70	20
4,5-Methylenephenanthrene	140	180	2,800	20	4	5	15	20	10
4- and/or 9-Methylphenanthrene	150	150	1,900	10	3	3	5	20	10
1-Methylphenanthrene	670	220	2,400	10	3	5	20	10	10
Fluoranthene	2,070	1,930	21,400	160	25	50	70	730	230
Benz[e]acenaphthylene	580	210	7,300	70	6	30	40	130	50
Pyrene	1,510	1,210	18,200	150	20	50	60	830	240
Ethylmethylenephenanthrene	450	360	3,600	20	10	20	10	60	30
Benzo[a]fluorene	360	170	4,100	25	3	5		5	30

Table 5. Continued

Compound	Boiler: A			B			C	D	
Sample:	1	2	3	4	5	6	7	8	9
Benzo[b]fluorene	190	120	2,800	10	3	5		5	20
Methylpyrenes		270	2,300	40		30	40	20	10
Benzo[ghi]fluoranthene	135	180	2,100	50	7	35	40	30	40
Cyclopenta[cd]pyrene		210	3,000	100	7	15	40	110	20
Benz[a]anthracene	250	240	3,700	90	6	40	50	60	20
Chrysene and triphenylene	510	320	5,100	100	10	35	40	140	50
Benzo[b,j, and k]fluoranthene	270	180	3,700	140	90	30	30	160	60
Benzo[e]pyrene	80	50	1,300	45	40	10	15	110	40
Benzo[a]pyrene	110	60	2,400	120	50	20	30	220	70
Perylene	10	5	260	30	10	4	10	20	10
Indeno[1,2,3-cd]pyrene	40	15	1,300	60	50	7	6	80	30
Benzo[ghi]perylene	30	15	1,600	60	40	4	10	150	60
Total	14,035	10,175	141,460	1,615	498	504	811	4,365	1,540

[a]Samples consisted of particulate matter, condensate, and volatiles trapped on XAD-2

Sources: Refs. 16 and 24.

Table 6. PAH in Emission from a 2 MW Boiler Burning Wood Chips and Peat Briquettes $(\mu g/m^3)$ [a]

Compound	Sample number, wood				Sample No. 6, peat
	3[b]	4	5	Mean	
Phenanthrene	51.1	220.4	58.4	110	1030
Anthracene	5.25	16.4	3.23	8.3	83.7
Fluoranthene	28.3	71.7	41.0	47	444
Pyrene	21.5	60.3	32.7	38	366
Cyclopenta[cd]pyrene	0.49	3.31	1.72	1.8	0.70
Benz[a]anthracene	5.75	9.36	5.23	6.8	108
Chrysene/triphenylene	8.21	11.9	8.13	9.4	121
Benzo[b and k]fluoranthenes	11.6	14.2	8.45	11	120
Benzo[e]pyrene	5.25	8.28	2.46	5.3	
Benzo[cd]pyrenone	6.67	12.0	5.51	8.1	19.7
Benzo[a]pyrene	2.39	4.69	1.32	2.8	9.76
Perylene	0.53	0.72	0.21	0.49	3.89
Indeno[1,2,3-cd]pyrene	0.92	2.97	0.53	1.5	35.0
Benzo[ghi]perylene	1.21	3.97	0.78	2.0	22.9
M_w 300	≤0.15	≤0.15	≤0.06	—	3.48
Coronene and M_w 302	≤0.11	≤0.21	≤0.11	—	24.0
Total	150	440	170	252	2400

[a]Samples consisted of particulate matter, condensate, and volatiles trapped on XAD-2.
[b]Line washings not included.
Source: Ref. 11. Reprinted with permission from Pergamon Press.

Similar and lower emission levels have been measured in a 35 kW chip-fired boiler with a precombustion chamber. Levels from 220 to less than 8 $\mu g/m^3$ were reported under normal burning conditions [10]. Wood chips are therefore preferred to whole wood as a fuel for central heating purposes. Other refined fuels, such a pellets and briquettes of wood and straw, are now being introduced on the market. With the good combustion equipment that exists for these fuels, the PAH emissions should be lower than those cited above. However, no emission studies have yet been reported.

Peat is another form of biomass fuel [25], although it belongs to the renewable resources only in the geological time scale [26]. Both peat briquettes and powder have been used as fuels in central heating boilers. The PAH emissions from the briquette combustion [11] are given in Table 6, and a typical PAH chromatogram is shown in Fig. 2. The emissions from this fuel were considerably higher than those of wood chip combustion. The reason for this is apparently the combustion conditions [11]. The combustion temperature varied to a greater extent when burning peat. This indicates a nonuniform combustion which may lead to increased PAH formation. There was also a greater difference between the combustion zone temperature and the flue gas temperature than with wood chips [11], which again may be favorable for PAH formation [27].

C. Larger Equipment for Biomass Combustion

The emissions from larger biomass combustion facilities, such as a bark-fired 4 MW district heating plant and a fluidized-bed combustor burning peat

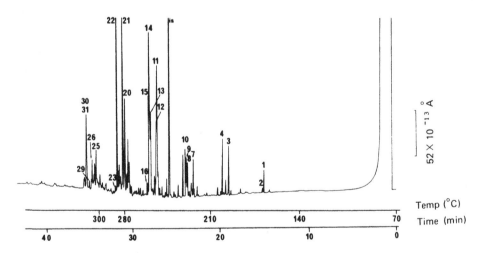

Figure 2. Chromatogram of the PAH fraction of filter-collected particles from combustion of peat briquettes in a central heating boiler. Peak identities: 1, phenanthrene; 2, anthracene; 3, fluoranthene; 4, pyrene; 7, benzo[ghi]fluoranthene; 8, cyclopenta[cd]pyrene; 9, benz[a]anthracene; 10, chrysen/triphenylene; 13, benzo[e]pyrene; 14, 6H-benzo[cd]pyren-6-one; 15, benzo[a]pyrene; 16, perylene; 20, indeno[1,2,3-cd]fluoranthene; 21, indeno[1,2,3-cd]pyrene/dibenz[a,h]anthracene; 22, benzo[ghi]perylene; 30, coronene; IS, internal standard. (Reprinted from Ref. 11 with permission from Pergamon Press.)

briquettes, was low, less than 50 $\mu g/m^3$ [28]. In the emissions from two Finnish 100 and 200 MW peat-powder power plants, no PAH were detected in the emissions (i.e., less than 100 ng/m^3 per individual compound) [28, 29]. This is due to more efficient combustion and flue gas cleaning [30], which is not possible (economical) on smaller combustion units. This shows that with finely distributed fuel, the emissions from biomass combustion may be as low as those from coal-fired power plants (see Chap. 2 and Ref. 31). As the combustion equipment size increase, they must be treated more or less as any other large combustion plant for coal or oil with respect to flue gas cleaning, particle removal, and so on.

V. PAH EMISSIONS FROM OPEN BURNING OF BIOMASS

The PAH emissions from open burning of biomass may also be considerable. Sources include forest fires, prescribed burning, and agricultural waste burning.

A. Forest Fires

Forest fires are both natural and man-made (accidental and intensional) events that apparently contribute significantly to the PAH content of the atmosphere. Large areas of forest are consumed every year by wild fires [32-34]. PAH emissions from forest fires varies widely, due to the variety of fuels, fire type (backing fires, with the fire line moving against the wind, or heading fires, with the fire line moving with the wind), fire intensity, and combustion phase (smoldering or flaming). McMahon and Tsoukalas [34] have performed several laboratory experiments to study PAH formation when burning slash pine needle litter as fuel, and their results are given in Table 7. The PAH present on the particles show that the heading fire produced higher amounts of particles but smaller total amounts of PAH.

B. Agricultural Burning

Large quantities of agricultural wastes, such as sugarcane, orchard prunings, and grain straws, are typically disposed of by open burning in every country around the world. In addition, backyard burning of biomass wastes occurs frequently [9]. The organic constituents of rice straw smoke from open burning has been studied by Mast et al. [35], and they found considerable amounts of PAH in the smoke. The high PAH content of barley straw smoke from an incinerator reported by Ramdahl and Møller [16] further supports the belief that agricultural burning may be a significant contributor to PAH in the atmosphere. An emission factor for this source is given in Chap. 1.

VI. POSSIBLE SOURCE-SPECIFIC PAH COMPOUNDS FROM WOOD COMBUSTION

Source-specific PAH compounds have been proposed for a number of sources, such as cyclopenta[cd]pyrene for automobile exhaust [36], benzo[b]naphtho-[2,1-d]thiophene for hard coal combustion [36], and picene derivatives for brown coal combustion [37].

On the basis of the study of several analyses of PAH in the emissions from wood combustion, a significant and unique peak has been identified in the

Table 7. PAH from Burning Pine Needles by Fire Type, Simulation of Forest Fire (ng/g of Fuel Burned, Dry Weight Basis)[a,b]

Compound	Backing fires			Heading fires		
	0.1 lb/ft^2	0.3 lb/ft^2	0.5 lb/ft^2	0.1 lb/ft^2	0.3 lb/ft^2	0.5 lb/ft^2
Anthracene/phenanthrene	12,181	2,189	584	2,525	5,542	6,768
Methylanthracene	9,400	1,147	449	1,057	4,965	7,611
Fluoranthene	14,563	2,140	687	733	974	1,051
Pyrene	20,407	3,102	1,084	1,121	979	1,133
Methylpyrene/fluoranthene	18,580	2,466	1,229	730	1,648	2,453
Benzo[c]phenanthrene	8,845	1,808	468	244	142	175
Chrysene/benz[a]anthracene	28,724	5,228	2,033	581	543	836
Methylchrysene	17,753	1,891	877	282	1,287	1,559
Benzofluoranthenes	12,835	1,216	818	164	129	241

Benzo[a]pyrene	3,454	555	238	38	40	97
Benzo[e]pyrene	5,836	1,172	680	61	78	152
Perylene	2,128	198	134	33	24	46
Methylbenzopyrenes	6,582	963	384	65	198	665
Indeno[1,2,3-cd]pyrene	4,282	655	169	—	—	—
Benzo[ghi]perylene	6,181	1,009	419	—	—	—
Total PAH	171,750	25,735	10,249	7,632	16,549	22,787
Total suspended particulate matter TSP (lb/ton)	21	9	5	20	73	118
Benzene-soluble organic substances (%)	55	50	45	44	73	75

[a] Moisture content for all fires ranged between 18 and 27%.
[b] Samples consisted of particulate matter sampled on a glass fiber filter using a modified Hi-Vol sampler.
Source: Ref. 34. Reprinted with permission from Raven Press, New York

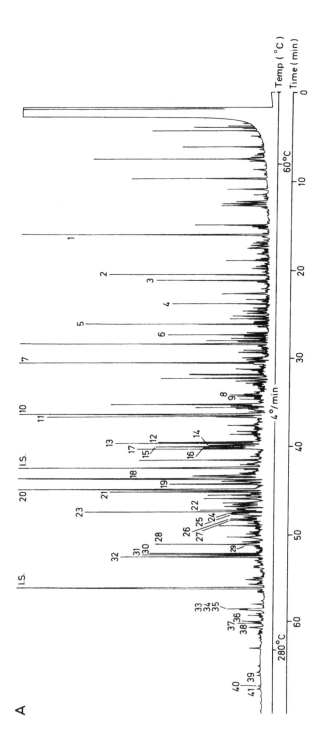

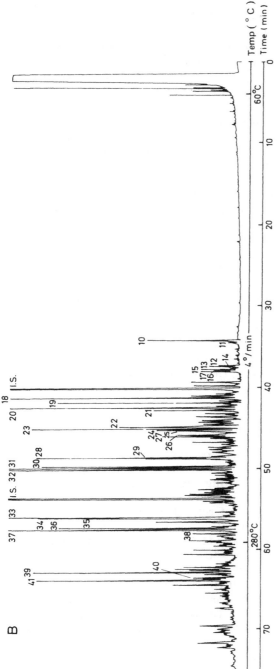

Figure 3. Chromatograms of the PAH fractions from (A) a wood (spruce, *Picea abies*) combustion emission sample from a 50 kW hot water boiler, and (B) an ambient air sample from Elverum, Norway [44]. Peak identities: 1, naphthalene; 2, 2-methylnaphthalene; 3, 1-methylnaphthalene; 4, biphenyl; 5, acenaphthylene; 6, acenaphthene; 7, fluorene; 8, 2-methylfluorene; 9, 1-methylfluorene; 10, phenanthrene; 11, anthracene; 12, 3-methylphenanthrene; 13, 2-methylphenanthrene; 14, 2-methylanthracene; 15, 4,5-methylenephenanthrene; 16, 9- and/or 4-methylphenanthrene; 17, 1-methylphenanthrene; 18, fluoranthene; 19, acephenanthrylene; 20, pyrene; 21, ethylmethylenephenanthrene; 22, benzo[a]fluorene; 23, 1-methyl-7-isopropylphenanthrene (retene); 24, benzo[b]fluorene; 25, 4-methylpyrene; 26, 2-methylpyrene; 27, 1-methylpyrene; 28, benzo-[ghi]fluoranthene; 29, benzo[c]phenanthrene; 30, cyclopenta[cd]pyrene; 31, benz[a]anthracene; 32, chrysene and triphenylene; 33, benzo[b]fluoranthene; 34, benzo[j]fluoranthene; 35, benzo[k]fluoranthene; 36, benzo[e]pyrene; 37, benzo[a]pyrene; 38, perylene; 39, indeno[1,2,3-cd]pyrene; 40, dibenz[a,c and/or a, h]anthracenes; 41, benzo[ghi]perylene; IS, internal standards. (Reprinted by permission from Nature 306:580–582. Copyright © 1983 Macmillan Journals Limited [37].)

PAH chromatograms from combustion samples of spruce (*Picea abies*) [38]. This is shown in Fig. 3A. This compound (peak 23) was identified by gas chromatography/mass spectroscopy (GC/MS) in comparison with an authentic standard as 1-methyl-7-isopropylphenanthrene (retene), a compound first isolated from fossil pine in 1837 [39]. The compound has been identified in sediments from all over the world [40-42] and in low levels in aerosols over the rural United States [43]. It is reported to originate from abietic acid, a diterpenoid in the resin of coniferous wood, by natural degradation. Ramdahl's study [38] showed the presence of retene in emissions from coniferous wood combustion, and that it is probably formed by thermal degradation of abietic and similar resin acids in the wood during combustion. It was shown that the formation of retene was very temperature dependent. At a low combustion temperature (260°C) inside the stove and a low air supply level, the thermal degradation of abietic acid was a dominant process and retene was the most abundant PAH formed. At a higher combustion temperature (i.e., higher than 500°C inside the stove) and a richer air supply, the abietic acid is more easily combusted and the yield of retene is lower. Retene is apparently not formed by the usual radical pyrosynthesis, which is the generally accepted mechanism for the formation of PAH from combustion of any carbonaceous material [44].

As combustion of coniferous wood seems to be the only source of retene in ambient air, this compound may serve as a molecular marker for this source. This has been demonstrated in the air of Elverum, Norway [45], a town with a high level of wood combustion (discussed further below). The peak representing retene is significant in the PAH chromatograms from ambient air samples from this town, clearly indicating that wood combustion makes a significant contribution to air pollution (see Fig. 3B).

VII. AMBIENT IMPACT OF PAH FROM WOOD COMBUSTION

As discussed above, the PAH emissions from residential wood combustion may be considerable and may therefore affect the ambient air quality. Several attempts have been made at estimating these impacts [5,7,46-50]. Some of these are modeling efforts; others present chemical analyses of air in wood-heated areas but in most cases not related to an evaluation of source emissions. The most promising quantitative method of assessing the ambient impact of wood combustion is the carbon-14 (^{14}C) method [47]. Due to the 5730-year half-life of ^{14}C, the ratio of ^{14}C activity per gram of carbon to the ^{14}C activity per gram of carbon in a modern carbon standard is a measure of the fraction of contemporary ("living") carbon in the air particulate sample. It is important to study only small particles in this context, because the larger ones may be biogenic (e.g., fractions of plants, pollen, and insects). This should, however, be a small problem during winter measurements.

In a study by Ramdahl et al. [45] this technique has been applied in combination with detailed chemical analysis, including PAH and source emission inventory for the first time. The study performed in Elverum, Norway (10,000 inhabitants), during January-February 1982 showed that on the average wood carbon contributed 65% to the carbon in ambient fine particles. With knowledge of the fuel consumption of the different emission sources, PAH from wood combustion were calculated by PAH emission factors (see Chap. 1) to contribute 64% to the total ambient PAH, in good agreement with the ^{14}C data. The PAH concentrations measured in Elverum were comparable to those observed in Oslo, the capital of Norway (500,000 inhabitants). Considering that wood combustion contributes only about 5% to the heating demand in Elverum, the results of this study clearly demonstrate the polluting potential of residential wood combustion.

VIII. CONCLUSIONS

The combustion of biomass, especially wood, in small units may result in significant PAH emissions to the atmosphere. The information resulting from these facts is forming an essential part of the basis for regulatory strategies for residential wood heating in Scandinavia and the United States (e.g., in Oregon) [51]. In Sweden solid fuels are not recommended for primary heating of houses situated in the most densely populated areas. In less populated areas, solid-fuel boilers have to be certified. Based on extensive monitoring of the emissions from boilers of varying design and size, tar (proportional to PAH) has been suggested as the index substance to be controlled (measured as amount of dichloromethane-extractable material), and 10 mg of tar per MJ of fuel input has been defined as the emission limit for boilers with a capacity of less than 30 kW. A sampling method and an analytical procedure have been recommended [52]. For boilers used for secondary heating and for residential stoves, no limit is presently suggested, but it is recommended that all models marketed should be tested to provide a data base for future emission limits. It will also be demanded that stoves and boilers be constructed so as to prohibit pyrolytic burning (starved air conditions). For residential wood heating units to be used in rural districts, no certification will be required. It is hoped that this policy will stimulate further development of the more efficient and less polluting stoves which recently have been put on the market, thereby reducing the environmental impact of the use of biomass fuels for heating.

Larger units for biomass combustion show PAH emissions at the same level as corresponding fossil-fueled plants, and thus the same stack gas cleaning requirements apply. PAH emissions from these plants seem not to be a major source of atmospheric PAH.

REFERENCES

1. L. P. White and L. G. Plaskett, *Biomass as Fuel*, Academic Press, London, 1981.
2. J. A. Cooper and D. Malek (Eds.), *Residential Solid Fuels*, Oregon Graduate Center, Beaverton, Oreg., 1982.
3. E. R. Frederick (Ed.), *Residential Wood and Coal Combustion*, APCA SP-45, Louisville, Ky., 1982.
4. S. S. Butcher and E. M. Sorenson, J. Air Pollut. Control Assoc. *29*: 724 (1979).
5. J. A. Cooper, J. Air Pollut. Control Assoc. *30*:855 (1980).
6. S. S. Butcher and M. J. Ellenbecker, J. Air Pollut. Control Assoc. *32*:380 (1982).
7. J. M. Dasch, Environ. Sci. Technol. *16*:639 (1982).
8. M. A. K. Khalil, S. A. Edgerton, and R. A. Rasmussen, Environ. Sci. Technol. *17*:555 (1983).
9. S. A. Edgerton, M. A. K. Khalil, and R. A. Rasmussen, J. Air Pollut. Control Assoc. *34*:661 (1984).

10. L. Rudling, B. Ahling, and G. Löfroth, in *Residential Solid Fuels*,
    J. A. Cooper and D. Malek (Eds.), Oregon Graduate Center, Beaverton,
    Oreg., 1982, p. 34.

11. T. Alsberg and U. Stenberg, Chemosphere 7:487 (1979).

12. W. D. Snowden, D. A. Alguard, G. A. Swanson, and W. E. Stolberg,
    Source Sampling Residential Fireplaces for Emission Factor Development,
    EPA-450/3-76-010, Research Triangle Park, N. C., 1975.

13. D. G. DeAngelis, D. S. Ruffin, and R. B. Reznik, *Preliminary Charac-
    terization of Emissions from Woodfired Residential Combustion Equip-
    ment*, EPA-600/7-80-040, Washington, D.C., 1980.

14. B. R. Hubble, J. R. Stetter, E. Gebert, J. B. L. Harkness, and R. D.
    Flotard, in *Residential Solid Fuels*, J. A. Cooper and D. Malek (eds.),
    Oregon Graduate Center, Beaverton, Oreg., 1982, p. 79.

15. T. Ramdahl, I. Alfheim, S. Rustad, and T. Olsen, Chemosphere *11*:
    601 (1982).

16. T. Ramdahl and M. Møller, Chemosphere *12*:23 (1983).

17. F. P. Kollmann and W. A. Cote, Jr., *Principles of Wood Science and
    Technology*, Vol. I, Springer-Verlag, New York, 1968.

18. E. D. Pellizzari, J. E. Bunch, B. H. Carpenter, and E. B. H. Sawicki,
    Environ. Sci. Technol. *9*:552 (1975).

19. P. W. Jones, R. D. Giammar, P. E. Strup, and T. B. Stanford, Environ.
    Sci. Technol. *10*:806 (1976).

20. L. D. Johnson and R. G. Merril, Toxicol. Environ. Chem. *6*:109 (1983).

21. A. Bjørseth (Ed.), *Handbook of Polycyclic Aromatic Hydrocarbons*,
    Marcel Dekker, New York, 1983.

22. S. Rustad and T. Olsen, in *Residential Solid Fuels*, J. A. Cooper and
    D. Malek (Eds.), Oregon Graduate Center, Beaverton, Oreg., 1982,
    p. 984.

23. L. Mortensen, Gartenbauwissenschaft *47*:14 (1982).

24. T. Ramdahl, G. Tveten, and A. Osvik, in *Proceedings of Bioenergy '84*,
    A. Ellegård and H. Egnéus (Eds.), Applied Science Publishers, Barking,
    Essex, England, 1984, in press.

25. O. Lindström, Ambio *9*:309 (1980).

26. H. Sjörs, Ambio *9*:303 (1980).

27. M. Blumer, Sci. Am. *234*:34 (1976).

28. T. Ramdahl, Central Institute for Industrial Research, Oslo, unpub-
    lished results, 1983.

29. L. Rudling and G. Löfroth, *Emissions from Combustion of Peat and Wood
    Chips*, National Swedish Environmental Protection Board (SNV), PM
    1449, Solna, Sweden, 1981 (in Swedish).

30. K. L. Foster, R. C. Scherr, and R. E. Dickson, J. Air Pollut. Control
    Assoc. *32*:872 (1982).

31. I. Alfheim, J. G. T. Bergström, D. Jenssen, and M. Møller, Environ.
    Health Perspect. *47*:177 (1983).

32. G. Yamate, *Developement of Emission Factor for Estimating Atmospher-
    ic Emissions from Forest Fires*, EPA-450/3-73-009, Research Triangle
    Park, N.C., 1973.

33. P. J. Crutzeu, L. H. Heidt, J. P. Krasnec, W. H. Pollock, and W.
    Seiler, Nature *282*:253 (1979).

34. C. K. McMahon and S. N. Tsoukalas, in *Carcinogenesis*, Vol. 3: *Poly-
    nuclear Aromataic Hydrocarbons*, P. W. Jones and R. I. Freudenthal
    (Eds.), Raven Press, New York, 1978, p. 61.

35. T. J. Mast, D. H. Hsieh, and J. N. Seiber, Environ. Sci. Technol. *18*:
    338 (1984).

36. G. Grimmer, K. -W. Naujak, and D. Schneider, Z. Anal. Chem. *311*: 475 (1982).

37. G. Grimmer, J. Jacob, K. -W. Naujak, and G. Dettbam, Anal. Chem. *55*:892 (1983).

38. T. Ramdahl, Nature *306*:580 (1983).

39. J. B. Trommsdorff, Liebigs Ann. Chem. *21*:126 (1837).

40. B. R. T. Simoneit, Geochim. Cosmochim. Acta *41*:463 (1977).

41. S. G. Wakeham, C. Schaffner, and W. Giger, Geochim. Cosmochim. Acta *44*:415 (1980).

42. R. E. Laflamme and R. A. Hites, Geochim. Cosmochim. Acta *42*:289 (1978).

43. B. R. T. Simoneit and M. A. Mazurek, Atmos. Environ. *16*:2139 (1982).

44. G. M. Badger, *Natl. Cancer Inst. Monogr. 9*:1 (1962).

45. T. Ramdahl, J. Schjoldager, L. A. Currie, J. E. Hanssen, M. Møller, G. Klouda, and I. Alfheim, Sci. Total. Environ. *36*:81 (1984).

46. J. A. Cooper and J. G. Watson, Jr., J. Air Pollut. Control Assoc. *30*:1116 (1980).

47. J. A. Cooper, L. A. Currie, and G. A. Klouda, Environ. Sci. Technol. *15*:1045 (1981).

48. J. F. Hornig, R. H. Soderberg, D. Larsen, and C. Parravano in *Residential Solid Fuels*, J. A. Cooper and D. Malek (Eds.), Oregon Graduate Center, Beaverton, Oreg., 1982, p. 506.

49. D. J. Murphy, R. M. Buchan, and D. G. Fox, in *Residential Solid Fuels*, J. A. Cooper and D. Malek (Eds.), Oregon Graduate Center, Beaverton, Oreg., 1982, p. 495.

50. R. E. Imhoff, J. A. Manning, W. M. Cooke, and T. L. Hayes, in *Residential Wood and Coal Combustion*, E. R. Frederick (Ed.), ACPA SP-45, Louisville, Ky., 1982, p. 161.

51. J. F. Kowalczyk and B. J. Tombleson, *A Potential Woodstove Certification Program*, Oregon Dept. of Environmental Quality, Portland, Oreg., 1982.

52. B. Timm, Swedish Environment Protection Board, Solna, Sweden, personal communication, 1983.

4
PAH Emissions from Automobiles

ULF R. STENBERG* / Department of Analytical Chemistry,
University of Stockholm, Stockholm, Sweden

I. Introduction   87

II. Sampling Methods   88

    A. Sampling from undiluted exhausts   88
    B. Proportional sampling of undiluted exhausts   91
    C. On-line measurements   93
    D. Sampling of diluted exhausts   93
    E. Gas-phase trapping of PAH   96
    F. Particle trapping   99

III. Factors Affecting Emissions   100

    A. Preconditioning of vehicle   100
    B. Starting conditions   100
    C. Ambient temperature   101
    D. Driving cycles   101
    E. Accumulation of PAH in oil   104
    F. Air/fuel ratio   104
    G. PAH content and aromaticity of fuel   106
    H. Comparison of PAH emission between differently fueled
       vehicles   107

IV. Conclusions   108

    References   108

I. INTRODUCTION

Motor traffic is a significant source of ambient polycyclic aromatic hydrocarbons (PAH) in urban areas, the emission coming from gasoline and diesel vehicles, as well as from small two-stroke engines such as mopeds and lawn mowers. The latter, however, are not regarded as important sources except on very rare occasions or locations. In this chapter we describe the development of sampling technology for PAH from mobile sources from both a historical point of view and in terms of the different strategies utilized. Parameters that influence emissions are also discussed.

The PAH include a large number of species representing a wide range of boiling points and volatility. Even if these compounds all have boiling points exceeding 200°C, they are distributed between the gas and particulate phase in emissions from an internal combustion engine, and the enrichment technique used has to be adjusted accordingly.

*Current affiliation: KabiVitrum AB, Stockholm, Sweden

II. SAMPLING METHODS

As early as the late 1920s some studies revealed that certain PAH were carcinogenic compounds [1]. At an early stage it was also shown that extracts from urban dust [2] showed carcinogenic properties when injected subcutaneously in mice, and that this effect could possibly be attributed to the PAH content [3]. However, Campbell's [4] animal experiments in the mid-1930s, with direct exposure to automobile exhausts, were negative concerning lung cancer-causing effects.

This direct method of trying to develop adverse effects in animals was hampered due to the concentration of carbon monoxide in the exhaust gases, which implied that the emissions must be diluted, and Campbell found that the amount of particles was very low in the exposure chamber. Since in previous experiments with dust from roads he had found an increased cancer rate in experimental animals, Campbell concluded that the particle emissions in vehicle experiments would be of importance. Later, in the beginning of the 1950s, when the emission from vehicles became a more widespread problem, new approaches were taken to assess the biological significance of this pollutant.

Sample collection and the validity of the sample obtained are two crucial factors when making calculations of emission factors from different sources. An advanced analytical procedure of high precision is of little use if the sample analyzed does not reflect the "true" composition of the emission of interest. The conditions that have to be fulfilled for a reliable sampling procedure are:

Collection must be quantitative.
No artifact should be formed.
The interface between collection and analysis must be free from interference and losses.

A. Sampling from Undiluted Exhausts

Studies published from the mid-1950s to the early 1970s often involved very bulky sample equipment which necessitated voluminous extraction and clean-up procedures. The reasons for this were that large sample amounts were required for existing analytical methods, and that all biological tests were performed with skin painting tests on intact animals.

The particle content of the emissions attracted most of the attention during early investigations in this field, although some attempts were made to enrich the gaseous part of the exhausts. Kotin et al. [5] constructed a sampling device, applicable on both diesel and gasoline engines, using a filter in combination with a cooled gas-trap device filled with carbon black. However, the system was not described in detail and no evaluation tests seem to have been performed. The system was used for the identification of various carcinogenic PAH, but no data were reported from the gas phase of PAH. Later, Lyons and Johnston [6] using hamp sacks connected directly to the tailpipe, collected at least some of the particle emission from "normal motoring." Mittler and Nicholson [7] enriched some of the gaseous PAH in both gasoline and diesel exhausts by condensation; however, no particulate emission was recovered.

These pioneer studies indicated that to be quantitative, trapping of PAH from vehicle exhausts must utilize a two-stage process. Thus in 1961, Stenburg et al. [8] presented a comparison of two sampling methods. One was called wet method and comprised washing flasks (impingers) connected prior to a filter (i.e., the emission was drawn through cooled ethanol before the particles were collected). In a second, dry method, the emission was cooled in a steel condenser before filtering. However, this method was discarded in favor of the first. In addition to evaluation experiments with impingers, they executed parallel sampling with different filter temperatures. Significant losses of benzo[a]pyrene were observed at a temperature of 150°C compared to 15°C, and temperatures below 35°C gave comparable results. Stenburg et al. omitted the dry process due to difficulties in recovery from the condenser and other drawbacks not specified. Later, however, this condensing technique was to become widely used for emission studies of automobile exhausts.

In 1962, Begeman [9] described a sampling system that enabled the collection of both gaseous PAH and particle-associated PAH, and the conclusion was that a large water condenser placed in front of the filter package should be sufficient to collect PAH quantitatively. The trapping system was dimensioned to enrich the total exhaust volume from a 364 in.^{3} (6 liter) V-8 engine operated in a chassis dynamometer. This system is shown in Fig. 1.

The steel condenser consisted of (286) tubes cooled by circulating water. Depending on the load (the driving cycle), two condensors could sometimes be connected in series. This type of condenser was reported to be very efficient, and the exhaust temperature when leaving the system was close to that of the circulating water (15-22°C), compared to an inlet temperature of 268-538°C. During the driving scheme used (80 hr with continuous cycles), the pressure drop in the condenser increased form 60 mmHg to 250 mmHg.

A large filter package (2.4 m^2) connected after the condensing system yielded only a small pressure drop, 4 mmHg, at the conditions given above. This large sample required a laborious recovery and extraction procedure. The condensed water was extracted with a countercurrent liquid-liquid extraction column (5 X 200 cm), using benzene as the extraction medium. The filters were Soxhlet-extracted (17 liters) with benzene-methanol (4:1) with approximately 20 siphoning actions in 30 hr.

To obtain the exhaust tar, the solvent (several gallons) was removed by vacuum distillation. The authors defined "tar" as residue that was benzene soluble and nonvolatile at 35°C and 0.1 mmHg. They also noted that the condensated water from the exhausts was acidic, pH 2-3. This type of sampling system with "total condensation" has also been used by other research groups [10-13].

A mobile sampling system in which the equipment was built into the vehicle and driving was performed on ordinary roads was presented by Hangebrauck et al. in 1966 [14]. However, this system had the same disadvantages as those described earlier, including laborious extraction procedures. The condenser could be immersed in a salt-ice mixture and a good cooling efficiency was reported; 3-27°C was the entrance temperature on the filter. The condensed water increased the back pressure in the system, and with the reported maximum resistance of 15 mmHg, approximately 5% of the engine power was lost. How this may have affected the emission of PAH was not reported. However, the possibility of obtaining samples under real driving conditions should be emphasized.

It is noteworthy that for the sampling systems described, no information is at hand concerning optimization arrangements or recovery data. In the early work almost no consideration was paid to the possibility of volatile losses from filter, artifact formation, or degradation during sampling. Some of these aspects were taken into account by Grimmer et al. when in 1972 they [15] presented a modified sampling system for the collection of the total

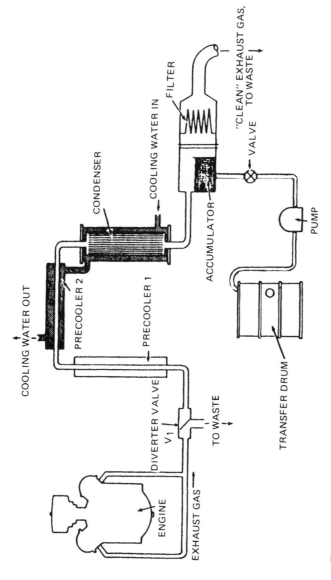

Figure 1. Sampling system for total exhaust flow. Condensation prior to filtering. Condenser: 1.3 m². (From Ref. 9.)

gasoline exhaust volume. They chose a vertical glass condenser with a cooling area of $0.25\ m^2$ (Figure 2). The reason for having a glass device was that it was easier to control the washing procedure, and that the catalytic effect on the condensed material was expected to be lower. An evaluation of the collecting arrangement was performed and some general conclusions drawn:

The entrance temperature on the filter should not exceed 35°C if quantitative trapping of three-ring PAH is sought.

Prolonged sampling, in this case repetitive sampling during ECE test procedure, will involve losses of certain species due to volatilization or degradation.

To increase filter efficiency, the filters could be impregnated with paraffin.

Paraffin-impregnated filters were used later by other investigators, especially for ambient air sampling. However, certain precautions must be taken when using this method since it is possible that the paraffin layer will adsorb other species from the gas stream, and these components might cause degradation of certain PAH [16].

Compared to earlier enrichment methods, this scaled-down equipment is less laborious to use and has been employed in a large series of analyses [17,18] for the assessment of the "total" PAH emission from gasoline vehicles. However, the dimensions of the condenser made it difficult to handle the emission from a vehicle operated in another driving cycle [e.g., federal test procedure (FTP) 1972], since the cooling capacity was too low.

An improved vertical glass condenser has recently been presented by Volkswagen in Germany for trapping the total exhaust volume [19]. The condenser, as illustrated in Fig. 3, has a cooling area of approximately $3\ m^2$. This large area permits sampling in both FTP and ECE-15 (see Sec. III. D). Extraction is performed by boiling acetone in a round-bottomed flask at the bottom of the system and using a condenser at the top, thus creating a continuous extraction with less solvent than that used in earlier methods. The system has been used for both diesel and gasoline vehicles.

Handa et al. [20] used a total condensation system, and ascertained that the filter temperature was below 15°C. By connecting additional cooling equipment at −196°C and extra filters, they concluded that their original construction was efficient to collect PAH with four rings or more. Furthermore, they found no significant breakdown of the PAH determined during sampling.

B. Proportional Sampling of Undiluted Exhausts

Collection of the total exhaust volume is an expensive method that is time consuming in terms of the subsequent analytical work. One way to circumvent the need for large sampling equipment is by the use of proportional sampling. This technique allow the use of a small sample, and has become an increasingly interesting alternative with the development of more sensitive analytical methods and biological tests in vitro.

A proportional sampling system has been described by Chipman and Massey for the sampling of gaseous pollutants such as HC, CO, and NO_x [21]. However, the same type of equipment can be used for the collection of PAH, thus decreasing the size of the condensing area required. A proportional sampling system uses a reference signal that should reflect the amount of emitted exhaust. This is obtained most easily by measuring the air passing

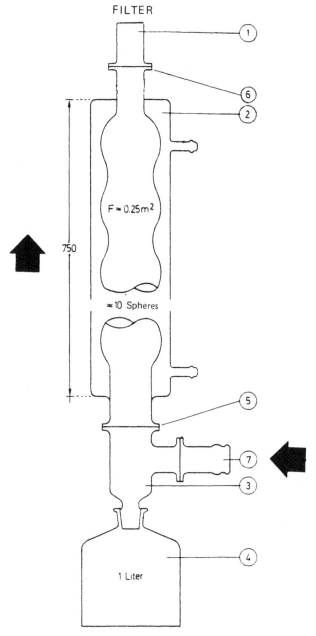

Figure 2. All-glass sampling system for total exhaust flow. Condensation prior to filtering. Condenser: $0.25 \, m^2$; filter: $0.6 \, m^2$. 1. Connection (80 mm) to filter; 2. condenser 750 mm; 3, connection, T-piece; 4, 1-liter bottle for condensate collection; 5 and 6, Teflon gaskets; 7, connection for exhaust pipe. (From Ref. 15.)

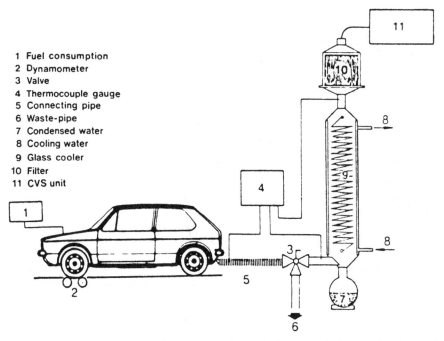

1 Fuel consumption
2 Dynamometer
3 Valve
4 Thermocouple gauge
5 Connecting pipe
6 Waste-pipe
7 Condensed water
8 Cooling water
9 Glass cooler
10 Filter
11 CVS unit

Figure 3. All-glass sampling system for total exhaust flow. Condensation prior to filtering. Condenser: $3\,m^2$; filter: $0.6\,m^2$. (From Ref. 19.)

through the carburetor of the engine, as shown in Fig. 4. This flow can be determined by a laminar flow element, thus creating a reference signal which is compared to a signal obtained from a similar element in the sample stream [22].

Spindt [23] applied this technique for the collection of diesel particles and gas phase trapping on Chromosorb 102 absorbent. Stenberg et al. [22, 24] used a proportional system connected to a glass condenser followed by a glass fiber filter. This equipment was used in conjunction with tests in different driving cycles with gasoline vehicles, and in spite of the small cooling area of the condenser ($0.1\,m^2$), it allowed the filtering temperature to be decreased to approximately 50°C. Typically, about 10% of the total exhaust volume was sampled and the PAH emission was determined from both FTP 1972 and ECE-15 driving cycles (see also Sec. III. D).

C. On-Line Measurements

Most PAH exhibit fluorescence. If they are illuminated with ultraviolet (UV) light, this incoming energy will be adsorbed and will transfer the molecule to an excited state. On returning to their normal energy level, transmission of visible light (fluorescence) will occur, which can be detected with a photo-multiplier. The advantage of using a fluorescence technique is the specificity, which can be used for on-line detection [25].

D. Sampling of Diluted Exhausts

The sampling techniques previously described all utilized raw exhausts (i.e., the emission is sampled directly at the tailpipe of the vehicle prior to dilution

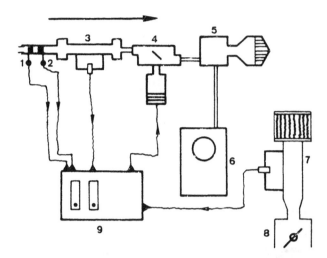

Figure 4. Proportional pump and feedback system. 1, temperature sensor; 2, pressure sensor; 3, laminar flow element; 4, valve; 5, vacuum pump; 6, dry gas meter; 7, laminar flow element including air filter; 8, carburetor; 9, electronics. (From Ref. 22.)

with ambient air), and it has been questioned whether this sampling point really reflects the true exposure to the environment. Thus it is feasible to assume that certain components will react immediately when dispersed into the air, and that the particle composition will be altered during the dilution process. This supposed behavior was one of the factors that stimulated construction of the first dilution tunnel.

Originally, dilution tunnels were built to measure lead emissions from gasoline vehicles [26,27]. However, the use of such tunnels has been extended, and today it is used in almost all measurements of automotive emissions, especially for unregulated pollutants in diesel emissions. The equipment comprises a tube of stainless steel through which filtered air is drawn by means of a large pump. The exhausts are introduced at an orifice to achieve sufficient mixing and the diluted exhausts are sampled isokinetically with a probe. The sampling point must not be situated less than 10 tunnel diameters from the inlet point, and the temperature must not exceed 125°F (52°C). Specifications according to the U.S. *Federal Register* are given in Ref. 28, and technical aspects are available in Ref. 29. Some of the details concerning the use of a dilution tunnel are discussed in the following paragraphs [30].

Condensation of Water Vapor

Gasoline exhausts contain about 10% of water by weight, and during a sampling procedure that includes a temperature decrease, this water will condense. Upon filtering, the water droplets formed can cause problems by making the filter fragile and by creating anomalies as to physical particle characterization. Condensation is easily prevented by dilution, and further lowering of the temperature is achieved with reduced collection of water. The water content of the exhausts varies during the driving cycles, but a dilution factor of 4 will prevent the dew point of water from being reached at actual temperatures (52°C).

Proportional Sampling

To obtain an appropriate sample during a cyclic operation, the dilution tunnel is connected to a constant volume sampler (CVS). This means that when a small amount of gas is emitted from the vehicle, a large proportion of dilution air is used, and vice versa, thus achieving a constant flow through the tunnel. To obtain a true constant flow, both the mixing air and the sample stream can be thermostated.

The sample can be taken isokinetically; that is, the sample stream through the probe should have the same velocity as the total flow in the tunnel. This will prevent enrichment of different particle sizes. The importance of isokinetic sampling, however, is not fully shown in the literature concerning automobile exhausts.

There are two essential parameters that have to be fulfilled for the tunnel to function satisfactorily:

1. Mixing capacity
2. Velocity profile

Assistance is required to achieve mixing of exhaust and dilution air over the comparatively short transporting distance involved, and this must not disturb the velocity profile by creating too much turbulence. Turbulence will create deposits and therfore must be kept as low as possible.

To investigate the mixing characteristics, either a tracer gas can be injected (propane has been suggested) [27] or use can be made of CO_2 originating from a test vehicle. The concentration of the gas is then measured along the axis over the diameter at the sampling point. The same procedure can also be used for particle measurements both on a weight basis and for size distribution.

The size of the tunnel will restrict the types of vehicles that can be tested. However, this can be circumvented in part by diverting the raw gas before introducing it to the tunnel. This is necessary when testing large diesel engines [29]. With the employment of the dilution technique it is anticipated that gas-phase components, among these certain PAH, will adsorb on particles, or that certain chemical reactions will take place before sampling. This will give a sample of "truer" composition. The residence time within a tunnel differs, but is probably less than 2-3 sec in most cases. Very few data are available in the literature to support this hypothesis, however, and no results had been presented covering parallel sampling of undiluted and diluted exhausts prior to those published in Ref. 31. Figure 5 shows the increased adsorption during dilution. However, it can also be seen that this process by no means is quantitative, especially regarding three- and four-ring PAH. That leads to the conclusion that a particle sample from gasoline exhausts alone will lead to significant underestimation of the emission (see also Sec. II E).

Another approach to sampling PAH has been published by Newhall et al [32]. After dilution of the exhaust in a small tube (0.1 X 5 m), approximately, 10% of the exhaust-air mixture was taken through two condensers in series, followed by a scrubbing tower and finally, an absolute filter. Recovery studies of radioactive benzo[a]pyrene (BaP) showed a yield of better than 90% and the authors emphasized that the trapping at subambient temperatures should minimize degradation processes.

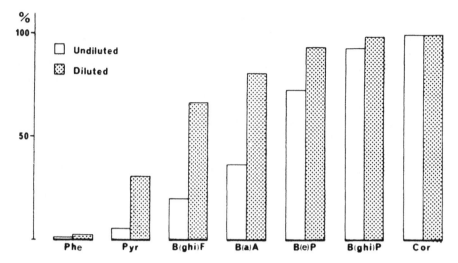

Figure 5. Comparison of percent particle associated PAH in gasoline exhausts.
Federal test procedure. Phe, phenanthrene; Pyr, pyrene; BghiF, benzo-
[ghi]fluoranthene; BaA, benz[a]anthracene; BeP, benzo[e]pyrene; BghiP,
benzo[ghi]perylene; Cor, coronene. Filter temperature: 45°C (diluted);
105°C (undiluted). (From Ref. 22.)

E. Gas-Phase Trapping of PAH

Enrichment of the gas phase after dilution is a problem that has been largely
overlooked; only a few papers deal with this question. The gas phase consists of compounds that are not retained by a Teflon filter at a temperature
not higher than 125°F. The enrichment methods employed have been either
adsorbent trapping [23,33] or a cryogenic technique [31,34]. Spindt [23]
used Chromosorb 102 and searched for BaP in diesel exhausts. Lee et al.
[33] used XAD-2 traps on gasoline emissions and found that a large proportion of three- and four-ring PAH was present in the gas phase. However,
they also reported problems with background from the adsorbent, probably
due to degradation.

A combination of impinger flasks filled with water, a Teflon filter, and an
adsorbent trap (XAD-2) has been used by Snow [35]. The conclusion was
that the flasks were not an effective collecting medium and that most of the
volatiles (>75%) were collected on the adsorbent. However, no special reference to PAH was given and the results were based on weight analysis.

Pedersen et al. [34] described briefly a cryogenic technique consisting
of two washing flasks immersed in liquid nitrogen. The authors concluded
that five-ring PAH and larger molecules were adsorbed to 100% on particles
from gasoline emissions. The other species analyzed (anthracene, fluoranthene, and pyrene) predominated in the gas phase. A cryogenic technique
has also been used on gasoline vehicles by Handa et al. [20], but no information is given of the relative distribution of gas-phase and particle-associated PAH, or of the construction of the system.

A gas-phase system that can be used on both diluted and undiluted exhausts has recently been described [31]. The system, illustrated in Fig. 6,
utilizes three condensers (ice-water, CO_2-ethanol, and liquid nitrogen).
This system has been demonstrated to have good trapping efficiency for com-

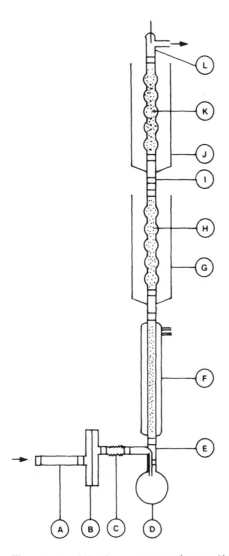

Figure 6. All glass cryogenic gradient sampling system for diluted exhausts or proportional flow of raw exhausts. A, Flexible metal tubing, 250 mm; B, filter holder; C, Teflon bellow; D, 1-liter flask; E, normal joint; F, ice/ water cooled condenser; G, cryogenic vessel, dry ice/ethanol; H, condenser cartridge; I, ball joint; J, cryogenic vessel, liquid nitrogen; K, condenser cartridge; L, normal joint, temperature sensor. (From Ref. 31.)

pounds more volatile than PAH. When only PAH are of interest in subsequent analyses, it is possible to exclude the condenser cooled with liquid nitrogen. The distribution of PAH between the gas and particle phases from a gasoline vehicle is presented in Table 1.

Stump et al. [36] have described a large-scale gas-phase trapping system for diluted gasoline exhausts. They used XAD-2 adsorbent, with mutagenicity as the quantitative tool, and concluded that only a negligible part of the mutagenic effect could be detected after the filter. These results are at variance with those obtained by Rannug [37] using the condensation technique described previously.

Table 1. Distribution (%) of Some PAH Between the Gas
Phase and Particles in Gasoline and Diesel Exhausts[a]

Compound	Gasoline		Diesel	
	Gas	Part.	Gas	Part.
Phenanthrene	95	5	53	47
Fluoranthene	36	64	9	91
Pyrene	25	75	6	94
Benzo[ghi]fluoranthene	6	94	<1	>99

[a]Samples taken in a dilution tunnel at approximately 30°C.

A comparison of cryogenic trapping and the adsorbent technique shows
that both methods have advantages. Handling is easier with adsorbents.
No cooling arrangements are necessary, and the sampling volume can be
altered by changing only the amount of adsorbent. High flow velocities
can be achieved due to the low pressure drop. Cryogenic trapping, on the
other hand, is restricted in sampling volume. If a large cooling area is
needed, the apparatus would be cumbersone and the demand for coolant
increases. Our experience with the sampling system described above indicates that larger equipment would be laborious to handle if the same cooling
technique were used. The maximum sampling limit is 100 liters/min with maintained flow during 20 min of diluted exhausts (1:10), but only 75 liters/min
has been used in most of the investigations.

Using a glass surface as a condensing area makes it easy to obtain a
good blank before sampling. In addition, it withstands reactions with NO/
NO_2, thus can be used several times and gives no artifact products. Adsorbents, on the other hand, have to be washed carefully prior to use.
More than 24 hr of continuous extraction with acetone or dichloromethane
is necessary if XAD-2 is used as the sampling matrix. XAD-2 is sensitive
to reactive gases (e.g., NO/NO_2) and degrades during sampling, giving
rise to artifacts. The products may be toxic, which makes subsequent
testing (e.g., for mutagenicity) uncertain or in some cases impossible.
Hanson et al. [38] have shown the occurrence of several benzene derivatives, such as phenols, acids, and aldehydes. They also point out that the
reaction product is mutagenic, and that some of these breakdown products
also result from combustion processes.

The trapping on a cold surface will give rise to an aqueous extract which
needs to be further treated before analysis. In addition, it cannot be excluded that this process per se forms reaction products, thus yielding anomolous results. Experimentally, we have found that a diluted (1:10) exhaust
volume of 200-300 liters gives approximately 10-20 ml of water with pH 3-5.
This precipitated water contains dissolved or dispersed gaseous compounds,
which for the subsequent analysis must be transferred into an organic solvent. PAH and alkyl derivatives are quantitatively extracted with DCM from
an aqueous solution that represents the "gas phase." The recovery of polar
derivatives of PAH from water solutions is not quantitative with DCM extraction, and thus has to be developed further.

With the use of this sampling system, a significant difference between gasoline- and diesel-fueled vehicles in the distribution of PAH between gas and particulate phase has been demonstrated (Table 1). For the latter, only 50% of three-ring PAH is in the gas phase, and four-ring PAH, such as fluoranthene and pyrene, comprise 5-10% in the gas phase. PAH with molecular weights $\geqslant 226$ are up to 99% particle-associated in diesel emissions.

Gas-phase enrichment has been discussed within the Coordinating European Council CEC/CF-11 working group, with the conclusion that the present literature does not support the selection of any specific method. Adsorbents such as XAD-2 react with nitrogen oxides and may produce artifacts; and condensation techniques may produce artifacts due to the acidic condensate. In general it was recommended that the efficiency of the total trapping system employed should be checked using a backup system of appropriate design.

F. Particle Trapping

Filtering is the technique most often utilized for trapping particles from automobile exhausts. Other methods are cyclones [34] or electrostatic precipitation [39], but these methods are much more complicated than conventional filtering and the advantages are not yet obvious. A comparison of parallel sampling on filters and electrostatic precipitation (ESP) has been presented [36]. The sample stream was undiluted diesel exhausts and the particles recovered were extracted and the extract fractionated. As a quantitative/qualitative tool, mutagenicity according to Ames was used. Small differences could be noted, but the authors were not able to distinguish if formation of more active species had occurred on the filter or if deactivation occurred in the ESP sample.

There are several factors that have to be taken into account when selecting filter media, including efficiency, pressure drop, background, vapor adsorption, and inertness toward the components of interest. In the following, the latter will be discussed further.

Glass fiber filters have the advantage of low pressure drop, good efficiency, and resistance toward reactive gases prevalent in automobile exhausts. They are also low in cost and easily obtainable from a number of manufacturers. However, their inertness has lately been questioned, especially for the sampling of PAH in automobile exhausts [40] and probably even for ambient air sampling [41].

In 1980, Lee et al. [40] reported an extensive investigation of filter media, focusing especially on PAH and their behavior during sampling. Recoveries of ^{14}C- labeled BaP and benz[a]anthracene were studied. The order of recovery for BaP was: Teflon membrane, Teflon coated, and glass/quartz fiber. Storage-dependent breakdown was also evaluated and it was concluded that Teflon filters were superior to glass fiber types. However, the test method was somewhat obscure since the radioactive component was adsorbed directly on the filter material, and probably the results are not transferable to real sampling conditions.

In another experiment, particle-loaded filters were used (ambient air partilces) and ^{14}C-labeled BaP was applied. Afterward, further ambient air was passed through and the recovery of BaP was calculated. It was found to be very low, 2-26%, and as the reaction products were hydroxy forms, diols of BaP were suggested.

For certification purposes (emission of diesel particles) the U.S. Environmental Protection Agency (EPA) has stipulated the use of Teflon-coated filters. These types of filters have also been prevalent in research concerning automobile emissions, especially for unregulated pollutants. An extensive investigation of the physical parameters of various filter media is presented in Ref. 42.

III. FACTORS AFFECTING EMISSIONS

A. Preconditioning of Vehicle

Prior to a test it is necessary to bring all vehicles to the same status (i.e., they must be conditioned). For routine measurements on gaseous emissions such as HC and CO there are standardized methods for this purpose. For the measurement of PAH, however, no procedure has been set up.

PAH emission is in the form of particles, which may be deposited on valves, the manifold, the exhaust system, and other areas. To circumvent this carryover effect, it has been suggested that the vehicle should be driven under heavy load for a certain amount of time before testing. This can be accomplished either on the road or on a chassis dynamometer. Because of heat formation, it is preferable to perform the conditioning on the road.

The importance of the deposits has been investigated [10] and the following conclusions can be drawn from these data:

The use of a high-PAH fuel increases the deposits of these substances.
Newly formed deposits give higher PAH emissions, which are continuously
     reduced when using a fuel with a low PAH content.

In another investigation [18] an attempt was made to compare the total emission with the PAH deposits (including what was found in the lube oil), and the conclusion was that only a minor part (about 1%) is deposited in the engine, compared to what is found in the emission and the lube oil. However, the authors emphasized the need for the vehicle to achieve steady state after changing the fuel (i.e., it is necessary to have a substantial driving period with the new fuel if this is a parameter of interest).

During discussions in the CEC/CF-11 working group, no proposal has been submitted for a particular preconditioning routine. However, it has been recommended that the oil should not be changed, and that the vehicle should have at least a 3000 km running-in period. It was also suggested that prior to testing there should be a driving period under heavy load, to circumvent carryover effects. The time and conditions should be stated in the report for the specific work performed. To date no accurate information is at hand concerning the driving period required, but 45-60 min at approximately 90 km/hr could be used. Sampling should be performed after at least 6 hr of standing at 22-25°C.

B. Starting Conditions

PAH emissions are dependent on the starting temperature [10,17,22,24]. However, the differences between hot and cold start are discussed only for temperatures specified in the regulations [i.e., +23°C (cold start) and approximately +80°C (hot start)]. Consequently, very little is known about emissions from vehicles started below 0°C, but the few results available indicate that starting at -10°C involves three-fold higher PAH emissions than those associated with a standardized cold start [24].

C. Ambient Temperature

It is well documented that the PAH content in air increases during the winter [43-45]. This has been attributed to the lower level of UV radiation, which minimizes degradation [46], as well as increased emissions from heat and energy generating plants. However, the phenomenon can also be found in areas where heat and energy generation takes place far from city-center areas. In this case, the increase of PAH in the air must have another source.

Motor vehicles are a significant PAH source [47] and during the winter the cause of increased PAH emissions in urban areas. As discussed later, the air/fuel ratio of the engine is a significant parameter for this effect. Most data representing mass emission rates from gasoline-powered vehicles come from driving done for certification purposes [17,19,24], which means that the temperature conditions are favorable (about 23°C) and thus the vehicle quickly reaches its working temperature. To obtain information on temperature-dependent emission, driving has to be performed in a closed room where it is possible to maintain temperatures below 0°C during a complete driving cycle. Previous investigations [48] concerning the effect of temperature on diesel vheicle emissions have shown that this factor is of little importance as regards particle emission and the mutagenicity of particle extracts. This is also in good agreement with the general consensus that diesel vehicles are rather insensitive to ambient temperature as regards driveability, and that CO and HC emissions are comparatively unaffected by temperature variability.

Gasoline engines are sensitive to temperature, however, especially concerning CO emission. On the assumption that there is a relationship between CO and the PAH emission, the latter should be expected to increase when driving at subambient temperatures. Recently, such experiments have been conducted [49] and a short summary of the results is given below. In this investigation several subambient temperatures were used, but here the discussion will be confined to -5°C and with a comparison of the standard +23°C.

Six different cars were tested; four were standard carburetor-equipped vehicles, one was fueled injected, and one was fuel injected with lambdasond* and a three-way catalyst (this car was run on unleaded gasoline). The results are summarized in Fig. 7. It should be noted that there is a significant difference in mass emission rates between the catalyst-equipped vehicle and the highest-emitting standard vehicle. In some tests this difference was more than one order of magnitude (i.e., the catalyst car emitted much less PAH at -5°C than do some standard vehicles at +23°C). The results show that the increased PAH levels in ambient air during the winter could be an effect of low-temperature driving conditions.

Another effect of low temperature driving conditions was that the distribution of gas and particle phases shifted toward higher particle levels. The relative amount of PAH per particle emission also increased with decreased temperature.

D. Driving Cycles

All measurements of automobile emissions have to be performed at conditions as close as possible to those of real driving. To achieve this, a chassis dyna-

---

*Oxygen-sensitive feedback system to achieve an optimal air/fuel ratio.

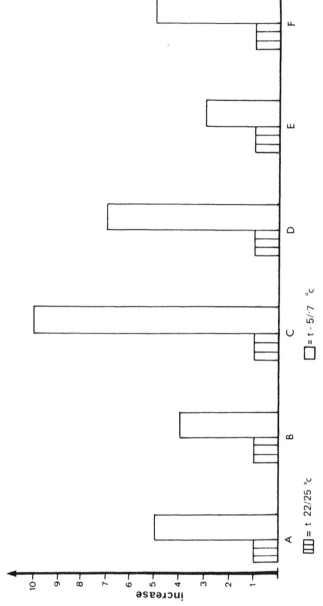

Figure 7. Normalized emission of particle associated PAH at different driving temperatures using Federal Test Procedure, FTP-1972. A-D, carburetor vehicles; E, fuel-injected vehicle; F, fuel-injected, three-way catalyst. (From Ref. 49.)

Driving Schedule (Cycle)   4 times repeated   Distance: 4 052 km

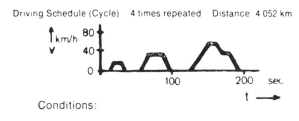

Conditions:

1. Cold start (6 h preconditioning min)
2. The three phases are repeated four times

Figure 8. European test cycle, 1970.

mometer is utilized. The test vehicle is kept in a room and the driving is performed on rollers. Due to different emission standards and driving habits, different driving modes are used in several countries. In Europe (except for Sweden and Switzerland), the ECE-15 test procedure is used (Fig. 8). Sweden and United States use the same driving mode (Fig. 9), and Japan has its own mode (Fig. 10). The latter two involve higher speeds and higher loads than those of the European test procedure.

There are also different regulations for hot and cold start, although by definition cold start is when the vehicle has been standing for at least 12 hr in a room at approximately 22-25°C. There are few publications that compare the various driving modes in terms of PAH emissions. Kraft and Lies [19] did not find any consistency in the results for gasoline engines, but did find some indication that emissions (µg/km) were lower at cold start in the FTP than in the ECE. For diesel vehicles the difference was more pronounced, which was reported in a later publication [50]. Stenberg [24] reported, concerning gasoline vehicles, higher emissions in the FTP than in the ECE when using hot start.

From a comparison of the data available it appears that the driving cycle does not play an important role in the emission level. PAH emissions from different investigations are shown in Table 2. The good agreement among the results, despite differences in driving cycles, sampling methods, analytical procedures, and emission control devices, is noteworthy.

Driving Schedule   Distance: 11.09 mi

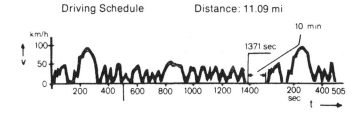

Conditions:

1. Cold start (12 h preconditioning min)

Figure 9. Federal test cycle, 1975.

11 Mode-cycle

Driving Schedule   4 times repeated
Distance: 1.02 km/schedule

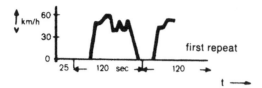

first repeat

Conditions:

1. Cold start
2. Driving cycle is repeated 4 times

Figure 10. Japanese test cycle, 1976.

E. Accumulation of PAH in Oil

PAH from gasoline and from combustion will be enriched in the motor oil
[10,18,51]. Experiments have shown that fresh oil with virtually no PAH at
the start will have its PAH content increased threefold during normal use
for 10,000 km [51]. However, there was no correlation between PAH emission with the exhausts and the amount of PAH in the lubricating oil (i.e.,
the accumulation of PAH in the oil during use did not influence the emissions).
On the other hand, Handa et al. [52] showed a correlation between PAH e-
missions and oil consumption; emissions increased with mileage.

F. Air/Fuel Ratio

The air/fuel ratio (λ) has a significant influence on PAH emissions [10,24,
34,53,54]. When running the engine on a stochiometric level of air and fuel,
this is defined as $\lambda = 1$. In most cases, PAH emissions will increase with
richer mixtures (>1), as will CO emissions. Furthermore, Pedersen et al.
[34] found that when λ reached 0.8, bad ignition led to a sharp increase
in PAH emissions, while small variations around $\lambda = 1$ did not affect the emissions.

Attempts have also been made to correlate PAH emission with CO or HC
(unburned gaseous hydrocarbons), but they do not present a consistent
picture. A high emission of CO [18] will not always be followed by a high
emission of PAH. Choking at cold start will increase CO emissions during
the first portion of the driving cycle, but over the entire driving period
the contribution is of less importance. PAH emissions (Fig. 11) are affected
much more by the cold start procedure (See also Sec. III C); thus the increase is more significant. This effect could possibly be attributed to the
passage of unburned fuel through the combustion chamber.

Alcohol-blended fuels also seem to have some influence on PAH emissions.
This can in part be ascribed to the leaning-out effect that occurs when using
such fuels with ordinary carburetor tuning [22]. Other results show a significant reduction in PAH emissions when using alcoholic fuel at moderate
blend of 15-20% [18].

Table 2. Emission of Some Selected PAH from Gasoline Exhausts (μg/km)

	Investigation[a]							
	A	B	C	D	E	F	G	H
Pyrene	–	54	–	60	184	–	43	–
Benz[a]anthracene	3.5	9	–	–	–	9	–	–
Benzo[a]pyrene	1.5	2.5	3.2	6	10	7	5.3	14.5
Coronene	–	–	–	1.5	20	16	9.3	–

[a]A: Gross (1972) [10]: condensing prior to filtering; undiluted exhausts; highway driving: 5-7 tests
B: Handa et al. (1979) [20]: cryogenic condensing prior to filtering; undiluted exhausts; 10 modes per steady state; 5 vehicles.
C: Williams and Swarin (1979) [60]: filter sample; diluted exhausts; cold start FTP 1975; 7 vehicles.
D: Kraft and Lies (1981) [19]: condensing prior to filtering; undiluted exhausts; cold start ECE-15; 3 vehicles.
E: Grimmer and Hildebrandt (1975) [17]: condensing prior to filtering; undiluted exhausts; cold start ECE-15; 100 vehicles.
F: Stenberg (1979) [24]: condensing prior to filtering; undiluted exhausts; cold start FTP 1972; 13 vehicles.
G: Hangebrauck et al. (1966) [14]: condensing prior to filtering; undiluted exhausts; real driving; 4 vehicles.
H: Zweidinger (1981) [61]: filter sample; diluted exhausts; FTP 1975; 4 vehicles.

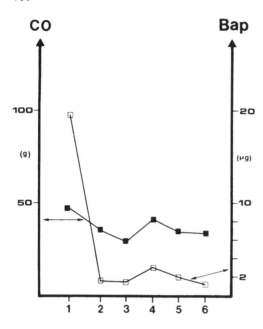

Figure 11. Temperature-dependent emissions. Repeated sampling in the European test cycle. 1, cold start (22-25°C); 2-6, hot start (80-85°C). Emission per cycle (1.013 km). (From Ref. 24.)

G. PAH Content and Aromaticity of Fuel

The origin of the PAH found in automobile exhausts has been investigated by several authors [10,13,34]. The results are not consistent but some conclusions can be drawn.

1. The input of PAH with the fuel far exceeds its emission in the exhaust.
2. Addition of C_9-C_{10} (or larger) aromatics in the form of catalytic reformate leads to increased PAH emissions. Whether this is a function of aromaticity as such, or of high levels of PAH in the reformate, is not clear.
3. Addition of a certain PAH to a test fuel will give an increased emission of this component.
4. High levels of PAH in the fuel will create larger deposits of these in the combustion chamber and exhaust system. This must be taken into account if measurements of different fuels are considered. Carryover effects can make it impossible to distinguish any difference in the emission between low- and high-PAH fuels.

In a recent compilation of data, [55] Candeli et al. state that fuel aromatics have a decisive influence on PAH emissions. This is especially pronounced for gasoline rich in C_9 or larger aromatic molecules. These compounds could constitute about 20-40% of the catalytic reformate, with remainder being lower aromatics.

Due either to carryover between the test runs or to unintentional adding of PAH with the catalytic reformate there is, however, still some doubt whether tests with gasoline containing different aromaticity really show a dependence between higher aromatics and PAH emission. Nunneman [56] has demonstrated the role of C_6-C_8 aromatics balanced with isooctane for the formation of PAH during combustion in a gasoline engine. His conclusion was that there is no correlation between PAH emission and these aromatics in the fuel. Nunneman emphasized the need to design good experimental conditions (i.e., to avoid carryover between tests) and pointed out that the analytical methods used possibly did not permit firm answers to submitted questions (e.g., the influence of aromaticity).

The fate of a single PAH during combustion has also been investigated [34], and apparently added PAH will resist passage in the combustion chamber. However, about a 20-fold increase in BaP in the fuel gave only about a 5-fold increase in emissions. All PAH do not follow the same reaction pattern in a gasoline engine. Calculation of fuel input shows that some compounds are degraded, that is, the input is greater than the output; this is true for BaP, benz[a]anthracene, and benzo[ghi]perylene. Others (e.g., cyclopenteno[cd]pyrene and coronene) have a higher output than input, thus are formed during combustion [57].

H. Comparison of PAH Emission Between Differently Fueled Vehicles

Results concerning PAH emissions from differently fueled vehicles are shown in Table 3 [58]. The data represent particle-associated PAH, and the emission was screened for mutagenicity [59]. The samples were taken in a dilution tunnel. The fuels investigated were diesel fuel; standard gasoline, leaded and unleaded; alcohol-blended fuels, 15% methanol and 23% ethanol, respectively; 95% methanol (5% isopentane), and LPG (liquified petroleum gas). In addition, some engine modifications, such as different compression ratios and "lean-burn combustion", were tested with emission control devices (e.g., different catalyst systems).

Table 3. PAH Emission ($\mu g/km \pm SD$) from Differently Fueled Vehicles[a]

Fuel[b]	PAH[c]	BaP
Diesel	960± 54	11± 0.5
Gasoline	170± 29	5± 1
M15	110± 27	3± 0.5
E23	43± 4	1± 0.1
M95	6± 1	<0.1
LPG	9	<0.5

[a]Particle samples from diluted exhausts. Federal test procedure.
[b]M15, gasoline with 15% methanol; E23, gasoline with 23% ethanol; M95, 95% methanol; LPG, liquefied petroleum gas.
[c]PAH, the sum of phenanthrene, fluoranthene, pyrene, benzo-[ghi]fluoranthene, cyclopenteno[cd]pyrene, benz[a]anthracene, chrysene, benzo[b]fluoranthene, benzo[k]fluoranthene, benzo-[e]pyrene, benzo[a]pyrene, indeno[1,2,3-cd]pyrene, benzo[ghi]-perylene, and coronene.

The emissions could be grouped into three classes: 1. diesel, 2. gasoline, and gasoline with methanol and ethanol respectively, 3. LPG, and gasoline used with a three-way catalyst. For vehicles fueled with LPG and neat methanol, the emission of PAH is drastically reduced compared to standard gasoline vehicles. Emissions when using these fuels are on the same order of magnitude as for catalyst-equipped vehicles.

The difference in the "total" PAH emission between gasoline and diesel exhausts is due mainly to a larger portion of lower molecular weight PAH from the diesels. These substances are substantially more particle-associated in the diesel emission then in gasoline exhausts. For compounds such as phenanthrene and pyrene (and equal boiling points), it is not sufficient to use only a particulate sample from gasoline emissions.

IV. CONCLUSIONS

Sampling and analysis of PAH from vehicle exhausts have attracted much attention during the past 30 years. The sampling procedures have been developed from direct sampling of raw exhausts to procedures involving a dilution of the emission. This lenient method has become widespread with the increasing knowledge that certain PAH are easily degraded during sampling. Furthermore, sampling artifacts have become even more important to avoid with the extended use of biological testing systems in emission research. Thus these improvements, in combination with modern high-resolving analytical methods, have yielded better knowledge of the PAH composition of both gasoline and diesel exhausts. In addition, new derivatives of PAH have recently become of interest to determine, and in the future more attention will probably be focused on PAH with functional groups attached to the rings than on the parent PAH.

REFERENCES

1. E. Clar, *Polycyclic Hydrocarbons,* Vol. 1, Academic Press, New York, 1968, pp. 131.
2. J. Leiter and M. J. Shear, J. Natl. Cancer Inst. 3:167 (1942).
3. R. E. Waller, Br. J. Cancer 6:8 (1952).
4. J. A. Campbell, Br. J. Exp. Pathol. 17:146 (1936).
5. P. Kotin, H. L. Falk, and M. Thomas, AMA Arch. Ind. Hyg., 9:164 (1954).
6. M. J. Lyons and H. Johnston, Br. J. Cancer 11:60 (1957).
7. S. Mittler and S. Nicholson, Ind. Med. Surg. 26:135 (1957).
8. R. L. Stenburg, D. J. von Lehmden, and R. P. Hangebrauck, Am. Ind. Hyg. Assoc. J. 22:271 (1961).
9. C. R. Begeman, SAE Paper 440C, Society of Automotive Engineers, Detroit. Mich., (1962).
10. G. P. Gross, CRC-APRAC Project CAPE-6-68, July 30, 1972.
11. A. Candeli, G. Morozzi, A. Paolacci, and L. Zoccolillo, Atmos. Environ. 9:843 (1975).
12. N. Zaghini, S. Mangolini, M. Arteconi, and F. Sezzi, SAE Paper 730836, Society of Automotive Engineers, Detroit, Mich., 1973.
13. A. Candeli, V. Mastrandrea, G. Morozzi, and S. Toccaceli, Atmos. Environ. 8:693 (1974).

14. R. P. Hangebrauck, R. P. Lauch, and J. E. Meeker, Am. Ind. Hyg. Assoc. J. *27*:693 (1966).

15. G. Grimmer, A. Hildebrandt, and H. Böhnke, Zentralbl. Bakteriol. Hyg. I Abt. Orig. *158*:22 (1973).

16. J. König and W. Funcke, in *Proceedings of COST 61 A bis Meeting,* Rome, Oct. 28-29, 1980.

17. G. Grimmer and A. Hildebrandt, Zentralbl. Bakteriol. Hyg. I Abt. Orig. *161*:104 (1975).

18. VW-Report UB-II, 2,334 62-10/75, Wolfsburg, West Germany, 1975.

19. J. Kraft and K. -H. Lies, SAE Paper 810082, Society of Automotive Engineers, Detroit, Mich., 1981.

20. T. Handa, T. Yamamura, Y. Kato, S. Saito, and T. Ishii, J. Japan Soc. Air Pollut. *14*:8 (1979).

21. J. C. Chipman and M. T. Massey, J. Air Polut. Control Assoc. *10*: 60 (1960).

22. U. Stenberg, T. Alsberg, and B. M. Bertilsson, SAE Paper 810441, Society of Automotive Engineers, Detroit, Mich., 1981.

23. R. S. Spindt, Report CAPE-24-72, July 1974.

24. U. Stenberg, *Emission of Polynuclear Aromatic Hydrocarbons from Gasoline Fueled Vehicles,* Report to the National Swedish Environment Protection Board, Feb. 1, 1979 (in Swedish).

25. J. L. Laity, M. D. Malbin, W. W. Haskel, and W. I. Doty, SAE Paper 730835, Society of Automotive Engineers, Detroit, Mich., 1973.

26. K. Habbi, Environ. Sci. Technol. *4*:239 (1970).

27. K. Habbi, Environ. Sci. Technol. *7*:223 (1973).

28. EPA, Fed. Regist. *45* (Mar. 15, 1980).

29. C. T. Hare, K. J. Springer, and R. L. Bradow, SAE Paper 760130, Society of Automotive Engineers, Detroit, Mich., 1976.

30. W. H. Lipkea, J. H. Johnson, and C. T. Vuk, SAE Paper 780108, Society of Automotive Engineers, Detroit, Mich., 1978.

31. U. Stenberg, R. Westerholm, T. Alsberg, U. Rannug, and A. Sundvall, in *Polynuclear Aromatic Hydrocarbons: Physical and Biological Chemistry,* M. Cooke, A. J. Dennis, and G. L. Fisher (Eds.), Battelle Press, Columbus, Ohio, 1982, p. 765.

32. H. K. Newhall, R. E. Jentoft, and P. R. Ballinger, SAE Paper 730834, Society of Automotive Engineers, Detroit, Mich., 1973.

33. F. S. -C. Lee, T. J. Prater, and F. Ferris, in *Polynuclear Aromatic Hydrocarbons,* P. W. Jones and P. Leber (Eds), Ann Arbor Science, Ann Arbor, Mich., 1979, p. 83.

34. P. S. Pedersen, J. Ingversen, T. Nielsen, and E. Larsen, Environ. Sci. Technol. *14*:71 (1980).

35. F. Snow, *Collection of Gas Phase Organics from the Raw Exhausts of Diesel and Gasoline Fueled Motor Vehicles,* Report Tr 81-14, Northrop Services, Inc., Environmental Sciences, Research Triangle Park, N.C., June 1981.

36. F. Stump, R. Bradow, W. Ray, D. Dropkin, R. Zweidinger, J. Sigsby, and R. Snow, *Trapping Gaseous Hydrocarbons for Mutagenic Testing,* EPA, Research Triangle Park, N.C. draft copy, Oct. 1981.

37. U. Rannug, Environ. Health Perspect. *47*:161 (1983).

38. R. L. Hanson, C. R. Clark, R. L. Carpenter, and C. H. Hobbs, Environ. Sci. Technol. *15*:701 (1981).

39. T. L. Chan, P. S. Lee, and J. S. Siak, in *Health Effects of Diesel Engine Emissions,* EPA 600/9-80-057a, 1980, pp. 230.

40. F. S. -C. Lee, W. R. Pierson, and J. Ezike, in *Polynuclear Aromatic Hydrocarbons: Chemistry and Biological Effects*, A. Björseth and A. J. Dennis (Eds), Battelle Press, Columbus, Ohio, 1980, p. 543.

41. J. N. Pitts, K. A. van Cauwenberghe, D. Grosjean, J. P. Schmid, D. R. Fitz, W. L. Belser, G. B. Knudson, and P. M. Hynds, Science *202*:515 (1978).

42. F. Black and L. Doberstein, *Filtermedia for Collecting Diesel Particulate Matter*, EPA 600-09, draft copy, 1981.

43. H. Yamasaki, K. Kuwata, and H. Miyamoto, Environ. Sci. Technol. *16*:189 (1982).

44. G. Broddin, W. Cautreels, and K. van Cauwenberghe, Atmos. Environ. *14*:895 (1980).

45. K. E. Thrane and A. Mikalsen, Atmos. Environ. *15*:909 (1981).

46. K. van Cauwenberge, Paper presented at the OECD Work Shop on PAH, Paris, Oct. 19-21, 1981.

47. M Möller, I. Alfheim, S. Larssen, and A. Mikalsen, Environ. Sci. Technol. *16*:221 (1982).

48. D. E. Seizinger, T. M. Naman, W. F. Marshall, C. R. Clark, and R. O. McClellan, SAE Paper 820818, Society of Automotive Engineers, Detroit, Mich., 1982.

49. U. Stenberg, *Report to the Swedish Motor Fuel Technology Company*, Stockholm, Nov. 1983 (in Swedish).

50. K. -H. Lies, in *Proceedings from Nato Advanced Research Workshop on Polycyclic Organic Matter from Exhaust Gases*, Liege, Belgium, Aug. 30-Sept. 2, 1982.

51. *Einflüsse der Bertriebszeit des Motors sowie des Motorschmieröles auf die Emission von polyzyklischer aromatischer Kohlenwasserstoffe aus Kraftfahrzeugen mit Ottomotoren*, DGMK Berichte 110, Hamburg, West Germany, 1977.

52. T. Handa, T. Yamamura, Y. Kato, S. Saito, and T. Ishii, Environ. Sci. Technol. *13*:1077 (1979).

53. E. S. Jacobs, P. J. Brandt, C. S. Hoffman, G. H. Patterson, and R. L. Willis, Polynuclear aromatic hydrocarbons from vehicles. Dupont Inc., Petroleum Laboratory, Wilmington Del. Presented to American Chemical Society, Los Angeles, Mar. 31, 1971.

54. G. Lepperhoff, in *Proceedings from Nato Advanced Research Workshop on Polycyclic Organic Matter from Exhaust Gases*, Liege, Belgium, Aug. 30-Sept. 2, 1982.

55. A. Candeli, G. Morozzi, and M. A. Shapiro, in *Proceedings from Nato Advanced Research Workshop on Polycyclic Organic Matter from Exhaust Gases*, Liege, Belgium, Aug. 30-Sept. 2, 1982.

56. F. Nunneman, in *Proceedings from Nato Advanced Research Workshop on Polycyclic Organic Matter from Exhaust Gases*, Liege, Belgium, Aug. 30-Sept. 2, 1982.

57. U. Stenberg, T. Alsberg, and R. Westerholm, Environ. Health Perspect. *47*:53 (1983).

58. T. Alsberg, R. Westerholm, U. Stenberg, and L. S. Elfver, in *Chemical and Biological Characterization of Exhausts Emissions from Vehicles Fueled with Gasoline, Alcohol, LPG and Diesel*, (K. E. Egebäck and B. M. Bertilsson (Eds), SNV-PM 1635, Jan. 1983.

59. A. Sundvall and U. Rannug, in *Chemical and Biological Characterization of Exhausts Emissions from Vehicles Fueled with Gasoline, Alcohol, LPG and Diesel*, (K. E. Egebäck and B. M. Bertilsson (Eds), SNV-PM 1635, Jan. 1983.

60. R. L. Williams and S. J. Swarin, SAE Paper 790419, Society of Automotive Engineers, Detroit, Mich., 1979.
61. R. B. Zweidinger, in *Proceedings of the 1981 EPA Diesel Emission Symposium,* Raleigh-Durham, N.C., 1981.

5

Recent Progress in the Determination of PAH by High Performance Liquid Chromatography

STEPHEN A. WISE / Organic Analytical Research Division, National Bureau of Standards, Gaithersburg, Maryland

I.   Introduction   113

II. Reversed-Phase Liquid Chromatography   114

III. Normal-Phase Liquid Chromatography   128

IV. Liquid Chromatographic Separations of Polycyclic Aromatic Sulfur Heterocycles   129

V.   High-Resolution Liquid Chromatography   136

VI. Analysis of Complex PAH Mixtures   137

   A. Selectivity in detection   137
   B. Multidimensional liquid chromatographic techniques   147

VII. Applications of Liquid Chromatography for the Determination of PAH   176

   A. Standard reference materials for the determination of PAH   176
   B. Review of liquid chromatographic applications for the determination of PAH   187

   References   187

I. INTRODUCTION

The use of liquid chromatography (LC) for the determination of polycyclic aromatic hydrocarbons (PAH) has increased significantly in recent years. A detailed review of the use of LC in the separation of PAH was included in the first volume of this handbook [1]. This previous review described the various modes of separation (e.g., reversed-phase and normal-phase LC), detection, and applications of LC for the measurement of PAH. This report reviewed the use of LC in PAH determinations from its introduction in the early 1970s through 1980. A recent book by Lee et al. [2] on the analytical chemistry of PAH contains a chapter on LC which primarily describes classical LC for PAH separations. Recent reviews by Bartle et al. [3] and Futoma et al. [4] provide more information concerning modern LC for PAH determinations.

This chapter will describe recent progress (since 1980) in the use of LC for the separation, identification, and quantification of PAH. Since 1980, LC has gained greater acceptance as a useful method for the determination of PAH in complex mixtures. Generally, these LC measurements utilize selective fluorescence or multiwavelength ultraviolet (UV) detection. In addition, normal-phase LC has found widespread use for the isolation and fractionation of specific groups of PAH prior to analysis by other chromatographic techniques (i.e., multidimensional chromatography). Applications of LC in either one or both modes in a multidimensional chromatographic separation of PAH will be described. As in the previous review [1], an updated summary of the recent literature for applications of LC in the analysis of PAH mixtures is provided in Table 17.

II. REVERSED-PHASE LIQUID CHROMATOGRAPHY

Reversed-phase LC on chemically bonded octadecylsilane (C_{18}) stationary phases is by far the most popular LC mode for the separation of PAH. Reversed-phase LC on C_{18} phases provides excellent selectivity for the separation of PAH isomers and alkyl-substituted PAH isomers [5]. Recent studies by Wise et al. [6], Ogan and Katz [7,8], Colmsjö and MacDonald [9], and Amos [10] found that C_{18} columns from various manufacturers provided not only different separation efficiencies, but often provided different selectivities and retention characteristics for PAH. Wise et al. [6] compared the selectivity for 14 PAH on seven different columns; Ogan and Katz [7] evaluated the selectivity of seven PAH on eight different columns; the retention characteristics of 11 PAH were studied on three columns by Colmsjö and MacDonald [9]; Amos [10] investigated the retention of 15 PAH on 10 different C_{18} columns. In these studies, attention was focused on three groups of PAH isomers which are difficult to separate: benz[a]anthracene and chrysene; benzo[e]pyrene, benzo[a]pyrene (BAP), benzo[b]fluoranthene, and benzo[k]fluoranthene; and benzo[ghi]perylene and indeno[1,2,3-cd]pyrene. All of these studies found that different C_{18} columns provided somewhat different selectivities for PAH and that one type of polymeric C_{18} material in these studies, Vydac 201TP/HC-ODS, was successful in resolving all of these difficult to separate PAH isomers. In addition to C_{18} columns, Amos [10] also studied 11 other alkyl reversed-phase columns (e.g., C_2, C_6, C_8, C_{22}, and phenyl), but concluded that the C_{18} phases provided better PAH separations. The differences in selectivity for C_{18} columns from different manufacturers described in these studies [6-10] (except that of Amos [10]) have been discussed in detail in the previous review [1]. Goldberg [11] compared k' values for anthracene and selectivity factors α ($\alpha = k_1'/k_2'$ for two solutes) for anthracene/naphthalene on 31 different commercial reversed-phase columns. Values for k' for anthracene ranged from about 0.5 to 6.5 on the different C_{18} columns, whereas α values for anthracene/naphthalene ranged from about 1 to 3.

Atwood and Goldstein [12] studied the variations in k' values for anthracene and variations in the separation factor α for anthracene/phenanthrene for 24 different manufacturer's lots of Vydac 201TP material. The mean value of k' for anthracene for all lots was 3.93 with a coefficient of variation of 8%. The separation factor for anthracene/phenanthrene had a mean value of 1.31 for all lots with a standard deviation of 0.07. They also compared the α values for anthracene/phenanthrene on eight other commercial C_{18} columns from different manufacturers and found that the Vydac 201TP material had an average value of 1.32 compared to a range of 1.09-1.17 for the eight other columns. Atwood and Goldstein [12] concluded that manufacturers of LC packing material should have quality control programs which characterize the variations in selectivity as fully as possible.

One important characteristic of the C_{18}-bonded material is the nature of the chemically bonded organic layer (i.e., whether the layer is monomeric or polymeric). Retention data for 100 PAH on both monomeric and polymeric phase have been reported by Wise et al. [1,5,6]. Several PAH (and particularly nonplanar PAH such as polyphenyl arenes) exhibit significantly different selectivities and retention characteristics on monomeric versus polymeric stationary phases. These differences in chromatographic retention and selectivity are the result of the use of different silica materials (i.e., different pore size and surface area) as supports and a variety of reagents (i.e., mono-, di-, or trichlorosilanes) and procedures to produce the C_{18}-bonded phase.

Recently, Wise and May [13] investigated the origin of these selectivity differences by studying the effect of the C_{18} surface coverage of the bonded material on the selectivity of various PAH, polyphenyl arenes, and polycyclic aromatic sulfur heterocycles (PASH) on different columns. The C_{18} surface concentrations ($\mu mol/m^2$) for the seven C_{18} columns previously compared by Wise et al. [6] were determined from the carbon content and specific surface area of the chemically modified silicas. These results, which were discussed briefly in the previous review [1], indicated that the selectivity was related to the C_{18} surface coverage of the silica [13]. To further evaluate the effect of surface coverage on selectivity, seven different lots of 5 μm polymeric C_{18} material from one manufacturer were studied. The results of the percent carbon and surface area measurements for these seven lots are summarized in Table 1. The manufacturer indicated that lots 10 and 11 had low carbon loadings and that lots 14, 15, and 16 had a high carbon loading. All five of these lots were rejected by the manufacturer's normal criteria for PAH selectivity. The remaining two lots, 12 and 13, were representative of those commercially available as 5 μm polymeric C_{18} material from this manufacturer.

As shown in Table 1, k' for BaP increases with increasing surface coverage. The selectivity factors, α, for a number of PAH (relative to BaP) versus the k'_{BaP} on the various polymeric C_{18} columns are shown in Figs. 1 and 2. The selectivity factors on a monomeric column (from a different manufacturer and with high-surface-area silica) are indicated on the far-left side to illustrate the differences in selectivity on a monomeric C_{18} phase. The position of the x-axis point for the monomeric phase does not prepresent the k'_{BaP} on the monomeric column (which is actually similar to lot 15, of the polymeric C_{18} material in Table 1, i.e., k' = 5.5), but is only representative of a maximum-surface-coverage monomeric material. In a similar plot using a monomeric column prepared on the same low-surface-area silica (about 90 m^2/g), lot 17 in Table 1 [14], it was found that the linear trends observed for the polymeric columns in Figs. 1 and 2 could generally be extrapolated back to the monomeric column.

Some interesting trends are observed in Fig. 1, particularly between monomeric and polymeric phases, and these results clarify some conflicting literature reports of PAH elution sequence on various C_{18} columns. The resolution of benz[a]anthracene (no. 3) and chrysene (no. 4) has often been viewed as a difficult separation on many columns. As illustrated in Fig. 1, these isomers were unresolved on the monomeric column and α values increase as the surface concentration values (related to k'_{BaP}) increase on the polymeric materials. Benzo[c]phenanthrene (no. 1) and triphenylene (no. 2), isomers of molecular weight 228, have reversed elution order on the monomeric column compared to the polymeric material.

Table 1. Physical Characteristics for Different Lots of 5 μm C_{18} Material[a]

Column (lot)	Percent carbon	Surface area (m^2/g)[b]	Surface coverage $(\mu mol/m^2)$	k'(BaP)[c]
17 (monomeric)	4.0 ± 0.1[d]	53.4	3.1 ± 0.1	0.9
10	4.5 ± 0.3[e]	48.5	4.3 ± 0.3	2.7
11	7.5 ± 0.2[d]	61.0	5.8 ± 0.5	3.3
12	7.7 ± 0.4[e]	60.0	5.9 ± 0.3	3.5
13	8.0 ± 0.1[d]	61.6	5.6 ± 0.4	4.0
14	8.0 ± 0.3[e]	50.1	7.4 ± 0.3	5.0
15	9.3 ± 0.1[d]	52.4	8.2 ± 0.1	5.5
16	8.1 ± 0.3[e]	48.3	7.8 ± 0.3	6.3

[a]Vydac 201TP 5 μm, 330 Å pore dia, 90 m^2/g surface area of underivatized silica; all lots except lot 17 are polymeric phases.

[b]Specific surface area of the C_{18}-modified material determined by a modified single-point B.E.T. [S. Branauer, P. H. Emmett, and E. Teller, J. Amer. Chem. Soc. 60:309 (1938)] using nitrogen as the adsorbate.

[c]Acetonitrile/water 85:15 (v/v).

[d]Four Samples analyzed, uncertainty is ±1σ (sample weights determined more accurately than samples in footnote e).

[e]Triplicate analyses, uncertainty is ±1σ.
Source: Refs. 13 and 14.

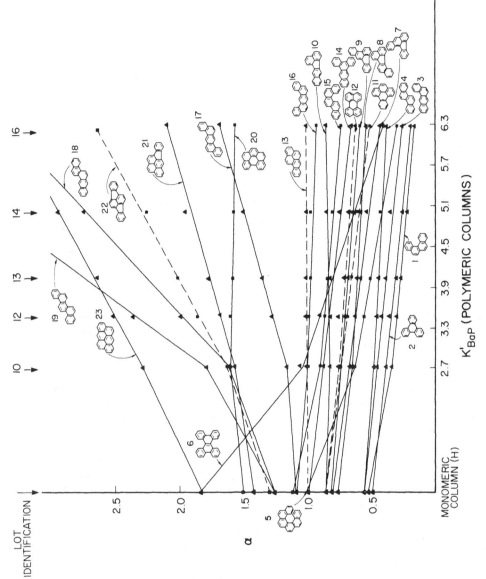

Figure 1. Selectivity factors (α), relative to benzo[a]pyrene, for PAH on five different polymeric C₁₈ columns and a monomeric C₁₈ column. (From Ref. 13.)

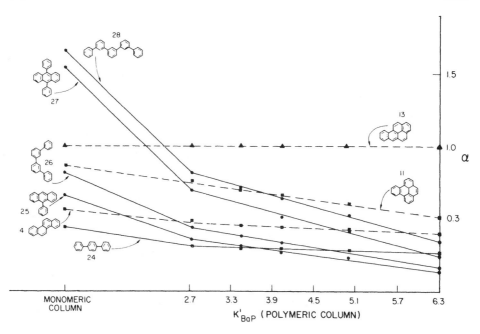

Figure 2. Selectivity factors (α), relative to benzo[a]pyrene, for poly-phenyl arenes on five different polymeric C_{18} columns and a monomeric C_{18} column. (From Ref. 13.)

Of particular interest in the determination of PAH in environmental mix-tures is the separation of the isomers of molecular weight 252 (compounds 7-10 and 11-13). As reported previously [5], PAH elution order on C_{18} columns is related to the shape [i.e., length-to-breadth ratio (L/B)] of the molecule. On the high-surface-coverage material the four benzofluoranthene isomers (nos. 7-10) elute in order of increasing L/B (i.e., L/B ratios of 1.16, 1.39, 1.40, and 1.48 for compounds 7-10, respectively), as do the three nonfluor-anthenic isomers, benzo[e]pyrene (no. 11), perylene (no. 12), and BaP (no. 13), with L/B ratios of 1.12, 1.27, and 1.50, respectively. On the high-coverage column, all seven isomers are at least partially resolved. On the low loading (lot 10) and on the monomeric packing, benzo[a]fluoranthene (no. 7) and benzo[j]fluoranthene (no. 8) reverse elution sequence and sev-eral of the isomers coelute.

The next isomers of interest are those of molecular weight 278 (nos. 14-19). Again, these isomers elute in the order of increasing length-to-breadth ratio (i.e., 1.24, 1.47, 1.47, 1.79, 1.84, and 1.99 for nos. 14-19, respectively) with the greatest resolution on the high-surface-coverage column (lot 16). On the monomeric column benzo[c]chrysene (no. 15) elutes after dibenz[a,h]-anthracene (no. 17), and benzo[b]chrysene (no. 18) and picene (no. 19) co-elute. The longer retention of benzo[c]chrysene is indicative of the behavior of slightly nonplanar PAH. For the three isomers with length-to-breadth ratios greater than 1.5, α values increase with increasing surface coverage; whereas for the other three isomers, α values decrease with increasing sur-face coverage. In general, α values increase much more rapidly as the length-to-breadth ratios increase (e.g., nos. 18 and 19).

The separation of dibenz[a,h]anthracene (no. 17) and benzo[ghi]perylene (no. 20) is of particular interest. These two compounds are easily resolved

on monomeric columns as shown in Fig. 1 and reported in the literature [6-
10]. However, conflicting reports on the elution order of this PAH pair on
this polymeric C_{18} material have appeared in the literature [6,9,10,15]. In
the LC method of Ogan et al. [15] for the separation of the 16 priority pollutant PAH and in the study of Amos [10], dibenz[a,h]anthracene eluted
prior to benzo[ghi]perylene, whereas Wise et al. [6] and Colmsjö and Mac-
Donald [9] showed chromatograms in which benzo[ghi]perylene eluted first.
Presumably, Wise et al. [6] and Colmsjö and MacDonald [9] used columns from
lots with high surface coverage, whereas Ogan et al. [15] and Amos [10]
used material with low surface coverage (see Fig. 1). As mentioned previously, the work of Ogan and Katz [7,8] was used to identify selected lots
of this polymeric material with specific selectivity for PAH. These selected
lots are now available commercially. However, the earlier columns of this
10 µm material were not screened for selectivity; thus these reports of different selectivity have appeared in the literature. The 5 µm columns of
this material are screened to provide columns with selectivity similar to that
of lots 12 and 13 in Table 1.

The unusual behavior of nonplanar PAH such as phenanthro[3,4-c]-
phenanthrene (no. 5) and tetrabenzo[a,c,f,h]naphthalene (no. 6) is illustrated in Fig. 1. For these two compounds, the α values decrease as the surface
coverage increases and the α values increase significantly on the monomeric
phase; for example, tetrabenzonaphthalene (no. 6) elutes prior to benzo[a]-
fluoranthene (no. 7) on the high-surface-coverage column and has the longest
retention of all the PAH in this study on the monomeric material. In addition,
other compounds which have slight nonplanarity owing to steric hindrance
[i.e., benzo[c]phenanthrene (no. 1) and benzo[c]-chrysene (no. 15)] have
higher α values on the monomeric than on the polymeric phases when compared to their structural isomers.

The effect of solute planarity is also illustrated in Fig. 2 for several
polyphenyl arenes. As the surface coverage decreases, the α values for the
polyphenyl arenes increase and the retention (as indicated by α) on the
monomeric column is significantly longer when compared to the fused-ring
PAH. The retention characterisitics of 13 polyphenyl arenes on a monomeric
and a polymeric material have been reported [5]. The polyphenyl arenes
are generally retained longer, relative to the fused-ring PAH, on the monomeric than on the polymeric materials; for example, m-quaterphenyl, 1,2'-
binaphthyl, 9,9'-bianthryl, 9-phenylanthracen, 1-phenylanthracene, and
9-phenylphenanthrene all elute prior to benz[a]anthracene on the polymeric
material, but after it on the monomeric material [5].

Ogan and Katz [15a] observed the anomolous behavior of the polyphenyl
arene, 9,10-diphenylanthracene, on C_{18} columns from various manufacturers,
but these data were not reported in their study [8]. However, the data
of Ogan and Katz [8,15a] for this compound and other PAH were combined with
similar data from another laboratory for a comparison of nine different columns
and published in a previous review on liquid chromatography of PAH [1].
As shown in Fig. 2, 9,10-diphenylanthracene (no. 27) elutes after BaP and
dibenz[a,h]anthracene on the monomeric column, but prior to BaP and benzo-
[e]pyrene on the polymeric columns, and it elutes even prior to chrysene on
the highest-surface-coverage material.

The long retention of polyphenyl arenes on the monomeric material is similar to that obtained on silica. Popl et al. [16] reported the following retention indices for silica: p-terphenyl (4.57), m-quaterphenyl (5.98), 9-phenyl-
anthracene (4.02), and 9,10-diphenylanthracene (5.13) [retention indices

are relative to benz[a]anthracene (4.00) and benzo[b]chrysene (5.00)].
The similarity of retention on silica, alumina, and reversed-phase C_{18} packings has been discussed previously [5]. Chmielowiec and Sawatzky [17]
reported that polyphenyl arenes behave in an "irregular" manner with respect to temperature-induced selectivity on a C_{18} phase. Snyder [18] suggested that these differences correlate with differences in the molecular
shape of the solute molecule and also with the "ordered' nature of the stationary phase.

Reversed-phase LC on C_{18} materials provides good selectivity for the
separation of methyl-substituted PAH [5]. The selectivity factors for the
methyl-substituted chrysene isomers on different polymeric columns have
been reported [13]. As with the parent PAH, the longer isomers have increasing α values with increasing surface coverage. As reported previously
[5], polymeric C_{18} materials generally provide more resolution of individual
methyl isomers than do the monomeric materials. However, even though the
resolution of individual methyl isomers improves as the surface coverage increases, the parent compound elutes closer to the methyl isomers as the surface coverage increases. These same trends were observed also for methylbenzo[a]pyrene and methylbenzo[b]naphtho[2,1-d]thiophene isomers.

The data in Figs. 1 and 2 are useful in the selection of the appropriate
polymeric C_{18} column (i.e., with appropriate surface coverage) to separate
selected PAH. The selectivity characteristics of a particular lot of this polymeric material can be predicted based on the k' value of BaP on the column.
In addition, a column with selectivity intermediate to that of two lots can be
prepared by mixing various portions of the two lots. Three "mixed"-phase
columns were prepared by mixing material from lots 11 (low coverage) and 15
(high coverage) in Table 1 in 70:30, 50:50, and 30:30 (w/w) ratios [14].
The selectivity factors for selected PAH on the columns from lots 11 and 15
and on the three mixed-lot columns are shown in Fig. 3. The efficiencies of
mixed-lot columns were equivalent to the columns from the individual lots
(i.e., approximately 20,000 plates based on anthracene).

Differences in column selectivity for seven PAH solutes are illustrated in
Fig. 4 on columns from lots 11, 15, and 17 and the three mixed-phase columns.
The columns from lots 11 and 15 did not separate all seven solutes. However,
a mixed-phase column of 70% lot 11 and 30% lot 15 provided the appropriate
selectivity to achieve separation of these selected solutes. Selectivities similar to those obtained with the physically mixed phases were also obtained by
coupling short columns of appropriate lengths, one containing material from
lot 11 and one containing material from lot 15. Thus the chromatographer
could have a collection of short C_{18} columns of different selectivities which
could be coupled together in various combinations to achieve the necessary
selectivities for a particular separation.

Sander and Wise [19] further investigated the origin of selectivity differences for PAH on different C_{18} columns. They described the synthesis
and characterization of monomeric, polymeric, and "oligomeric" C_{18} alkyl
phases on a series of wide-pore (300 Å) silica substrates. The surface
coverage ranges for these three types of phases were monomeric (2.23 to
3.02 μmol/m^2), oligomeric (3.29 to 4.11 μmol/m^2), and polymeric (4.73 to
6.41 μmol/m^2). The selectivity characteristics of these phases were evaluated by using a three-component mixture of phenanthro[3,4-c]phenanthrene
(PhPh), tetrabenzo[a,c,f,h]napthalene (TBN), and BaP. The selectivity
of these three PAH was observed to be strongly dependent on the type
of phase and the surface coverage (see Fig. 1). Depending on the elution

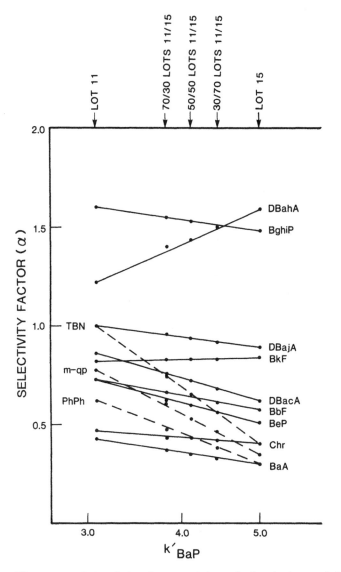

Figure 3. Selectivity factors (α), relative to benzo[a]pyrene, for selected PAH on polymeric C_{18} columns from mixtures of two different lots (see Table 1). BaA, benz[a]anthracene; Chr, chrysene; BeP, benzo[e]pyrene; BbF, benzo[b]fluoranthene; DBacA, dibenz[a,c]anthracene; BkF, benzo[k]fluoranthene; DBajA, dibenz[a,j]anthracene; BghiP, benzo[ghi]perylene; DBahA, dibenz[a,h]anthracene; TBN, tetrabenzo[a,c,g,h]napththalene; PhPh, phenanthro[3,4-c]phenanthrene; m-qp, m-quinquephenyl. (From Ref. 14.)

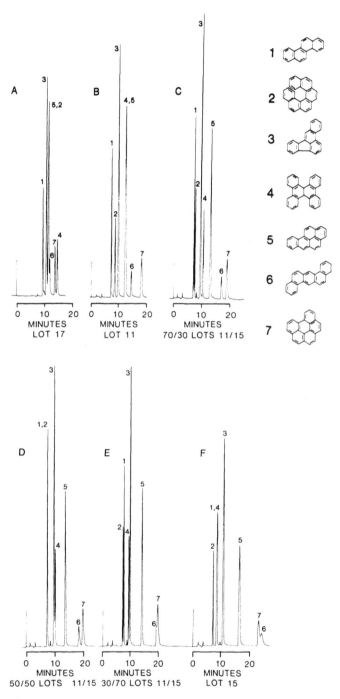

Figure 4. Reversed-phase LC separation of selected PAH on a monomeric C_{18} column from lot 17 (A) and on polymeric C_{18} columns from lot 11 (B) and lot 15 (F) and mixtures of these two lots [70:30 (C), 50:50 (D), and 30:70 lot 11/15 (E)]. Compound identification: 1, chrysene; 2, phenanthro-[3,4-c]phenanthrene; 3, benzo[b]fluoranthene; 4, tetrabenzo[a,c,f,h]naphthalene; 5, benzo[b]fluoranthene; 6, dibenz[a,h]anthracene; 7, benzo[ghi]-perylene. (From Ref. 14.)

order of this mixture, the selectivity of a C_{18} phase toward more complex PAH mixtures could be rapidly classified. The separations of this test mixture and a 16-component PAH mixture on a monomeric, an oligomeric, and a polymeric C_{18} phase are shown in Fig. 5. Sander and Wise [19] concluded that the differences in selectivity between these three phase types were a matter of degree of surface coverage rather than some fundamental difference in the phases.

In the work of Sander and Wise [19], it was found that the excellent selectivity reported previously for PAH separations on Vydac 201TP material [6-10] could be reproduced on other wide-pore silica materials. Polymeric C_{18} phases were synthesized on five different 300 Å silicas (Zorbax, LiChrosphere, Vydac TP, Hypersil, and Protosil) with surface areas ranging from 45 to 250 m^2/g. The surface coverages on these five materials were extremely reproducible (5.1 $\mu mol/m^2$ ± 5% relative standard deviation) and the selectivities toward PAH were very similar.

Sander and Wise [20] also investigated the differences in selectivity of PAH in polymeric and monomeric C_{18} phases prepared on silica substrates with different pore sizes. As shown in Fig. 6, the selectivity of polymeric C_{18} phases on small-pore silica (60 Å) was similar to that of a monomeric C_{18} phase (see Fig. 5), whereas the wide-pore silicas (150 and 300 Å) provided the selectivity necessary to separate the 16-component PAH mixture. Thus the combination of a polymeric C_{18} phase on a wide-pore silica provides the best selectivity for separation of this PAH mixture.

Regarding the mechanism of retention in reversed-phase LC, Wise et al. [5] described a relationship between the shape of PAH solutes (the length-to-breadth ratio, L/B) and the reversed-phase retention on C_{18} bonded phases. In this study the length-to-breadth ratios for more than 100 unsubstituted and methyl-substituted PAH were compared to LC retention characteristics on both a monomeric and a polymeric C_{18} stationary phase. In nearly all cases, the ratio was successful in predicting the elution order of isomeric PAH (i.e., retention increases with increasing length-to-breadth ratio). Two particularly interesting groups of isomers are the benzofluoranthenes of molecular weight 252 and the cata-condensed five-ring isomers of molecular weight 278. The benzofluoranthene isomers—benzo[a]fluoranthene, benzo[j]fluoranthene, benzo[b]fluoranthene, and benzo[k]fluoranthene—have L/B ratios of 1.16, 1.39, 1.40, and 1.48, respectively, and have the same reversed-phase elution order on C_{18} polymeric materials with high surface coverage (see Fig. 1). The L/B ratios for five isomers of molecular weight of 278 were reported in the initial study of Wise et al. [5] [i.e., dibenz[a,c]anthracene (1.24), dibenz[a,j]anthracene (1.47), dibenz[a,h]anthracene (1.79), benzo[b]chrysene (1.84), and picene (1.99)]. The reversed-phase LC separation of 11 of the 12 possible cata-condensed PAH isomers of molecular weight 278 is shown in Fig. 7. The elution order of these compounds is in good agreement with that predicted by the length-to-breadth ratios [21].

The effect of position of methyl substitution on the retention of a PAH solute is illustrated in Fig. 8 for 10 methylbenzo[a]pyrene isomers. A correlation coefficient of 0.890 was found for the correlation between the retention indices and the length-to-breadth ratios for these 10 methylbenzo[a]pyrene isomers. Issaq et al. [22] compared LC separation of these same methylbenzo[a]pyrene isomers with GC separation on two nematic liquid-crystal stationary phases. The GC elution order of the methylbenzo[a]pyrenes was similar to the elution order in LC with the exception of the 6-methyl isomer

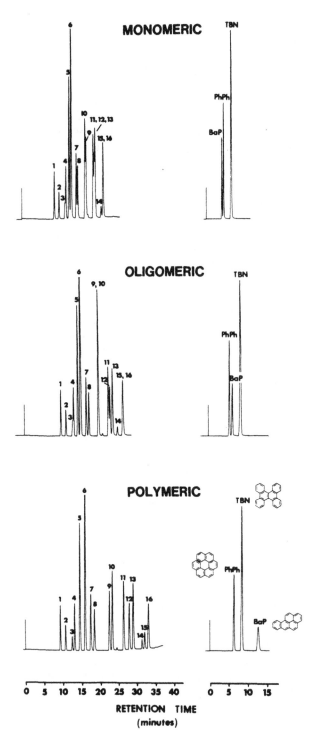

Figure 5. Reversed-phase LC separation of 16 PAH (NBS SRM 1647) and a three-component test mixture on representative monomeric, oligomeric, and polymeric C_{18} phases. Peak identification: 1, naphthalene; 2, acenaphthylene; 3, acenaphthene; 4, fluorene; 5, phenanthrene; 6, anthracene; 7, fluoranthene; 8, pyrene; 9, benz[a]anthracene; 10, chrysene; 11, benzo[b]fluoranthene; 12, benzo[k]fluoranthene; 13, benzo[a]pyrene; 14, dibenz[a,h]anthracene; 15, benzo[ghi]perylene; 16, indeno[1,2,3-cd]pyrene. Conditions: SRM 1647—linear gradient from 40 to 100% CH_3CN in H_2O in 30 min at 2ml/min; test mixture—isocratic at 85% CH_3CN in H_2O. UV detection at 254 nm. (From Ref. 19.)

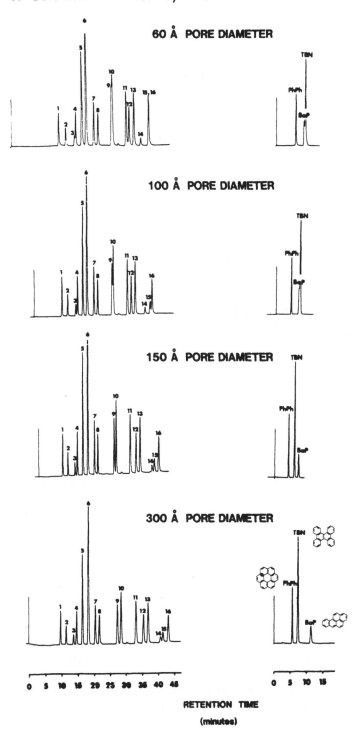

Figure 6. Reversed-phase LC separation of 16 PAH (NBS SRM 1647) and a three-component test mixture on polymeric C_{18} phase prepared on silica substrates with different pore sizes. Peak identifications: see Fig. 5. Conditions: same as Fig. 5. (From Ref. 20.)

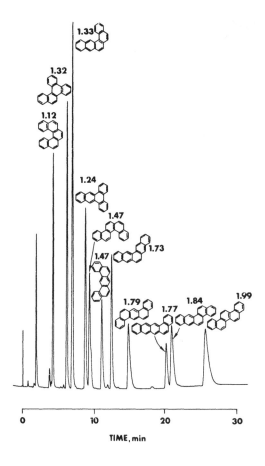

Figure 7. Reversed-phase LC separation of 11 PAH isomers of molecular weight 278. Numbers for each peak correspond to length-to-breadth ratios. Column: Vydac 201TP C_{18} 5 μm. Mobile phase: linear gradient from 85 to 100% CH_3CN in H_2O at 1%/min at 1.5 ml/min. UV detection at 254 nm. (From Ref. 21.)

eluting after the 1-methyl isomer and the 3-methyl isomer eluting after the 8-methyl isomer. The chromatograms in Figs. 7 and 8 illustrate the excellent selectivity of reversed-phase LC, particularly on polymeric C_{18} phases, for the separation of PAH isomers.

Hasan and Jurs [23] used the reversed-phase retention data set from the studies of Wise et al. [5] to predict LC retention based on PAH structural features. They generated numerical descriptors from the PAH molecular structures (e.g., fragments, molecular connectivity, substructure environment, geometric and calculated physical property descriptors) which were used as predictor variables in multiple linear regression analysis. Their analysis indicated that the two columns, monomeric and polymeric, provided different selectivity toward the compounds in the data set. Thus two linear equations (i.e., one for each type of column) with four and five variables were developed. The variables included the following descriptors: number of aromatic rings, molecular volume, cluster three molecular connectivity,

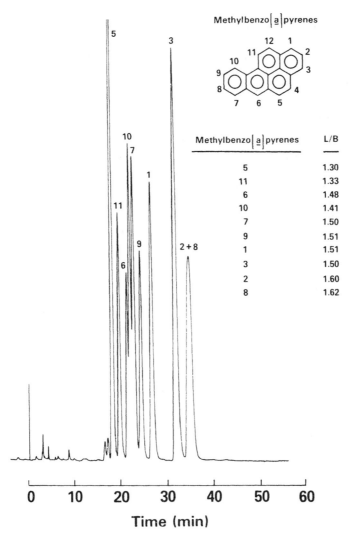

Figure 8. Reversed-phase LC separation of methylbenzo[a]pyrene isomers on polymeric C_{18} column. Column: Vydac 201TP C_{18} 5 μm. Mobile phase: 90% CH_3CN in H_2O at 1 ml/min. (From Ref. 5.)

largest and smallest principal axis, smallest principal moment, and X/Y principal moments. The type of descriptors selected to represent the PAH structures supports the results of previous studies on the influence of the solute structure on chromatographic retention. The number of aromatic rings and the molecular volume are directly related to the size of the molecules. The cluster three molecular connectivity index indicates the degree of branching in the molecule and the presence and position of a methyl group in the molecule. The presence of the principal axis and the principal moment descriptors in the equations strongly supports the importance of shape of the molecules on their retention as previously reported [5], since the magnitude of the three principal axes can be associated with the length, breadth, and thickness of the molecule. The third principal axis and the principal moment descriptors (i.e., the thickness) serve as correction factors for the nonplanar PAH. The anomalous LC retention behavior of nonplanar PAH has been pointed out by Wise et al. [5]. As shown in Figs. 1 and 2, the nonplanar PAH have shorter retention on polymeric than on monomeric C_{18} phases. This observation was supported by the larger negative correction in the retention equation for the polymeric column than for the monomeric column [23]. Hasan and Jurs [23] demonstrated that the two retention equations could be used to predict the retention of a number of PAH which were not included in the original data set.

Hurtubise et al. (24) investigated the correlation of molecular connectivity of PAH hydroaromatics with reversed-phase LC retention. They measured the retention of 34 compounds on a µBondapak C_{18} column with a methanol-water mobile phase. Log capacity factor (k') was plotted as a function of molecular connectivity and compared with graphs of log k' as a function of F (F is a chromatographic correlation factor developed by Schabron et al. [25] which relates solute structure to k'). They found less scatter with the correlation of F and log k' than with molecular connectivity.

III. NORMAL-PHASE LIQUID CHROMATOGRAPHY

Even though reversed-phase LC is the most widely used mode of LC for the determination of PAH, investigations of normal-phase LC on polar chemically bonded phases for the separation of PAH have also been reported. As described in the previous review [1], normal-phase LC on polar bonded phases, particularly amine and diamine phases, has been used extensively for the separation of PAH based on the number of aromatic carbon atoms. Recent applications of the amine and similar phases for the isolation of particular fractions from complex PAH mixtures will be described in detail later.

Matsunaga [26] recently compared the separation of PAH and polar compounds on several polar chemically bonded phases (i.e., nitro, amine, cyano, and sulfonic acid), silica, alumina, and porous polystyrene gel. These materials were evaluated for the separation of aromatic compounds by the number of aromatic rings, separation of polar compounds (nitrogen and oxygen containing compounds) from PAH, and the resolution between polar compounds by functional group. The nitro phase was found to give the largest capacity factor (k') values for selected PAH. The amine phase had selectivity similar to the nitro phase for PAH but smaller k' values. Both the nitro and amine phases had better selectivity and greater retention than silica. Matsunaga [26] evaluated these columns for the separation of aromatic compounds in fossil fuel liquids.

In addition to investigating the retention of 16 PAH on an amine phase, Liphard [27] also studied the retention characteristics of 27 partially hydrogenated PAH. The capacity factors for the hydrogenated PAH were found to decrease with increasing hydrogen content (i.e., the elution order is perhydroanthracene, octahydroanthracene, tetrahydroanthracene, dihydroanthracene, and anthracene). Compounds with the same number of double bonds eluted as a group (e.g., tetralin and octahydrophenanthrene, naphthalene and tetrahydrophenanthrene). However, within these groups with the same number of double bonds, a minor dependence on molecular mass was observed (i.e., capacity factors increased with increasing molecular weight).

Charge-transfer or donor-acceptor complex chromatography has been investigated by several workers [28-30] for the separation of PAH in complex fuel materials. Holstein [28,29] described the preparation of a new tetrachlorophthalimido phase for LC and reported retention data for about 40 aromatics and 20 nitrogen-containing aromatics. The aromatics were found to elute according to the number of rings, with only slight influence on retention for the addition of alkyl substituents. The performance of this phase was demonstrated for the separation of coal liquids [28] and compared with the separation of the same sample on the amine phase [29].

A comparison of alumina and chemically bonded diamine and 2,4-dinitro-anilinopropyl (DNAP) phases was performed by Grizzle and Thomson [30] for compound class separation of 86 model aromatic hydrocarbons expected or known to be present in petroleum, coal liquids, or shale oil. They compared relative retention strengths, grouping tendencies, and the observed structure and substituent effects. The silica-DNAP and alumina were determined to be more sensitive to molecular structure but less sensitive to alkyl substituent effects than the diamine phase. On the basis of retention strengths and grouping tendencies of model compounds, the silica-DNAP was considered superior to alumina and the diamine phase. However, chromatograms of a shale oil sample on silica-DNAP and alumina were similar. The most obvious difference in the shale oil chromatograms on the silica-DNAP and alumina colums versus the diamine column was the increased retention of compounds with less than three rings.

Ecknig et al. [31] studied group separations of crude oil distillates on perfluorocarbon (PFEA)-modified silica. Aromatic hydrocarbons eluted in order of increasing π-electron energy which was characterized by a linear correlation between ln k' and the resonance energy. In general, when compared to unmodified silica, the PFEA phase had reduced retention but alkyl substituents had little effect on the retention. Therefore, Ecknig et al. [31] suggested the use of this phase for separation of crude oil distillates into groups according to aromatic ring types.

Another novel polar phase, a phenylmercuric acetate phase, was described by Chmielowiec [32] and normal-phase LC retention data for over 60 PAH were reported on this phase. For PAH the phenylmercuric acetate phase provided normal-phase retention characteristics which were similar to other polar phases (e.g., amine, nitro, diamine, etc.), but some selectivity differences were observed. As with other polar phases, alkyl substitution was observed to have little effect on the normal-phase retention. Concerning possible selective retention of polycyclic aromatic sulfur heterocycles (PASH) relative to PAH, very few retention data were reported for thiophene-type sulfur-containing compounds. However, it appears that selectivity for PASH is not sufficient to separate PAH from PASH by normal-phase LC. No applications of the phenylmercuric acetate column to provide novel separations of PAH or PASH have been reported.

IV. LIQUID CHROMATOGRAPHIC SEPARATIONS OF POLYCYCLIC AROMATIC SULFUR HETEROCYCLES

Recently, interest in the determination of PASH in environmental samples has increased. However, very few data have been reported concerning the liquid chromatographic retention characteristics of PASH. Colmsjö et al. [33] reported GC and LC retention data for eight peri-condensed PASH. Recently, Wise et al. [34] investigated the normal- and reversed-phase LC retention characteristics for 27 cata-condensed four- and five-ring PASH (molecular weights of 234 and 284) and 10 peri-condensed five-ring PASH isomers (molecular weight of 258). These results are summarized in Table 2. Normal-

Table 2. LC Retention Data for Polycyclic Aromatic Sulfur Heterocycles[a]

Compound	Normal-phase LC[b] (NH$_2$)	Reversed-phase LC[c] (C$_{18}$)	
		Poly.	Mono.
Four ring cata-condensed			
Benzo[b]naphtho[1,2-d]thiophene	3.65	3.79	4.02
Benzo[b]naphtho[2,3-d]thiophene	3.78	4.05	4.05
Benzo[b]naphtho[2,1-d]thiophene	3.59	4.20	4.24
Anthra[2,3-b]thiophene	4.19	3.41	2.97
Phenanthro[3,4-b]thiophene	3.89	3.57	3.70
Phenanthro[2,3-b]thiophene	4.26	3.65	3.70
Phenanthro[3,2-b]thiophene	4.24	3.67	3.67
Phenanthro[9,10-b]thiophene	3.89	3.77	3.86
Anthra[2,1-b]thiophene	4.14	3.78	3.74
Phenanthro[4,3-b]thiophene	3.94	3.79	3.86
Phenanthro[2,1-b]thiophene	4.13	3.80	3.71
Anthra[1,2-b]thiophene	3.94	3.99	3.93
Phenanthro[1,2-b]thiophene	3.98	4.08	3.93
Five ring peri-condensed			
Triphenyleno[4,5-bcd]thiophene	4.09	4.30	4.61
Chryseno[4,5-bcd]thiophene	4.04	4.51	4.77
Benzo[2,3]phenanthro[4,5-bcd]thiophene	3.83	4.65	4.98
Benzo[1,2]phenaleno[4,3-bc]thiophene	—	4.16	4.81
Pyreno[4,5-b]thiophene	4.32	4.31	4.55
Benzo[1,2]phenaleno[3,4-bc]thiophene	—	4.35	4.62
Pyreno[2,1-b]thiophene	4.55	4.35	4.44
Benzo[4,5]phenaleno[9,1-bc]thiophene	—	4.36	4.51
Benzo[4,5]phenaleno[1,9-bc]thiophene	—	4.37	4.48
Pyreno[1,2-b]thiophene	4.32	4.45	4.62

Table 2. Continued

Compound	Normal-phase LC[b] (NH$_2$)	Reversed-phase LC[c] (C$_{18}$)	
		Poly.	Mono.
Five ring cata-condensed			
Benzo[b]phenanthro[4,3-d]-thiophene	4.00	4.18	4.80
Benzo[b]phenanthro[3,2-d] thiophene	4.91	4.52	4.84
Benzo[b]phenanthro[3,2-d]-thiophene	4.91	4.52	4.84
Benzo[b]phenanthro[9,10-d]-thiophene	4.49	4.68	5.17
Benzo[b]phenanthro[2,3-d]-thiophene	4.81	4.74	4.91
Benzo[b]phenanthro[3,4-d]-thiophene	4.49	4.81	5.21
Dinaphtho[2,3-b:2',3'-d]-thiophene	4.77	4.84	5.12
Dinaphtho[1,2-b:1',2'-d]-thiophene	4.33	4.95	5.27
Anthra[1,2-b]benzo[b]thiophene	4.52	>5(10.6)[d]	5.25
Dinaphtho[1,2-b:2',1'-d]thiophene	4.40	>5(11.7)[d]	5.51
Benzo[b]phenanthro[2,1-d]thiophene	4.58	>5(15.4)[d]	5.32
Dinaphtho[1,2-b:2',3'-d]thiophene	4.55	>5(15.7)[d]	5.38
Triphenyleno[2,1-b]thiophene	4.81	4.21	4.53
Triphenyleno[2,3-b]thiophene	4.16	4.30	4.61
Triphenyleno[1,2-b]thiophene	4.98	4.40	4.68

[a]Retention data reported as log I, where I = retention index (Ref. 6).
[b]Bondapak NH$_2$ column, n-hexane as the mobile phase
[c]Vydac 201TP 5 μm (poly.) and Zorbax ODS (mono.) columns, acetonitrile/water 85:15 (v/v) as the mobile phase.
[d]Log I > 5, values in parentheses are k'.
Source: Ref. 34.

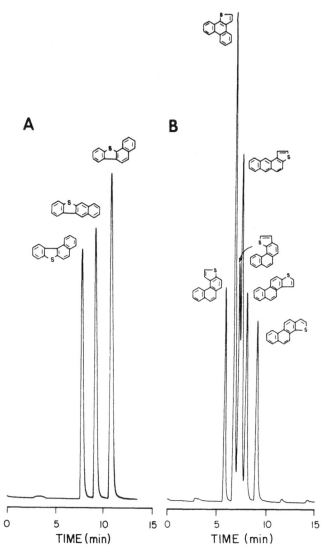

Figure 9. Reversed-phase LC separation of selected four-ring cata-condensed PASH of molecular weight 234 (A, B) and selected peri-condensed five-ring PASH of molecular weight 258 (C). Column: Vydac 201 TP C_{18} 5 μm. Mobile phase: linear gradient from 80 to 100% CH_3CN in H_2O at 1%/min at 1.5 ml/min. UV detection at 254 nm. (From Ref. 34.)

phase LC on a chemically bonded polar phase (i.e., aminosilane) can be used to separate PASH according to the number of aromatic carbons, as described previously for PAH [6].

In normal-phase LC, the cata-condensed PASH containing a thiophene ring with ring fusion on two sides elute in the same region as the corresponding PAH structure (e.g., benzofluorenes and the benzo[b]naphtho-thiophenes which both have 16 aromatic carbon atoms). The peri-condensed five-ring PASH with 18 aromatic carbons and ring fusion on three sides (e.g., triphenyleno[4,5-bcd] thiophene, chryseno[4,5-bcd]thiophene, and benzo-[2,3]phenanthro[4,5-bcd]thiophene) elute with the corresponding PAH with the same number of aromatic carbons [i.e., triphenylene (4.07), benz[a]-

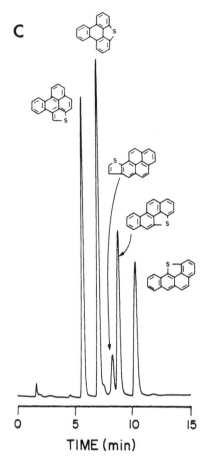

TIME (min)

Figure 9. Continued.

anthracene (4.00), and chrysene (4.01) (see Table 2)]. The five-ring peri-
condensed PASH elute as sulfur-substituted PAH rather than with the related
PAH structural analogues [i.e., BaP (4.38) and benzo[e]pyrene (4.46)]
(Table 2). Colmsjö et al. [33] also found that the peri-condensed PASH
produced fluorescence spectra similar to a sulfur-substituted PAH with the
same number of aromatic carbons. For the five-ring cata-condensed PASH
with ring fusion on two sides, the retention indices range from 4.33 to
4.91, excluding benzo[b]phenanthro[4,3-d]thiophene, which elutes earlier
due to its compact, nonplanar structure. In general, the PASH ring fusion
on only one side of the thiophene ring have longer normal-phase LC retention
than PASH with ring fusion on two sides of the thiophene ring (e.g., phen-
anthrothiophenes compared to benzo[b]naphthothiophenes).

Reversed-phase LC offers excellent selectivity for the separation of isomeric PASH, as found previously for PAH. Since monomeric and polymeric
C_{18} reversed-phase LC columns provide somewhat different selectivities,
PASH were studied on both types of columns [34]. In reversed-phase LC,
as in the case of the PAH, the polymeric C_{18} column provided better selectivity in the separation of PASH isomers than the monomeric column. Reversed-phase liquid chromatograms of selected four-ring cata-condensed
PASH isomers and five-ring peri-condensed PASH isomers are shown in
Fig. 9. The three benzo[b]naphthothiophenes (see Fig. 9A) are easily

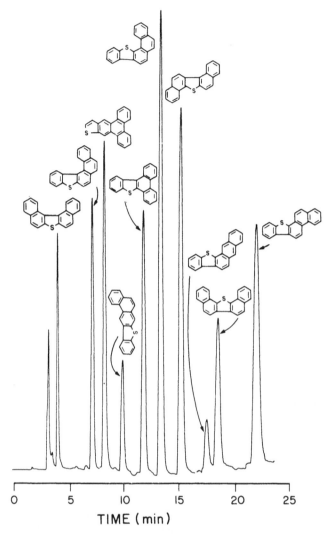

Figure 10. Reversed-phase LC separation of selected five-ring PASH of molecular weight 284. Conditions: same as Fig. 8, except linear gradient from 90 to 100% CH$_3$CN in H$_2$O at 1%/min. (From Ref. 34.)

separated as are triphenyleno[4,5-bcd]thiophene, chryseno[4,5-bcd]thiophene, and benzo[2,3]phenanthro[4,5-bcd]thiophene (see Fig. 9C). These two groups of isomers have been reported in environmental samples. The separation of 10 PASH isomers of molecular weight 284 is shown in Fig. 10. Several of these isomers are not generally separated using capillary GC (e.g., anthra[1,2-b]benzo[d]thiophene, benzo[b]phenanthro[2,1-d]thiophene, and benzo[b]phenanthro[3,4-d]thiophene). As in the case of PAH, the methyl-substituted PASH have normal-phase retention volumes similar to the parent PASH.

    The sequential combination of normal-phase and reversed-phase LC for the analysis of the PASH fraction isolated from an SRC-1 coal liquid has been

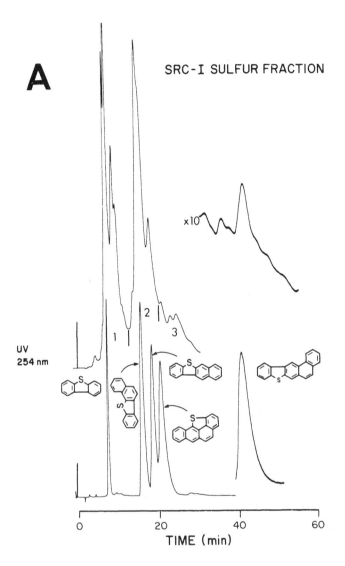

Figure 11. (A) Normal-phase LC fractionation of PASH fraction from SRC-I coal liquid and standard PASH compounds. Column: μBondapak NH_2. Mobile phase: n-hexane at 5 ml/min. UV detection at 254 nm. (B) Reversed-phase LC analysis of fraction 2 [see part (A)] of PASH fraction from SRC-I Coal Liquid. Conditions: same as Fig. 8, except linear gradient from 85 to 100% CH_3CN in H_2O at 1%/min. (From Ref. 34.)

reported [34]. The normal-phase prefractionation of the sample and the separation of selected PASH standards are shown in Fig. 11A. The reversed phase liquid chromatogram from the analysis of fraction 2 is shown in Fig. 11B. Peaks 1, 2, and 4 are the three benzo[b]naphthothiophene isomers as confirmed by LC retention data and fluorescence spectra. Peak 3 was tentatively identified as an alkyl-substituted benzo[b]naphtho[1,2-d]thiophene and peaks 6-11 were tentatively identified as alkyl-substituted benzo-[b]naphtho[2,1-d]thiophenes using fluorescence spectroscopy.

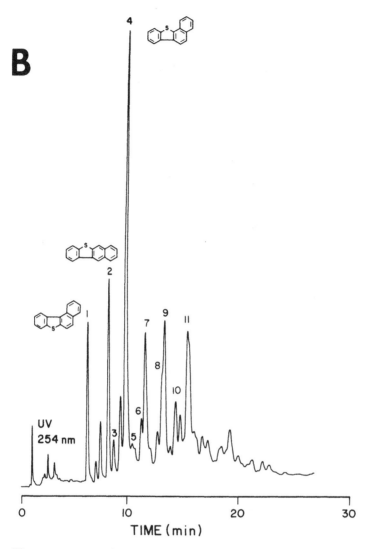

Figure 11. Continued.

V. HIGH-RESOLUTION LIQUID CHROMATOGRAPHY

Research directed toward increasing the efficiency of LC separations in general has focused on three approaches: (1) narrow-bore (capillary) microparticulate packed columns, (2) coupling conventional columns, and (3) short columns of conventional inner diameter packed with 3 to 5-μm particles. Although these efforts have not been directed toward PAH ananlyses specifically, many of these studies have used PAH mixtures to demonstrate the separation capabilities of these approaches.

The use of packed capillary columns to achieve high separation efficiencies has received considerable attention [35-37]. Yang [35] described a 2 m X 0.32 mm i.d. column packed with 3 μm C_{18} material with a total column plate count of 144,000 measured for pyrene (k' = 10). Unfortunately, about

6 hr was required for elution of pyrene from the column. He also reported the separation of the 16 priority pollutant PAH on a column with 92,000 plates and commented on the baseline resolution of benz[a]anthracene and chrysene. However, the separation of these two was probably due to the selectivity of the column packing material, as discussed previously, rather than the high column efficiency. Hirata and Jinno [38] illustrated the separation of the aromatic fraction of coal tar on a 1 m X 0.2 mm i.d. column packed with 3 µm material and a 5 m X 0.2 mm column packed with 10 µm material which required 3 and 14 hr, respectively, for completion.

As an alternative to the long analysis times required in capillary LC, Verzele and Dewaele [39] demonstrated that an improvement in efficiency of 10 over traditional LC could be achieved by coupling three conventional columns (25 cm X 0.46 cm i.d.) packed with 5 µm spherical silica. Separations with 20,000-100,000 plates were easily performed in less than 1 hr. This approach was illustrated by the separation of polychlorinated biphenyls (Arochlor 1260).

The use of short columns (100-125 mm) of conventional diameters (4-6 mm) packed with 3-5 µm particles to achieve separations in very short times has been described by DiCesare et al. [40] as "very high speed LC." Dong and DiCesare [41,42] compared the separation of PAH by conventional LC versus very high speed LC. Comparable separations of the 16-priority-pollutant PAH were achieved in less than 8 min using the very high speed LC (5 µm particles, 3.5 m./min) compared to 40 min on a conventional column (10 µm particles, 0.5 ml/min). At present, these later two approaches using conventional i.d. columns offer a more practical solution to the quest for high-resolution LC since capillary LC requires such long analysis times.

VI. ANALYSIS OF COMPLEX PAH MIXTURES

PAH mixtures encountered in environmental samples are extremely complex because of the presence of alkyl-substituted PAH as well as the numerous isomeric parent compounds. As a result of this complexity, the identification and quantification of individual PAH in such mixtures by LC requires the use of the following approaches: (1) column selectivity/multidimensional chromatography (LC/LC or LC/GC), (2) detector selectivity (ultraviolet or fluorescence), or (3) a combination of multidimensional chromatography with selective detection. The use of these three approaches to determine individual PAH in complex mixtures will be discussed.

A. Selectivity in Detection

Because of the complexity of PAH mixtures from environmental samples, selective fluorescence detection has found widespread use for the determination of individual PAH. By selection of the appropriate excitation and emission wavelengths, a high degree of specificity can be achieved. This spectral selectivity often permits the determination of individual PAH in the mixture even when complete liquid chromatographic resolution of the components is not achieved. The selectivity of LC analysis with fluorescence detection is illustrated in Fig. 12 for the analysis of an air particulate PAH fraction [43, 44]. In Fig. 12 the upper chromatogram (A) is the reversed-phase LC analysis of a total PAH fraction (obtained from normal-phase LC isolation procedure [6]) with ultraviolet (UV) detection. UV detection provides a nearly

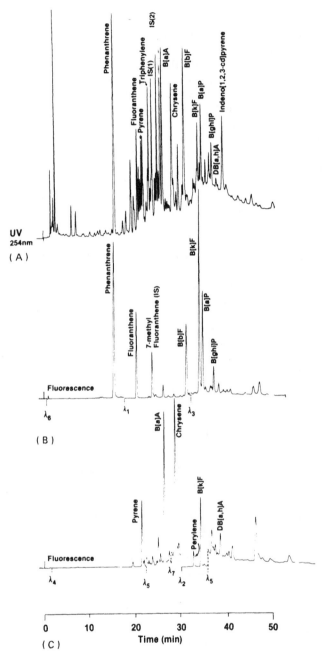

Figure 12. Reversed-phase LC analysis of PAH fraction from Washington, D.C., air particulate sample (SRM 1649) isolated by normal-phase LC clean-up: (A) UV detection at 254 nm; (B,C) fluorescence detection at conditions in Table 3. Column: Vydac 201TP C_{18} 5 μm. Mobile phase: linear gradient from 40 to 100% CH_3CN in H_2O at 1%/min at 1 ml/min.

"universal" LC detector for PAH. The major PAH are discernible in this chromatogram; however, accurate quantification of the majority of these components would be difficult due to incomplete chromatographic resolution. Chromatograms B and C in Fig. 12 are the LC analyses of the same total PAH fraction using fluorescence detection at the wavelength conditions described in Table 3. The selectivity of the fluorescence detection is enhanced by changing the fluorescence excitation and emission wavelengths during the liquid chromatographic separation (i.e., wavelength programming) in order to optimize the detection of each PAH measured.

Of particular interest is the extreme selectivity achievable for the determination of perylene. In Fig. 12A, perylene is present at very low levels and is not quantifiable from the UV chromatogram. However, with excitation at 400 nm and emission at 440 nm, even trace amounts of perylene are detectable in the presence of other coeluting PAH (see chromatogram C). Hellmann [45] utilized the specificity of fluorescence detection (excitation at 430 nm, emission at 466 nm) to measure perylene in extracts from water and soil samples.

The PAH fraction shown in the chromatograms in Fig. 12 was isolated by normal-phase LC prior to the reversed-phase LC analysis. However, fluorescence detection offers an extremely selective response for PAH, as illustrated in Fig. 13. The air particulate extract analyzed in Fig. 13 was not cleaned up by normal-phase LC as was the sample shown in Fig. 12 [44]; instead, the hexane soluble extract was only partitioned into nitromethane prior to analysis. The selectivity of fluorescence detection with wavelength programming is illustrated in Fig. 13 by comparing chromatograms C, D, and E with the UV chromatogram (A).

Using LC with fluorescence wavelength programming, May and Wise [44] quantified 13 PAH in an air particulate material which is now available as a Standard Reference Material (SRM 1649—Urban Dust/Organics) [46]. The results of three LC methods (i.e., different internal standards and/or different cleanup procedures) are summarized in Table 4. Results obtained on the same SRM material using capillary gas chromatography (GC) are also shown in Table 4. Perylene-d_{12} was found to be an excellent internal standard for these analyses because of the selective excitation/emission characteristics. (Additional advantages of using deuterated PAH as internal standards in complex mixtures are described later.) This LC method described for the analysis of the air particulate material was also used for the determination of PAH in diesel particulate samples [44].

Stöber and Reupert [47] have reported a similar LC method using fluorescence wavelength programming for the determination of PAH in river water samples. A total of nine PAH were determined in two different chromatographic runs each with three different sets of excitation and emission wavelength conditions. The detection limits as reported by Stöber and Reupert for the nine PAH under the six different wavelength conditions are summarized in Table 5. Values for each PAH were obtained in each of the two chromatographic runs, except for pyrene, perylene, and indeno[1,2,3-cd]pyrene, which had a low response in run A. Because of the simplicity, precision, and accuracy of this LC/fluorescence method, Stöber and Reupert [47] suggested that it be considered as a reference method for the determination of PAH in water samples.

In the LC method described by May and Wise [44] and Stöber and Reupert [47], the fluorescence detector could be programmed to change excitation and emission wavelength conditions a maximum of three times during a chro-

Table 3. Fluorescence Conditions for the LC Determination of Selected PAH in Particulate Matter

	Wavelengths (nm)		PAH quantified
	Excitation	Emission	
λ_1	285	450	Fluoranthene and 7-methylfluoranthene (I.S.)[a]
λ_2	400	440	Perylene-d_{12} (I.S.) and perylene
λ_3	295	405	Benzo[k]fluoranthene, benzo[a]pyrene, and benzo[ghi]perylene
λ_4	330	385	Pyrene
λ_5	285	385	Benz[a]anthracene, dibenz[a,h]anthracene, and benzo[ghi]perylene
λ_6	290	360	Phenanthrene
λ_7	270	360	Chrysene
λ_8	300	500	Indeno[1,2,3-cd]pyrene
λ_9	250	400	Anthracene[b]

[a]I.S., internal standard.
[b]Not measured in the certification of SRM 1649.
Source: Ref. 44.

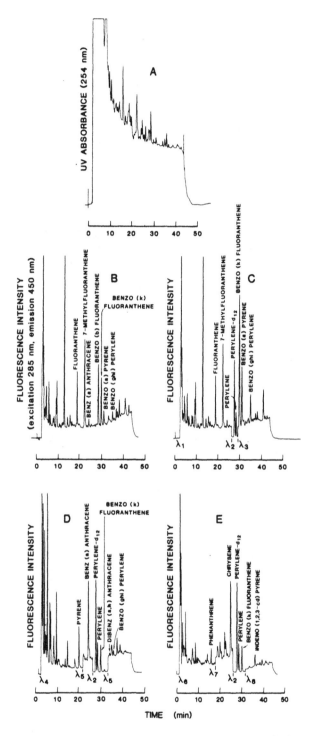

Figure 13. Reversed-phase LC analysis of air particulate extract (SRM 1649). No normal-phase LC cleanup. Column: Vydac 201TP C_{18} 5 μm. Mobile phase: linear gradient from 40 to 100% in 45 min at 1.5 ml/min. Detection conditions: (A) UV absorption at 254 nm; (B) fluorescence detection at excitation 285 nm and emission at 450 nm; (C-E) fluorescence detection at conditions listed in Table 3. (From Ref. 44.)

Table 4. Summary of Results by the Various Analytical Methods for the Analysis of the Washington, D.C., Urban Air Particulate Material (SRM 1649)

Compound	Analytical methods[a,b]			
	GC [4] (μg/g)	LC-I [18] (μg/g)	LC-II [9] (μg/g)	LC-III [3] (μg/g)
Phenanthrene	—	—	4.5 ± 0.3[c]	4.7 ± 0.1
Fluoranthene[d]	7.3 ± 0.2[c]	7.0 ± 0.5[c] (24)	6.8 ± 0.4	7.1 ± 0.5
Pyrene	7.2 ± 0.2	6.3 ± 0.4 (17)	6.2 ± 0.2	6.0 ± 0.2
Chrysene	4.6 ± 0.2[e]	3.5 ± 0.1 (5)	3.7 ± 0.2 3.7 ± 0.1[f] (3)	3.5 ± 0.1
Benz[a]anthracene[d]	2.4 ± 0.1	2.8 ± 0.3	2.4 ± 0.1 (3)	2.7 ± 0.1
Triphenylene	—	—	1.7 ± 0.1[f] (3)	—
Perylene	0.84 ± 0.09	0.80 ± 0.004 (17)	0.65 ± 0.02	0.74 ± 0.005
Benzo[e]pyrene	3.3 ± 0.2	—	—	3.9 ± 0.3

Benzo[a]pyrene[d]	3.0 ± 0.3	2.6 ± 0.4	2.6 ± 0.1	2.4 ± 0.2
Benzo[b]fluoranthene	—	6.2 ± 0.3	—	6.0 ± 0.3
Benzo[k]fluoranthene	—	2.0 ± 0.1	2.1 ± 0.1	2.0 ± 0.1
Benzo[ghi]perylene[d]	4.7 ± 0.2	3.9 ± 0.8 (12)	5.2 ± 0.6	4.1 ± 0.1
Dibenz[a,h]anthracene	—	—	0.41 ± 0.07	0.45 ± 0.04
Indeno[1,2,3-cd]pyrene[d]	3.3 ± 0.3	3.4 ± 0.4 (16)	3.6 ± 0.2	3.4 ± 0.1

[a]LC-I, 7-methylfluoranthene used as internal standard; sample cleanup consisted of partitioning into nitromethane; LC-II, perylene-d$_{12}$ used as internal standard; sample preparation same as LC-I; LC-III, perylene-d$_{12}$ and 7-methyl-fluoranthene used as internal standards; sample cleanup consisted of isolation of total PAH fraction by normal-phase LC on aminosilane column.

[b]Numbers in brackets indicate number of samples extracted; numbers in parentheses indicate number of measurements, if different from number in brackets.

[c]Uncertainty is 1 standard deviation.

[d]Indicates compounds with certified values.

[e]GC determination includes triphenylene, which coelutes with chrysene.

[f]Triphenylene and chrysene determined by LC analysis of 18-aromatic-carbon PAH fraction isolated by normal-phase LC on aminosilane column; benz[a]anthracene (2.4 μg/g) used as internal standard.

Source: Ref. 44.

Table 5. Detection Limits (ng) for PAH Using Different Wavelength Combinations ($\lambda_{ex}/\lambda_{em} = \lambda_i$)

| | Run A | | | | Run B | |
| | λ_1 | λ_2 | λ_3 | λ_4 | λ_5 | λ_6 |
Compound	355/425	350/415	355/415	335/420	385/455	300/502
Anthracene	0.038			0.055		
Fluoranthene	0.090			0.130		
Pyrene	4			0.062		
Perylene		4			0.024	
Benzo[b]fluoranthene		0.075			1.6	
Benzo[k]fluoranthene		0.040			0.050	
Benzo[a]pyrene		0.038			0.064	
Benzo[ghi]perylene			0.09			2.2
Indeno[1,2,3-cd]pyrene			12			0.150

Source: Ref. 47.

matographic run. Thus it was necessary to make several chromatographic runs (e.g., three as shown in Fig. 13) to quantitate the 13 PAH of interest. However, a fluorescence detector is now available which is capable of changing wavelength conditions 15 times in one chromatographic run, thereby making it possible to quantify all of the major PAH during one chromatographic run rather than three as shown in Fig. 13. Using this fluorescence detector, the PAH isolated from an air particulate extract (SRM 1648) were determined in one chromatographic run as shown in Fig. 14.

Other workers [48-52] have reported the use of LC with fluorescence detection for determination of PAH in environmental samples. Matsushita et al. [48] measured levels of BaP in air particulate samples with fluorescence detection at an excitation wavelength of 370 nm and an emission wavelength of 406 nm. Obana et al. [49] used a variable-wavelength fluorescence detector at three sets of conditions (excitation/emission wavelengths: 365/430, 334/384, and 300/393) to monitor nine PAH in sediments, oysters, and seaweed. Using the same LC/fluorescence methodology, these workers [50] also quantified parts per trillion (ppt) levels of PAH in human fat and liver tissues. A filter fluorometer was used by Joe et al. [51] to measure PAH in barley malt. Kodama et al. [52[determined BaP in air particulate extracts after cleanup on alumina using fluorescence at excitation 368 nm and emission at 406 nm.

In the past several years, videofluorometry for LC detection of PAH in complex mixtures has been developed particularly by Warner and co-workers [53,54]. Using liquid chromatography/videofluorometry (LC/VF), they described the qualitative analysis of a crude oil ash residue sample. Using LC/VF, excitation and emission spectral data were collected during the chromatographic run. Following the chromatographic analysis, these data were reconstructed as emission/excitation matrices (EEMS); spectral deconvolution techniques were used and the resulting spectra compared to standard PAH spectra compiled in a reference library to identify the components in the chromatogram. As illustrated by Warner and co-workers [53, 54], LC/VF offers the potential of real-time aquisition of fluorescence spectral information on chromatographic effluents.

Even though UV spectra generally provide less definitive information on the identity of a PAH, rapid-scanning UV detection has also been used for the monitoring of the LC effluents containing PAH. Choudhury and Bush [55] confirmed the presence of 15 PAH in various air particulate samples using LC with rapid-scanning UV detection and gas chromatography/mass spectrometry (GC/MS). However, accurate quantification using only UV detection is generally difficult, as mentioned previously, since many PAH and alkyl-substituted PAH coelute and more selective detection (e.g., fluorescence detection) is required. Variable-wavelength UV detection and stop-flow scanning were used by Readman et al. [56] for the determination of PAH extracted from an estuarine sediment.

Several novel LC detection systems have been reported for possible use in the detection of PAH. Schaper [57] described a detector based on thin-layer electrochemiluminescence (ECL) which allowed monitoring of compounds both by their ECL emission and by the electrochemical activity. Using this system 25 pg of rubrene was detected. Locke et al. [58] described a direct liquid-phase photoionization detector with picogram sensitivity to PAH. However, this detector could not be used with the common reversed-phase mobile phases due to high background currents. A spiked mobile-phase fluorescence detector, in which aniline was added to the mobile

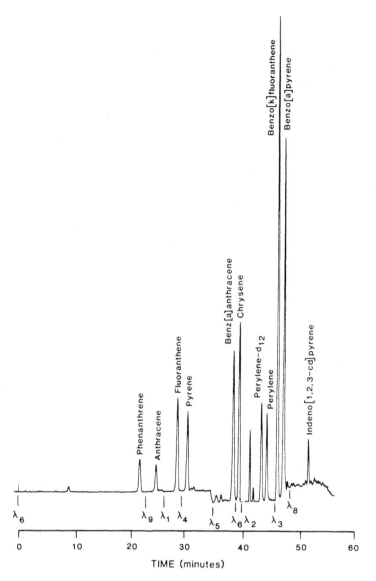

Figure 14. Reversed-phase LC analysis of total PAH fraction from St. Louis, Mo., air particulate sample (SRM 1648) isolated by normal-phase LC clean-up. Conditions: same as Fig. 12. (From Ref. 44.)

phase to enhance fluorescence, was described by Su et al. [59]. Armstrong et al. [60] reported an enhanced fluorescence response for PAH using a mobile phase containing sodium dodecyl sulfate. For example, pyrene was enhanced by a factor of 10, benz[a]anthracene by a factor of 4, and biphenyl by a factor of 4. Even though all of these novel detection systems [57-60] were evaluated using PAH as test solutes, no applications to the determination of PAH in real samples were reported.

B. Multidimensional Liquid Chromatographic Techniques

The second approach to the analysis of complex mixtures is the use of multi-dimensional chromatographic techniques (i.e., the sequential use of different chromatographic separation modes). In this section the use and advantages of the multidimensional liquid chromatographic approach to isolate and measure individual components will be presented. Multidimensional chromatography is used in two ways: (1) for the detailed characterization of a complex mixture, and (2) for the determination of a few selected components in the complex mixture. For applications involving extremely complex matrices (e.g., oil, coal liquids, coal tar, etc.), a combination of multidimensional chromatography with selective detection is often required to achieve accurate quantification.

An excellent example illustrating multidimensional chromatography is the characterization of PAH from an air particulate sample using normal-phase LC to isolate specific PAH fractions followed by GC, GC/MS, or reversed-phase LC analysis [43,44]. The PAH mixture from the air particulate extract was separated into fractions using normal-phase LC on an aminosilane column. The normal-phase LC procedure separates the PAH according to the number of aromatic carbon atoms in the PAH as shown in Table 6. In addition, alkyl-substituted PAH have normal-phase LC retention characteristics on the aminosilane column similar to those of the unsubstituted PAH (see Table 6). Thus the separation of PAH mixtures using this normal-phase LC procedure results in fractions that contain isomeric PAH and their alkyl-substituted homologs. The reversed-phase LC separations of fractions isolated in this manner are shown in Fig. 15. Each fraction contains PAH with the same number of aromatic carbon atoms: for example, fraction 2, phenanthrene and anthracene (14 aromatic carbons); fraction 3, pyrene and fluoranthene (16 aromatic carbons); fraction 4, triphenylene, chrysene, and benz-[a]anthracene (18 aromatic carbons); fraction 5, benzo[b]fluoranthene perylene, benzo[k]fluoranthene, and benzo[a]pyrene (20 aromatic carbons); and fraction 6, benzo[ghi]perylene and indeno[1,2,3-cd]pyrene (22 aromatic carbons). These chromatograms illustrate the usefulness of this approach for the analysis of complex mixtures. Using UV absorption detection at 254 nm (generally considered as a "universal" detector for PAH), quantification of even the major components in the total PAH fraction would be difficult owing to the complexity of the mixture. However, the normal-phase preseparation provides fractions suitable for LC analysis even with the universal UV detector. This method was used to obtain the value for triplenylene in Table 4.

For most analyses of environmental samples by GC or LC, only the 10-15 major PAH components are determined. However, for a detailed characterization of such a complex PAH mixture, a normal-phase LC preseparation provides fractions in which many of the minor components can be separated from the major components and identified. This multidimensional approach (i.e., normal-phase LC followed by GC and/or reversed-phase LC) has been described for the detailed characterization of the PAH from two air particulate samples [61,62]. The normal-phase LC separation of one air particulate extract is shown in Fig. 16. The individual fractions were analyzed by GC/MS and the components were identified based on GC retention times and mass spectra. Retention times and fluorescence spectra obtained from the LC analyses of these same fractions were also used for identifications. The GC analyses of these fractions are shown in Fig. 17 and the results are summarized in Table 7. Quantification of the numerous minor components in each

Table 6. Liquid Chromatographic Retention of Polycyclic Aromatic Hydrocarbons[a]

Compound	Mol. wt.	Number of aromatic C	NH$_2$ column retention index[b]	C$_{18}$ column retention index[c]
Naphthalene	128	10	2.00	2.00
Fluorene	166	12	2.55	2.70
Anthracene	178	14	2.94	3.20
2-Methylanthracene	192	14	3.01	3.71
9-Methylanthracene	192	14	3.02	3.39
Phenanthrene	178	14	3.00	3.00
Phenanthrene-d$_{10}$	188	14	3.03	2.91
1-Methylphenanthrene	192	14	3.02	3.38
2-Methylphenanthrene	192	14	3.00	3.72
3-Methylphenanthrene	192	14	3.12	3.32
9-Methylphenanthrene	192	14	3.02	3.31
4H-Cyclopenta[def]phenanthrene	190	14	3.10	3.16
Benzo[a]fluorene	216	16	3.51	3.72
Benzo[b]fluorene	216	16	3.54	3.84
Pyrene	202	16	3.37	3.48
Pyrene-d$_{10}$	212	16	3.44	3.40
1-Methylpyrene	216	16	3.46	3.90
Fluoranthene	202	16	3.51	3.37

Compound			
Fluoranthene-d$_{10}$	212	3.55	3.31
Benz[a]anthracene	228	4.00	4.00
Benz[a]anthracene-d$_{12}$	240	4.05	3.91
1-Methylbenz[a]anthracene	242	3.90	4.14
5-Methylbenz[a]anthracene	242	4.04	4.28
6-Methylbenz[a]anthracene	242	4.03	4.10
9-Methylbenz[a]anthracene	242	4.08	4.39
Chrysene	228	4.01	4.10
Chrysene-d$_{12}$	240	4.06	4.03
1-Methylchrysene	242	4.07	4.43
3-Methylchrysene	242	4.12	4.29
5-Methylchrysene	242	3.94	4.35
Triphenylene	228	4.07	3.70
Benzo[c]phenanthrene	228	3.64	3.64
Benzo[ghi]fluoranthene	226	3.84	3.95
Benzo[b]fluoranthene	252	4.48	4.29
Benzo[j]fluoranthene	252	4.56	4.24
Benzo[k]fluoranthene	252	4.45	4.42
Benzo[a]pyrene	252	4.38	4.68
Benzo[a]pyrene-d$_{12}$	264	4.49	4.61
Benzo[e]pyrene	252	4.46	4.28

Table 6. Continued

Compound	Mol. wt.	Number of aromatic C	NH$_2$ column retention index[b]	C$_{18}$ column retention index[c]
Perylene-d$_{12}$	264	20	—	4.27
Perylene	252	20	4.61	4.33
Benzo[ghi]perylene	276	22	4.83	4.73
Anthanthrene	276	22	4.80	4.93
Indeno[1,2,3-cd]pyrene	276	22	4.90	4.83
Dibenz[a,c]anthracene	278	22	4.93	4.40
Dibenz[a,h]anthracene	278	22	4.94	4.72
Benzo[b]chrysene	278	22	5.00	5.00
Picene	278	22	5.03	5.10

[a]Retention reported as logarithm of the retention index (see Ref. 6).
[b]n-Hexane as the mobile phase (normal phase).
[c]Mixtures of 70 or 85% acetonitrile in water as mobile phase (reversed phase).
Source: Refs. 6 and 44.

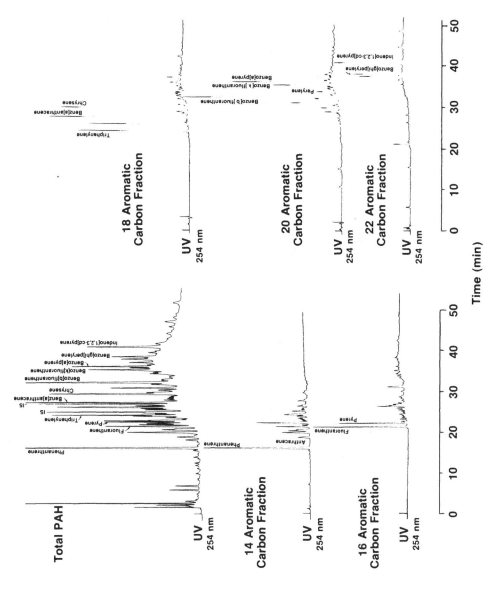

Figure 15. Reversed-phase LC analysis of fractions obtained from normal-phase LC fractionation of extract of SRM 1649. Conditions: same as Fig. 11, UV detection at 254 nm. (From Ref. 43.)

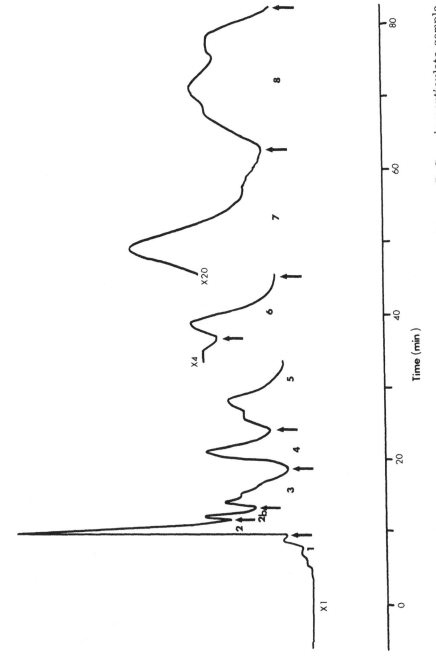

Figure 16. Normal-phase LC separation of PAH isolated from Washington, D.C., air particulate sample (SRM 1649). Column: μBondapak NH$_2$. Mobile phase: 0.5% CH$_2$Cl$_2$ in n-C$_5$H$_{12}$. UV detection at 254 nm. (From Ref. 61.)

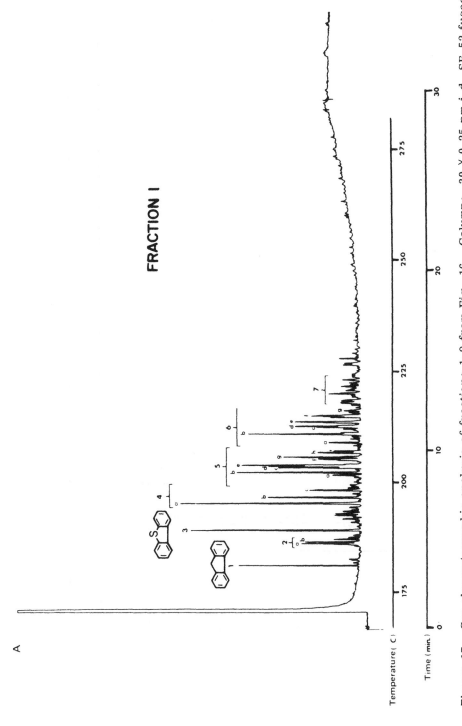

Figure 17. Gas chromatographic analysis of fractions 1-8 from Fig. 16. Column: 30 X 0.25 nm i.d. SE-52 fused silica capillary. (A-E) Temperature programmed from 175°C at 4°C/min; (F-I) initial temperature 200°C. See Table 7 for compound identifications. (From Ref. 61.)

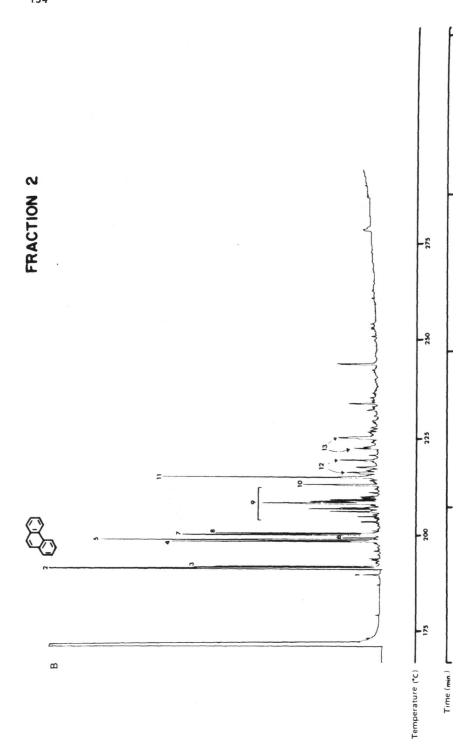

Figure 17.   Continued

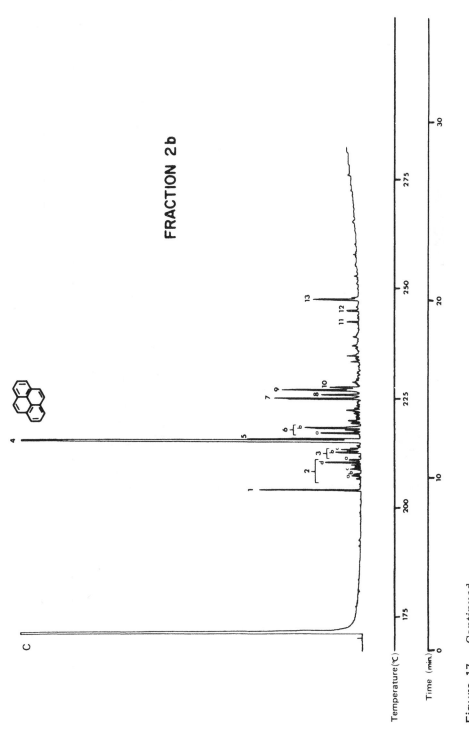

Figure 17.   Continued

FRACTION 3

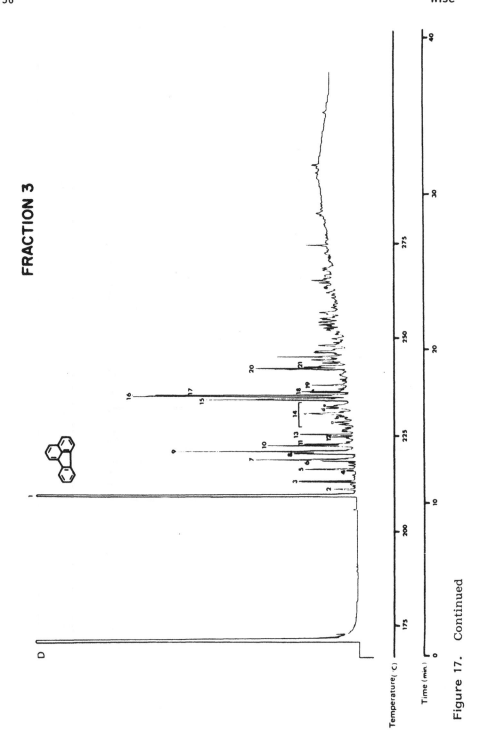

Figure 17.   Continued

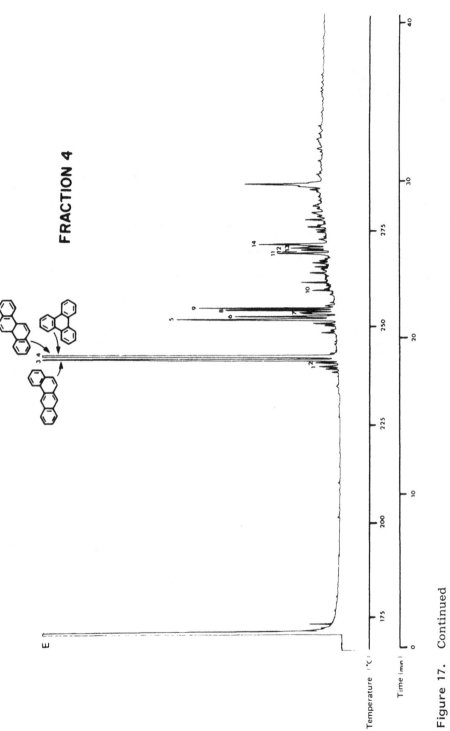

Figure 17. Continued

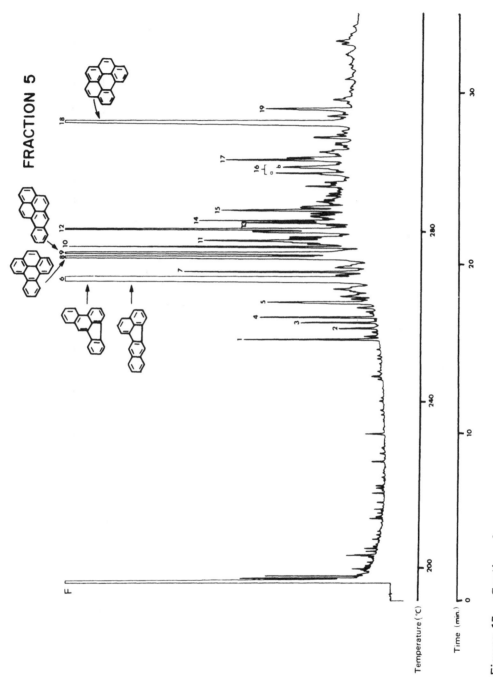

Figure 17. Continued

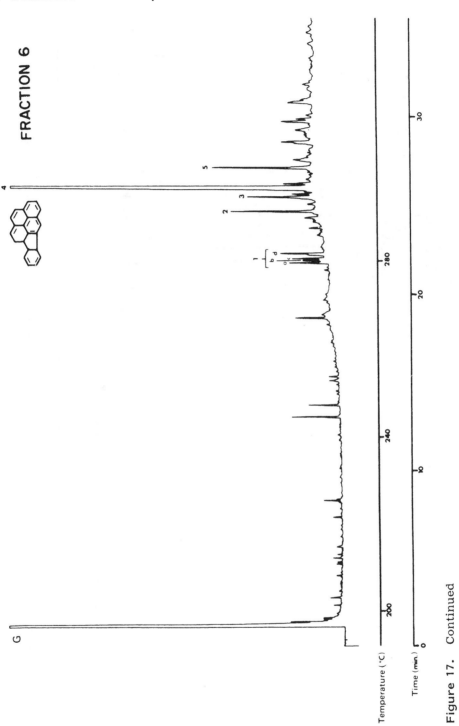

Figure 17.   Continued

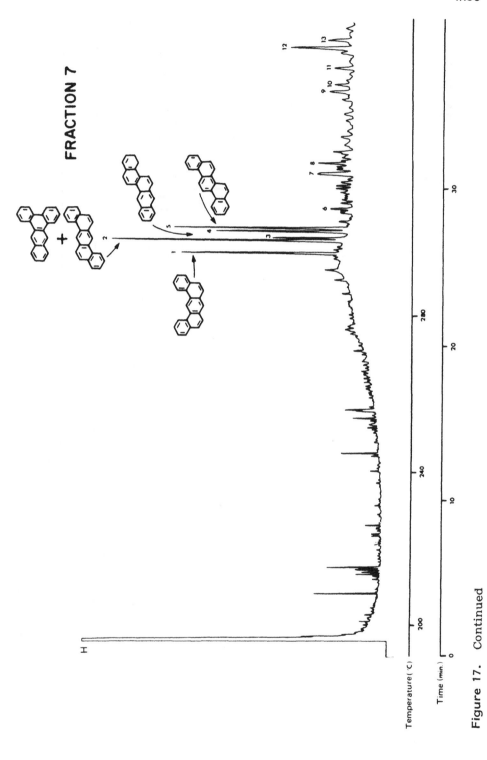

Figure 17. Continued

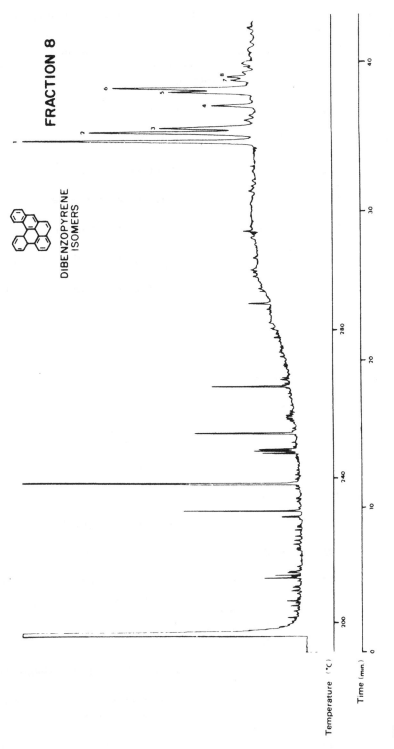

Figure 17. Continued

Table 7. Polycyclic Aromatic Hydrocarbons Identified in Washington, D.C.,
Urban Particulate Matter (SRM 1649)[a]

Peak	Mol. wt.	Conc. (μg/g)	Compound
		Fraction 1	
1	166	—	Fluorene[b,c]
2 (a,b)	180	—	Methylfluorene[b]
3	184	—	Dibenzothiophene[b,c]
4 (a-c)	198	—	Methyldibenzothiophene[b]
5 (a-i)	212	—	C_2-substituted dibenzothiophene[b]
6 (a-g)	226	—	C_3-substituted dibenzothiophene[b]
7	240	—	C_4-substituted dibenzothiophene[b]
		Fraction 2	
1	184	—	Dibenzothiophene[b,c]
2	178	4.5 ± 0.3	Phenanthrene[b,c,d]
3	178	0.49	Anthracene[b,c,d]
4	192	0.62	3-Methylphenanthrene[b,c]
5	192	0.80	2-Methylphenanthrene[b,c]
6	192	0.11	2-Methylanthracene[b,c]
7	190	1.0	4H-Cyclopenta[def]phenanthrene[b,c]
8	192	0.47	1-Methylphenanthrene[b,c]
9	206	0.06, 0.08, 0.19, 0.22, 0.12, 0.46, 0.33, 0.18, 0.11	C_2-substituted phenanthrene/anthracene[b]
10	208	0.29	Phenanthro[4,5-bcd]thiophene[b,c]
11	218	0.95	C_2-substituted 4H-cyclopenta[def]-phenanthrene[b]
12 (a,b)	218	0.15, 0.12	C_2-substituted 4H-cyclopenta[def]-phenanthrene[b]

Table 7. Continued

Peak	Mol. wt.	Conc. (μg/g)	Compound
13 (a,b)	232	0.15, 0.40	C_3-substituted 4*H*-cyclopenta[def]-phenanthrene[b]
		Fraction 2b	
1	204	0.30	C_1-substituted 4*H*-cyclopenta[def]-phenanthrene[b]
2	206	0.13 (d)	C_2-substituted phenanthrene/anthracene[b]
3	218	0.10 (b)	C_2-substituted 4*H*-cyclopenta[def]-phenanthrene[b]
4	202	7.2 ± 0.2	Pyrene[b,c,d]
5	204	0.33	C_1-substituted 4*H*-cyclopenta[def]-phenanthrene[b]
6 (a,b)	218	0.15, 0.23	C_2-substituted 4*H*-cyclopenta[def]-phenanthrene[b]
7	216	0.36	2-Methyl- or 4-methylpyrene[b,c]
8	218	0.15	C_2-substituted 4*H*-cyclopenta[def]-phenanthrene[b]
9	216	0.47	1-Methylpyrene[b,c]
10	218	0.12	C_2-substituted 4*H*-cyclopenta[def]-phenanthrene[b]
11	244	0.04	C_3-substituted pyrene/fluoranthene[b]
12	258	0.04	Unknown
13	242	0.21	Unknown
		Fraction 3	
1	202	7.1 ± 0.5	Fluoranthene[b,c,d]
2	202	0.09	Acephenanthylene[b]
3	202	0.19	Pyrene (overlap from fraction 2b)[b,c]
4	230	—	Unknown
5	218	0.17	C_2-substituted 4*H*-cyclopenta[def]-phenanthrene[b]
6	230	0.52	C_2-substituted fluoranthene/pyrene[b]
7	216		Methylfluoranthene[b,e]

Table 7. Continued

Peak	Mol. wt.	Conc. (μg/g)	Compound
8	216	0.16	Methylfluoranthene[b,e]
9	216	0.78	Benzo[a]fluorene[b,c]
10	216	0.32	Benzo[b]fluorene[b,c]
11	216	0.13	Methylfluoranthene[b,e]
12	216	0.08	Methylfluoranthene[b,e]
13	216	0.19	Methylfluoranthene[b,e]
14 (a-e)	230	0.18 (d)	C_2-substituted fluoranthene/pyrene[b]
15	234	0.77	Benzo[b]naphtho[2,1-d]thiophene[b,c]
16	226		Benzo[ghi]fluoranthene[b,c]
		1.51	
17	228		Benzo[c]phenanthrene[b,c]
18	234	0.22	Benzo[b]naphtho[1,2-d]thiophene[b,c]
19	234	0.16	Benzo[b]naphtho[2,3-d]thiophene[b,c]
20	228		Benz[a]anthracene (overlap with fraction 4)[b,c]
	226	0.41	Cyclopenta[cd]pyrene[b,c]
21	228	0.16	Chrysene (overlap with fraction 4)[b,c]

Fraction 4

Peak	Mol. wt.	Conc. (μg/g)	Compound
1	234	0.10	Benzo[b]naphtho[1,2-d]thiophene isomer[b]
2	234	0.10	Benzo[b]naphtho[1,2-d]thiophene isomer[b]
3	228	2.6 ± 0.3	Benz[a]anthracene[b,c]
4	228		Chrysene[b,c]
	228	5.5	Triphenylene[b,c]
5	242	1.0	Methylchrysene/benz[a]anthracene/triphenylene[b]
6	242	0.52	Methylchrysene/benz[a]anthracene/triphenylene[b]

Table 7. Continued

Peak	Mol. wt.	Conc. (μg/g)	Compound
7	242	0.26	Methylchrysene/benz[a]anthracene/ triphenylene[b]
8	240	0.52	Methylbenzo[ghi]fluoranthene[b]
9	242	0.61	Methylchrysene/benz[a]anthracene triphenylene[b]
10	256	0.12	C_2-substituted 228 isomer[b]
11	268	0.31	Unknown
12	258	0.12	Unknown
	268		Unknown
13	268	0.25	Unknown
14	258	0.46	Unknown
	268		Unknown

Fraction 5

Peak	Mol. wt.	Conc. (μg/g)	Compound
1	242	0.24	Methyl-substituted 228 isomer[b]
2	254	0.07	Binaphthyl[b]
3	254	0.12	Binaphthyl[b]
4	254	0.22	Binaphthyl[b]
5	256	0.47	C_2-substituted 228 isomer[b]
6	252	8.0	Benzofluoranthenes[b,c]
7	252	0.32	Unknown
8	252	2.8	Benzo[e]pyrene[b,c,d]
9	252	2.9 ± 0.5	Benzo[a]pyrene[b,c,d]
10	252	0.49	Perylene[b,c,d]
11	266	0.52	Methyl-substituted 252 isomer or dibenzofluorene isomer[b]
12	306	0.52	Unknown
13	266	0.21	Methyl-substituted 252 isomer or dibenzofluorene isomer[b]
14	266	0.30	Methyl-substituted 252 isomer or dibenzofluorene isomer[b]
15	264	0.37	Unknown

Table 7. Continued

Peak	Mol. wt.	Conc. (μg/g)	Compound
16 (a,b)	284	0.37, 0.34	C_2-substituted 252 isomer[b]
17	276	0.65[f]	Unknown
18	276	4.5 ± 1.1[f]	Benzo[ghi]perylene[b,c,d]
19	276	0.50[f]	Anthanthrene[b,c]
		Fraction 6	
1 (a-d)	300	0.10, 0.08 0.09, 0.12	Unknown
2	276	0.28	Indeno[1,2,3-cd]fluoranthene[b,c]
3	276	0.31	Unknown
4	276	3.3 ± 0.5	Indeno[1,2,3-cd]pyrene[b,c,d]
5	276	0.36	Benzo[ghi]perylene (overlap from fraction 5)[b,c]
		Fraction 7	
1	278	—	Dibenz[a,j]anthracene[b,c,d]
2	278	—	Dibenz[a,c]anthracene and dibenz-[a,h]anthracene[b,c,d]
3	278	—	Pentaphene[b,c,d]
4	278	—	Benzo[b]chrysene[b,c,d]
5	278	—	Picene[b,c,d]
6	290	—	Unknown
7	292	—	Methyl-substituted 278 isomer[b]
8	304	—	Unknown
9	302	—	Dibenzopyrene[b]
10	302	—	Dibenzopyrene[b]
11	302	—	Dibenzopyrene[b]
12	300	—	Unknown
13	300	—	Unknown
		Fraction 8	
1	302	—	Dibenzopyrene isomer[b]

Table 7. Continued

Peak	Mol. wt.	Conc. (µg/g)	Compound
2	302	—	Dibenzopyrene isomer[b]
3	302	—	Dibenzopyrene isomer[b]
4	302	—	Dibenzopyrene isomer[b]
5	302	—	Dibenzopyrene isomer[b]
6	302	—	Dibenzopyrene isomer[b]

[a]Identifications are best estimates based on mass spectral data, GC and LC retention data compared to standard compounds, and fluorescence spectra compared to standard compounds. Quantitation is based on using the certified and information values for selected PAH as internal standards in a particular fraction, i.e., fraction 2, phenanthrene; fraction 2b, pyrene; fraction 3, fluoranthene; fraction 4, benz[a]anthracene; fraction 5, benzo[a]pyrene; fraction 6, indeno[1,2,3-cd]pyrene (see Table 4).
[b]Identification based on MS.
[c]Identification based on GC retention.
[d]Identification based on LC retention and fluorescence spectra.
[e]Identification as methylfluoranthene rather than methylpyrene based on normal-phase LC retention.
[f]Values based on benzo[ghi]perylene as internal standard rather than benzo[a]pyrene.
Source: Refs. 61 and 62.

fraction was achieved by using one of the major peaks in each fraction as an internal standard. The values for these internal standards were the values determined previously in the unfractionated samples (see Table 4).

Fraction 1 contains fluorene and dibenzothiophene as the major parent PAH and PASH. Methylfluorenes and C_1- through C_4-substituted dibenzothiophenes are the remaining constituents. Both fluorene and dibenzothiophene have 12 aromatic carbons and elute prior to phenanthrene in the normal-phase LC separation (retention indices of 2.55 and 2.87, respectively). Phenanthrene is the major peak in fraction 2. Anthracene is present, but only at about the same level as several of the methylphenanthrenes (see Table 4). 4H-Cyclopenta[def]phenanthrene and the sulfur-containing analog of molecular weight 208 are also included in fraction 2, as would be expected by the number of aromatic carbons.

The isomeric pyrene and fluoranthene were isolated into separate LC fractions, 2b and 3 (LC retention indices of 3.37 and 3.51, respectively) for the GC analysis, whereas they were combined in one fraction for the reversed-phase LC analysis shown in Fig. 15 (fraction 3). Minor components in the pyrene fraction include alkyl-substituted 4H-cyclopenta[def]phenanthrene isomers and methylpyrenes. The fluoranthene fraction (no. 3) contains methylfluoranthenes, benzofluorenes, benzo[ghi]fluoranthene, benzo[c]phenanthrene and three PASH of molecular weight 234 (i.e., benzo[b]-

naphtho[1,2-d]thiophene, benzo[b]naphtho[2,1-d]thiophene, and benzo[b]-
naphtho[2,3-d]thiophene). All of the PAH and PASH in this fraction have
16 aromatic carbons except benzo[c]phenanthrene and benzo[ghi]fluoran-
thene. These two PAH elute earlier than expected based on aromatic carbon
number because of their compact and slightly nonplanar structure.

The four-ring cata condensed PAH found in fraction 4 include benz[a]-
anthracene, chrysene, triphenylene, C_1- and C_2-substituted isomers of
molecular weight 228, methylbenzo[ghi]fluoranthene, and possible PASH of
molecular weight 258. The major peaks in fraction 5 are the five-ring peri-
condensed isomers of molecular weight 252 (i.e., benzo[b]fluoranthene,
benzo[k]fluoranthene, benzo[e]pyrene, BaP, and perylene), three isomers
of molecular weight 276, and a compound of molecular weight 306. Benzo[ghi]-
perylene and anthanthrene (six-ring peri-condensed PAH appeared in fraction 5, but could have been fractionated into fraction 6 (as in Fig. 15) or
a separate fraction based on the retention indices of these compounds
(i.e., 4.61 for perylene compared to 4.80 and 4.83 for anthanthrene and
benzo[ghi]perylene). Minor components identified in fraction 5 included
binaphthyl isomers and methyl-substituted 252 isomers (or dibenzofluorenes).
Fraction 6 contains four isomers of molecular weight 276 (indeno[1,2,3-cd]-
pyrene is the major peak and traces of benzo[ghi]perylene overlap from
fraction 5) and four unknown constituents of molecular weight 300.

Of particular interest are fractions 7 and 8, which contained isomers
of molecular weight 278 and 302, respectively. Six isomers of molecular
weight 278 were the major components found in fraction 7. Three compounds of molecular weight 300 and four dibenzopyrene isomers (MW 302)
were the minor components. Additional dibenzopyrene isomers (six peaks)
were the major components of fraction 8. The reversed-phase LC analyses
of these two fractions using both UV absorption and fluorescence detection
are shown in Figs. 18 and 19. The unique capability of reversed-phase LC
to separate isomers is illustrated by the separation of these two groups of
isomers. Using fluorescence spectra and retention data, two previously
unidentified PAH were confirmed in these samples (i.e., pentaphene in
fraction 7 and dibenzo[b,k]fluoranthene in fraction 8).

The chromatograms in Figs. 18 and 19 also illustrate the third approach
to analyzing complex mixtures (i.e., the combination of multidimensional
chromatography and detector selectivity). For the analysis of these particular fractions from the air particulate extract, the increased selectivity
of the fluorescence detection was not essential for the measurement of the
individual components. However, this combination is often required for
the analysis of complex mixtures such as petroleum, shale oil, and coal
liquids.

The use of multidimensional LC with selective fluorescence detection for
the quantification of individual PAH in shale oil was described by Hertz et
al. [63] and May et al. [64] as part of the process of certifying a shale oil
sample as a Standard Reference Material (SRM) (see later discussion of SRMs
for determination of PAH). In this procedure two specific fractions, one
containing pyrene and fluoranthene (16 aromatic carbons) and the other
containing BaP, benzo[e]pyrene, and perylene (20 aromatic carbons), were
isolated by normal-phase LC and analyzed by reversed-phase LC with fluorescence detection. The reversed-phase LC analysis of the BaP fraction is
shown in Fig. 20. The results of these analyses are summarized in Table
8 and compared with the results obtained by GC/MS using a direct injection

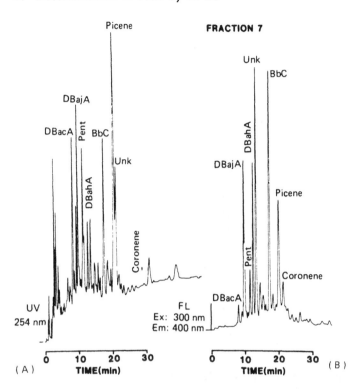

Figure 18. Reversed-phase LC analysis of fraction 7 (Fig. 16). Column: Vydac 201 TP C$_{18}$ 5 μm. Mobile phase: linear gradient from 85 to 100% CH$_3$CN in H$_2$O at 1%/min at 1.5 ml/min. (A) UV detection at 254 nm; (B) fluorescence detection, excitation 300 nm and emission 400 nm. (From Ref. 62.)

Table 8. Summary of Results of Analytical Methods Used in Certification of SRM 1580—Organics in Shale Oil

Compound	Conc.[a] (μg/g)		Sample preparation technique	Analytical technique
Fluoranthene	55	± 5	Direct injection	GC/MS
	53	± 2	HPLC	HPLC
Pyrene	101	± 5	Direct injection	GC/MS
	107	± 8	HPLC	HPLC
Benzo[a]pyrene	20	± 1	Direct injection	GC/MS
	23	± 1	HPLC	HPLC
Benzo[e]pyrene	17	± 1	Direct injection	GC/MS
	20	± 3	HPLC	HPLC
Perylene	2.8	± 0.5	Direct injection	GC/MS
	3.9	± 0.6	HPLC	HPLC

[a]Uncertainty is the standard deviation of a single measurement.
Source: Ref. 64.

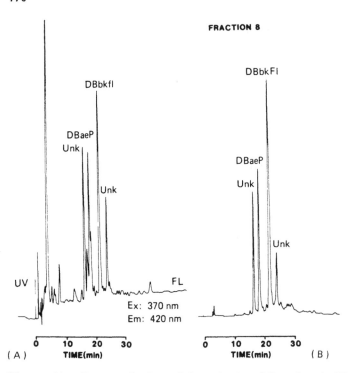

Figure 19. Reversed-phase LC analysis of fraction 8 (Fig. 16). Column: Vydac 201TP C_{18} 5 μm. Mobile phase: linear gradient from 80 to 100% CH_3-CN in H_2O at 1%/min at 1.5 ml/min. (A) UV detection at 254 nm; (B) fluorescence detection at excitation 370 nm and emission 420 nm. Peak identification: Unk = unknown, DBaeP = dibenzo[a,e]pyrene, and DBbkFl = dibenzo-[b,k]fluoranthene.

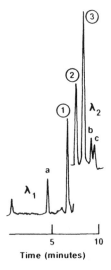

Figure 20. Reversed-phase LC analysis of benzo[a]pyrene fraction isolated from shale oil (SRM 1580) by normal-phase LC. Peak identification: (1), perylene; 2, benzo[a]pyrene; 3, benzo[k]fluoranthene; a-c, unknown. Column: Vydac 201TP C_{18} 5 μm. Mobile phase: fluorescence detection: λ_1, excitation 405 nm, emission 440 nm; λ_2, excitation 295 nm, emission 403 nm. (From Ref. 64.)

technique (i.e., no sample preparation). The agreement of the results obtained by these two different techniques confirms the utility of LC with selective fluorescence detections for the accurate quantification of individual PAH. Information values for benzo[k]fluoranthene (5 ± 1) and benzo-[b]fluoranthene (12 ± 2) have also been reported using an LC technique with quenchofluorimetric detection [65].

A similar multidimensional LC approach has also been employed in the certification analysis of a crude oil sample [66]. This sample presented somewhat different analytical problems than the shale oil sample (i.e., lower concentrations for many PAH and significant amounts of alkyl-substituted PAH). In these LC analyses, perdeuterated PAH were used as internal standards for quantification. Perdeuterated PAH are excellent internal standards for use in multidimensional LC analysis since the perdeuterated analog elutes in the same fraction as the undeuterated PAH in the normal-phase fractionation, and in the reversed-phase LC analysis it elutes immediately prior to the undeuterated PAH (see retention data in Table 6). In the analysis of the crude oil samples, four fractions (containing 14, 16, 18, and 20 aromatic carbons) were collected and analyzed; each fraction contained a perdeuterated PAH previously added to the oil as an internal standard. The compounds measured and the internal standards and fluorescence conditions used are summarized in Table 9. The reversed-phase LC analysis of the pyrene/fluoranthene fraction from a shale oil sample is shown in Fig. 21 to illustrate this method. The results of the analyses of these four fractions from the crude oil sample are summarized in Table 10 and compared with the results obtained from GC/MS analyses. Except for pyrene, the LC and GC/MS results were in excellent agreement. Compared to the shale oil analyses (Table 8), the LC results for the crude oil were more precise, indicating an improvement in quantification using the perdeuterated PAH as internal standards. The analyses of the shale oil and crude oil SRMs illustrate the selectivity achievable by using multidimensional LC and selective fluorescence detection to quantify individual PAH.

The multidimensional LC method employed for the analyses of the shale oil and crude oil SRMs were accomplished off-line (i.e., the fractions collected from the aminosilane column were concentrated and aliquots reinjected onto the reversed-phase C_{18} column). Sonnefeld et al. [67,68] developed an on-line multidimensional system using the same approach (i.e., normal-phase LC followed by reversed-phase LC with fluorescence detection). This system employed a "concentrator column" (a diaminosilane column) for interfacing the normal- and reversed-phase systems and overcoming the incompatibility of the mobile phases in the two systems. In this on-line multidimensional system, the desired fraction from the aminosilane column was valved through and trapped on the "concentrator column." The normal-phase solvent (pentane) was evaporated by using an inert gas pruge. The concentrator column was then valved into the reversed-phase analytical system and the fraction transferred to the head of the C_{18} column by gradient elution focusing. This on-line system was validated by analyzing several fuel samples and comparing the results to the off-line procedure (see Table 11). This on-line multidimensional system offers several advantages when compared to the off-line procedure: (1) sample handling is minimized, (2) sensitivity is enhanced since the whole fraction is transferred to the C_{18} column rather than an aliquot, and (3) the system is readily automated for routine analyses.

Table 9. Analytical Parameters for the LC Determination of PAH in Crude Oil (SRM 1582)

Measured compound	Internal standard	Number of aromatic carbons	Fluorescence wavelength conditions	
			Excitation	Emission
Phenanthrene	Phenanthrene-d_{10}	14	250	360
Fluoranthene	Fluoranthene-d_{10}	16	285	450
Pyrene	Pyrene-d_{10}	16	335	385
Benz[a]anthracene	Benz[a]anthracene-d_{12}	18	285	385
Benzo[a]pyrene	Perylene-d_{12}	20	295	405[a]
Perylene	Perylene-d_{12}	20	405	440

[a]Perylene-d_{12} monitored at Ex 405 nm; Em 500.
Source: Ref. 66.

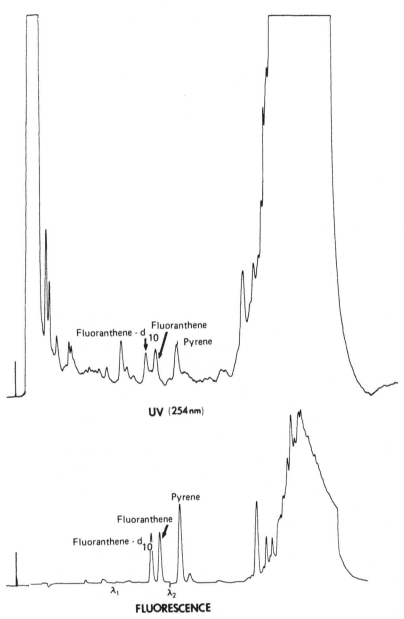

UV (254 nm)

FLUORESCENCE

Figure 21. Reversed-phase LC analysis of pyrene/fluoranthene fraction isolated from shale oil (SRM 1580) by normal-phase LC. Column: Vydac 201TP C_{18} 5 μm. Mobile phase: isocratic at 65% CH_3CN in H_2O at 1 ml/min until after elution of pyrene, then rapid gradient to 100% CH_3CN. (From Ref. 66.)

Table 10. Concentrations (μg/g) of Selected PAH in Wilmington Crude Oil (SRM 1582)

	LC	GC/MS
Phenanthrene	103 $\pm$ 3[a]	100 $\pm$ 6
Fluoranthene	2.4 $\pm$ 0.2	2.6 $\pm$ 0.3
Pyrene	6.6 $\pm$ 0.6	7.6 $\pm$ 0.3
Benz[a]anthracene	3.0 $\pm$ 0.3	2.8 $\pm$ 0.2
Benzo[a]pyrene	1.3 $\pm$ 0.1	1.4 $\pm$ 0.3
Perylene	33 $\pm$ 1	31 $\pm$ 1

[a]Uncertainty is one standard deviation from the mean value.
Source: Ref. 66.

Tomkins et al. [69] used a multidimensional LC procedure similar to that of Hertz et al. [63] and May et al. [64] for the determination of BaP in a number of natural, synthetic, and refined crudes. In their procedure the BaP-enriched fraction was isolated from an aminocyano-phase column and than analyzed on a C_{18} column with fluorescence detection (filter fluorimeter, excitation 360 nm and emission 425 nm). This procedure was applicable to samples ranging in concentration from 0.02 to 500 μg/g with a nominal precision of $\pm$6% (relative standard deviation). This procedure was used to measure BaP in a number of samples and the results were compared to a more tedious reference procedure involving the use of ^{14}C-labeled BaP. Results obtained by this procedure were also reported for BaP in several National Bureau of Standards (NBS) reference samples [i.e., SRM 1580, shale oil (31 ± 2.4), a coal liquid (SRC-II) (179 ± 4), and the Wilmington crude oil (1.2 ± 0.2)]. When compared to the NBS values for these same samples, the results reported by Tomkins et al. [69] for the shale oil and coal liquid were about 30% higher than the NBS values.

Chmielowiec et al. [70] used normal-phase LC on a diamine phase to isolate PAH fractions from two Canadian crude oils. These fractions were than analyzed by GC, MS, ^{13}C nuclear magnetic resonance (NMR), and spectrofluorimetry. Katoh et al. [71] used synchronous fluorescence spectrometry to characterize PAH fractions isolated from a coal liquid sample first by normal-phase LC and then by reversed-phase LC.

Ogan and Katz [72] coupled reversed-phase LC with size-exclusion chromatography to provide a multidimensional chromatographic method for the analysis of complex mixtures. This procedure included three chromatographic steps; the first two steps served to isolate the sample into fractions which were more readily suitable for analysis by the final high-resolution analytical LC step. The first step was a low-resolution version of the final analytical step (i.e., reversed-phase C_{18}) and was used to select components with retention times comparable to those compounds of analytical interest. The second step, size-exclusion chromatography, isolated compounds of appropriate molecular weight. These first two steps were coupled on-line to provide rapid sample preparation. The final high-resolution reversed-phase C_{18}

Table 11. Determination of Selected PAH in Three Oil Samples by Multidimensional LC

Compound	Multidimensional LC (on-line)[a]	Multidimensional LC (off-line)[a]	Certified values[b]
	Shale oil (SRM 1580) (concentration in μg/g)		
Fluoranthene	53.8 ± 3.5 (3)[c]	53 ± 2 (9)	54 ± 10
Pyrene	109.7 ± 7.1 (3)	107 ± 8 (10)	104 ± 18
Perylene	2.32 ± 0.15 (5)	3.9 ± 0.6 (11)	3.4 ± 2.2
Benzo[a]pyrene	20.7 ± 2.2 (8)	23 ± 1 (8)	21 ± 6
	SRC II coal liquid (concentration in mg/g)		
Fluoranthene	3.15 ± 0.04 (3)	3.3 ± 0.16 (3)	
Pyrene	6.65 ± 0.05 (3)	6.0 ± 0.2 (3)	
Perylene	0.024 ± 0.002 (3)		
Benzo[a]pyrene	0.133 ± 0.008 (5)	0.134 ± 0.007 (3)	
Benzo[e]pyrene	0.153 ± 0.006 (3)	0.143 ± 0.005 (3)	
Benzo[k]fluoranthene	0.062 ± 0.001 (5)		
	Wilmington crude oil (SRM 1582) (concentration in μg/g)		
Fluoranthene	2.4 ± 0.3 (3)	3 ± 1 (3)	
Pyrene	7.4 ± 1.1 (3)	7 ± 1 (3)	

[a]Uncertainty is 1 standard deviation.
[b]Certified values derived from a combination of LC and GC/MS values.
[c](n) indicates the number of replicate measurements.
Source: Ref. 67.

analysis was used to provide identification and quantification of the individual components of interest. Ogan and Katz [72] applied this multidimensional method to the determination of PAH in several coal liquid and oil samples.

Several groups have reported the use of low-temperature fluorimetry in Shpol'skii matrices to characterize PAH fractions after multidimensional LC analysis [73-77]. Colin et al. [73] isolated PAH fractions from medicinal white oils using a sequence of frontal elution on silica gel, adsorption chromatography on alumina, partition chromatography on Sephadex LH-20 impregnated with dimethylformamide and water, and reversed-phase LC on C_{18} bonded silica. The reversed-phase LC analysis utilized fluorescence detection and peaks were collected and analyzed by low-temperature fluorimetry in Shpol'skii matrices to confirm the identity of the PAH. High-resolution spectrofluorimetry at 15 K (Shpol'skii effect) was also used by Ewald et al. [74] to identify the five methylphenanthrene isomers in petroleum extracts. The methylphenanthrenes were extracted from the crude petroleum by a three-step chromatographic approach: adsorption chromatography on Florisil to remove aliphatics, resins, and asphaltenes; normal-phase LC on the amine phase to isolate the phenanthrene fraction; and reversed-phase LC on C_{18} phase. The final analysis of the peaks collected from the C_{18} column was performed with low-temperature fluorimetry to identify the methyphenanthrene isomers. Colmsjö and co-workers [76,77] used reversed-phase LC to separate PAH and PASH isolated from carbon black and coal tar and then used low-temperature spectrofluorimetry to identify the chromatographic peaks.

VII. APPLICATIONS OF LIQUID CHROMATOGRAPHY FOR THE DETERMINATION OF PAH

A. Standard Reference Materials for Determination of PAH

Until recently, Standard Reference Materials (SRMs) for use in the validation of analytical methods for trace organic analysis were unavailable. However, since 1980 several SRMs have been issued by the National Bureau of Standards with certified values for a number of PAH (see Table 12) [78].

Table 12. Standard Reference Materials for Determination of PAH

1580	Organics in Shale Oil
1582	Organics in Wilmington Crude Oil
1644	Generator columns for Polynuclear Aromatic Hydrocarbons
1647	Priority Pollutant Polynuclear Aromatic Hydrocarbons (in Acetonitrile)
1648[a]	Urban Particulate Matter
1649	Urban Dust/Organics
1650	Diesel Particulate Matter

[a]Certified for trace element constituents only; information values available for selected PAH.

Table 13. Certified Values for Selected PAH in Natural Matrix SRMs (μg/g)

	SRM 1580	SRM 1582	SRM 1649
Phenanthrene	—	101 ± 6	—
Pyrene	104 ± 9	7.1 ± 0.8	—
Fluoranthene	54 ± 10	2.5 ± 0.3	7.1 ± 0.5
Benz[a]anthracene	—	2.9 ± 0.3	2.6 ± 0.3
Benzo[a]pyrene	21 ± 6	1.3 ± 0.3	2.9 ± 0.5
Benzo[e]pyrene	18 ± 8	—	—
Perylene	3.3 ± 1.1	32 ± 2	—
Benzo[ghi]perylene	—	—	4.5 ± 1.1
Indeno[1,2,3-cd]pyrene	—	—	3.3 ± 0.5

In the certification of an SRM, two independent and reliable analytical methods are used for the measurement of the concentration. When results are obtained from only one analytical method, the values are provided only as information values rather than as certified values. In the certification of all the SRMs listed in Table 12, LC with UV or fluorescence detection was used as one of the analytical methods.

Three natural matrix material SRMs are currently available from NBS with certified values for PAH: urban particulate material (SRM 1649), shale oil (SRM 1580), and crude oil (SRM 1582). The use of LC in the analysis of each of these materials has already been discussed and the LC analytical results are summarized in Tables 4, 8, and 10. The certified values for these three SRMs, which are a combination of the results from the two analytical methods (i.e., LC and GC or GC/MS), are summarized in Table 13. These SRMs have already found widespread use in serving as benchmark materials for the validation of analytical methods for the determination of PAH in alternative fuels and particulate matrices [69,79-81]. A fourth natural matrix material certified for PAH, diesel particulate material, is currently in preparation as an SRM (see Table 14). Another matrix SRM (1648), which has certified values for trace element content, has also been analyzed and the results for selected PAH are also summarized in Table 14.

The 16 PAH on the priority pollutant list of the EPA are certified in SRM 1647 (Priority Pollutant Polynuclear Aromatic Hydrocarbons in Acetonitile). This SRM was prepared specifically for use in the calibration of chromatographic systems, particularly for EPA method 610, which specified the use of LC for the measurement of PAH in water samples [82]. The certified values for SRM 1647 are summarized in Table 15 and the direct LC analysis of this

Table 14. Analytical Results for the Determination of Selected PAH in SRMs 1650 and 1648

Compound	SEM 1650 diesel particulate matter		SRM 1648 urban particulate matter	
	LC	GC/MS	LC	GC
Phenanthrene	63 ± 2	79 ± 1	4.6 ± 0.3	3.8 ± 0.2
Anthracene	—	—	0.36 ± 0.01	—
Fluoranthene	49.8 ± 0.3	48.5 ± 1	8.7 ± 0.4	9.2 ± 0.5
Pyrene	45.5 ± 1.7	49.0 ± 0.7	6.8 ± 0.2	7.3 ± 0.4
Chrysene	22 ± 1	—	6.6 ± 0.2	—
Benz[a]anthracene	7.1 ± 0.3	6.0 ± 0.1	3.2 ± 0.1	3.2 ± 0.2
Perylene	0.13 ± 0.02	—	0.69 ± 0.02	0.76 ± 0.07
Benzo[k]fluoranthene	2.1 ± 0.2	—	3.4 ± 0.05	3.5 ± 0.2
Benzo[e]pyrene	—	9.6 ± 0.3	—	5.2 ± 0.2
Benzo[a]pyrene	1.4 ± 0.1	1.3 ± 0.1	3.4 ± 0.2	3.1 ± 0.2
Benzo[ghi]perylene	2.4 ± 0.4	2.3 ± 0.1	—	4.7 ± 0.1
Indeno [1,2,3-cd] pyrene	3.2 ± 0.3	1.8 ± 0.1	4.7 ± 0.2	4.4 ± 0.15

Table 15. Analytical Results for the Certification of SRM 1647

	Concentration (µg/ml)			
	Gravimetry	HPLC	GC	Certified values[a]
1. Naphthalene	22.5	22.4 ± 0.5		22.5 ± 0.2
2. Acenaphthylene	19.0	19.2 ± 0.5		19.1 ± 0.2
3. Acenaphthene	20.8	21.2 ± 0.4		21.0 ± 0.4
4. Fluorene	4.89	4.96 ± 0.18		4.92 ± 0.10
5. Phenanthrene	5.00	5.12 ± 0.18		5.06 ± 0.10
6. Anthracene	3.25	3.33 ± 0.10		3.29 ± 0.10
7. Fluoranthene	9.99	10.3 ± 0.5		10.1 ± 0.2
8. Pyrene	9.82	9.85 ± 0.58		9.84 ± 0.10
9. Benz[a]anthracene	4.99	5.12 ± 0.14	4.97 ± 0.06	5.03 ± 0.10
10. Chrysene		4.69 ± 0.15	4.68 ± 0.06	4.68 ± 0.10
11. Benzo[b]fluoranthene	5.11	5.13 ± 0.21	5.09 ± 0.06	5.11 ± 0.10
12. Benzo[k]fluoranthene	5.00	5.06 ± 0.15	4.99 ± 0.10	5.02 ± 0.10
13. Benzo[a]pyrene	5.28	5.32 ± 0.13	5.31 ± 0.19	5.30 ± 0.10
14. Benzo[ghi]perylene	4.00	4.09 ± 0.30	3.99 ± 0.14	4.01 ± 0.10
15. Dibenz[a,h]anthracene		3.73 ± 0.12	3.63 ± 0.07	3.68 ± 0.10
16. Indeno[1,2,3-cd]pyrene	4.07	4.11 ± 0.15	4.02 ± 0.06	4.06 ± 0.10

[a]The estimated uncertainty for each compound is based on judgement, and represents an evaluation of the combined effects of method imprecision and possible systematic errors among methods.
Source: Ref. 78.

Table 16. Concentration of Aqueous Solutions from SRM 1644

Temperature (°C)	Concentration (μg/kg)[a]		
	Anthracene	Benz[a]anthracene	Benzo[a]pyrene
10	16.6 ± 0.7	3.38 ± 1.2	0.59 ± 0.06
12	18.7 ± 0.6	3.83 ± 0.91	0.67 ± 0.04
14	21.1 ± 0.5	4.36 ± 0.68	0.76 ± 0.03
16	23.8 ± 0.5	4.96 ± 0.54	0.87 ± 0.03
18	27.0 ± 0.5	5.65 ± 0.60	0.99 ± 0.03
20	30.7 ± 0.6	6.45 ± 0.77	1.13 ± 0.03
22	35.0 ± 0.6	7.38 ± 0.94	1.30 ± 0.04
24	39.9 ± 0.6	8.45 ± 1.0	1.49 ± 0.04
26	45.7 ± 0.8	9.69 ± 1.0	1.71 ± 0.04
28	52.4 ± 1.3	11.1 ± 1.2	1.96 ± 0.05
30	60.1 ± 2.1	12.8 ± 1.7	2.26 ± 0.09

[a]The uncertainties are 90% confidence bands for the entire regression curve. The difference between the true and certified concentration should be less than the stated uncertainty at the 99% confidence level.

Source: Ref. 78.

Table 17. Applications of LC for the Determination of PAH

Publication year [ref.]	Authors	Application	Column	Detection	Comments
1981 [4]	Futoma et al.	Review			Review of Chromatographic methods for PAH in water
1981 [47]	Stöber and Reupert	Water	LiChrosorb RP-18	Fluorescence	River water samples
1981 [48]	Matsushita et al.	Air particulate extracts	Nucleosil 7 C_{18}	Fluorescence: Ex, 370 nm; Em, 406 nm	
1981 [49]	Obana et al.	Human fat and liver	LiChrosorb RP-18	Fluorescence at excitation/emission 334/384, 365/430, and 300/393 nm	Measured ppt level PAH
1981 [54]	Shelly et al.	Crude oil ash residue	Aminosilane and Ultrasphere ODS	Videofluorometry	Reversed-phase LC analysis of normal-phase LC fractions
1981 [84]	Takeuchi and Ishii	Standards	Micro LC	UV	Liquified alkanes as mobile phase
1981 [85]	Takeuchi and Ishii	Standards	Micro LC, ODS-SC-01	UV	
1981 [104]	Alfredson	Petroleum products	Comparison of polar bonded phases		Hydrocarbon group separations
1982 [24]	Hurtubise et al.	Standards	µBondpak C_{18}		Correlation of retention and molecular connectivity

Table 17. Continued

Publication year [ref.]	Authors	Application	Column	Detection	Comments
1982 [30]	Grizzle and Thomson	Shale oil	2,4-Dinitroanilino-propylsilica (DNAP)	UV	Retention data for PAH compounds on DNAP, alumina, and diamine columns
1982 [31]	Ecknig et al.	Standards, crude oil	Silica modified with perfluoroethoxy-ethanol		Correlation of retention data for 37 PAH with resonance energy
1982 [33]	Colmsjö and Östman	PASH Standards	Vydac 201 TP C_{18}		Retention data for eight PASH
1982 [35,36] 1984	Yang	16 PAH standards	ODS	UV	Pack capillary column
1982 [39]	Verzele and Dewaele	PCBs	5-μm ROSIL-C_{18}		Coupled three conventional columns
1982 [41,42]	Dong and DiCesare	Standards, air particulate extracts, coal liquid, and waste water	HC-ODS HS-3C_{18} (3 μm)	UV and fluorescence	High-speed LC
1982 [51]	Joe et al.	Barley malt	Zorbax ODS	Fluorescence	
1982 [59]	Su et al.	Standards	Ultrasphere 5 μm ODS	Negative-peak fluorescent detector	
1982 [67,68]	Sonnefeld et al.	Fuel samples	μBondapak NH_2 Vydac 201 TP C_{18}	Fluorescence and UV	On-line multidimensional method

Year [Ref]	Author	Sample	Column	Detection	Notes
1982 [72]	Ogan and Katz	Fuel samples	SEC column and HC-ODS	Fluorescence	Multidimensional LC
1982 [77]	Colmsjö et al.	Carbon black	Vydac 201 TP C_{18}	UV	PASH identified by Shpol'skii fluorescence
1982 [78]	May et al.	Standard Reference Materials	Vydac 201TP C_{18} 5 μm	UV and fluorescence	Use of LC for certification of SRMs for PAH measurements
1982 [86]	Takeuchi and Ishii	Standards	Several columns, including C_{18} and NH_2 columns	UV at 254 nm	Micro LC
1982 [87]	Grimalt and Albaiges	Crude oils	3 μm C_{18}	Multiwavelength UV	Oil spill identification
1982 [88]	DiCesare and Dong	Standards	Comparison of several reversed-phase columns		
1982 [89]	Alexander et al.	Standards, fish, soot, and water	Radial PAK C_{18}	UV	
1982 [90]	Wang and Meresz	Vegetables	Vydac 201TP C_{18} 5 μm	Fluorescence	
1982 [91]	Black et al.	Water samples	Vydac 201TP C_{18} 10 μm	UV and fluorescence	
1982 [92]	Romanowski et al.	Air particulate extracts	Sephasorb HP methanol	UV	Cleanup of air particulate extracts prior to GC

Table 17. Continued

Publication year [ref.]	Authors	Application	Column	Detection	Comments
1982 [93]	Levine and Skewes	Diesel particulate extracts	Preparative column 10 μm Porasil	UV and fluorescence	
1983 [13]	Wise and May	Standards	Vydac 201TP C_{18} and other commercial columns	UV	Comparison of seven different C_{18} columns and columns from same manufacturer with different surface coverage
1983 [34]	Wise et al.	Standards and coal liquid	μBondapak NH_2 Vydac 201TP C_{18} Zorbax ODS	UV	Normal- and reversed-phase retention data for 40 PASH
1983 [38]	Hirata and Jinno	Coal tar aromatic fraction	Packed capillary Develosil ODS 3 μm	UV	
1983 [45]	Hellmann	Soil extracts	HC-ODS	Fluorescence	Comparison of PAH groups vs. individual PAH determined
1983 [52]	Kodama et al.	Air particulate extracts	LiChrosorb RP-18	Fluorescence: Ex, 368; Em, 406	Determined benzo[a]pyrene
1983 [74]	Ewald et al.	Petroleum extracts (monomethyl phenanthene isomers)	Spherisorb NH_2 Spherisorb C_{18}	UV detection peaks identified by Shpol'skii fluorescence	Multidimensional LC

184

Year [ref]	Author	Sample	Column	Detection	Comments
1983 [75]	Garrigues et al.	Petroleum extracts	Spherisorb NH_2 Spherisorb C_{18}	UV detection peaks identified by Shpol'skii fluorescence	Multidimensional LC
1983 [83]	Velapoldi et al.	SRM 1644	Vydac 201TP C_{18}	Fluorescence	
1983 [94]	Anderson et al.	Workplace atmosphere	PAH-ODS	Fluorescence (filter)	Determination of 16 PAH
1983 [95]	Grzybowski et al.	Sewage	Zorbax ODS	UV	
1983 [96]	Becher and Bjørseth	Urine	Vydac 201TP C_{18} 5 μm	Fluorescence	
1983 [97]	Karlesky et at.	Standards	Chromasorb LC-9 (amine)	UV	Normal-phase LC retention data for standards
1983 [98]	Haugen et al.	Coal gasification tar	Silica, Partisil 10 PAC, and Partisil 10 ODS-2	UV at 280 nm	Mutagenicity testing of LC effluent for identification of active components
1983 [99]	Sirota et al.	Lobsters	Vydac 201TP C_{18}	UV and fluorescence	
1983 [100]	Dark	Shale oil	μBondapak NH_2	UV	Separation of saturates, aromatics, and polars from oils
1983 [101]	Bylina et al.	Standards			Prediction of LC retention from structural features
1983 [102]	Ishii and Takeuchi	Standards PCBs	Silica ODS SC-01 5 μm	UV	

Table 17. Continued

1983 [103]	van der Wal and Young	Standards	Microbore with 5 μm Vydac 201TP C_{18} and 3 μm Varian SP C_{18}		Gradient elution system for microbore LC
1984 [14]	Wise et al.	Standards, air particulate fraction	Vydac 201TP C_{18}	UV	Columns contained mixtures of C_{18}; lots of varying selectivity
1984 [19]	Sander and Wise	Standards	C_{18} phases on various wide-pore silicas	254 nm	Compared monomeric and polymeric phases on five different 300 A silicas
1984 [20]	Sander and Wise	Standards	C_{18} phases on silicas with different pore sizes	254 nm	Compared monomeric and polymeric phases on silicas of different pore sizes
1984 [44]	May and Wise	Air particulate extracts	Vydac 201TP C_{18}	Fluorescence wavelength programming	Quantification of 15 PAH in air particulate extracts
1985 [66]	Kline et al.	Crude oil and shale oil	μBondapak NH_2 Vydac 201 TP C_{18}	Fluorescence	Multidimensional procedure for quantification of PAH in oil samples

SRM is shown in Figs. 5 and 6. Even though this SRM is a synthetic mixture rather than a natural matrix material, it has found widespread use in environmental analyses.

The final SRM related to PAH measurements is SRM 1644 (Generator Columns for Polynuclear Aromatic Hydrocarbons), which is intended to provide accurate concentrations of anthracene, benz[a]anthracene, and BaP in water. Since the aqueous solubilities of PAH in general, and the above-mentioned PAH in particular, are very low (43, 9.1, and 1.6 µg/kg, at 25°C, respectively), such solutions are difficult to prepare and to preserve for long periods of time. SRM 1644 allows the user to circumvent these problems by generating the solutions as needed [78,83]. The SRM consists of three separate generator columns (50 cm X 0.6 cm stainless steel columns) packed with 80-100 mesh sea sand coated with 0.5% (w/w) of the PAH of interest. Saturated aqueous solutions are produced by pumping HPLC-grade water through the column at flow rates between 0.1 and 5 ml/min. Since the aqueous solubility is a well-defined thermodynamic property, these saturated solutions are standard solutions. The concentrations of the aqueous solutions produced from SRM 1644 between 10 and 30°C are summarized in Table 16. In addition to providing accurate low-level concentrations of PAH in water, this SRM could also be used to generate constant concentrations of PAH for marine organism uptake or exposure studies.

B. Review of Liquid Chromatographic Applications for the Determination of PAH

Applications of LC for the determination of PAH are summarized in Table 16 for literature from 1981 through 1983. Only references from 1981 on, which were not included in the previous review [1], are listed in Table 17.

REFERENCES

1. S. A. Wise, in *Handbook of Polycyclic Aromatic Hydrocarbons*, A. Bjørseth (Ed.), Marcel Dekker, New York, 1983, p. 183.
2. M. L. Lee, M. Novotny, and K. D. Bartle, *Analytical Chemistry of Polycyclic Aromatic Compounds*, Academic Press, New York, 1981.
3. K. D. Bartle, M. L. Lee, and S. A. Wise, Chem. Soc. Rev. *10*:113 (1981).
4. D. J. Futoma, S. R. Smith, J. Tanaka, and T. E. Smith, CRC Crit. Rev. Anal. Chem. *10*:69 (1981).
5. S. A. Wise, W. J. Bonnett, F. R. Guenther, and W. E. May, J. Chromatogr. Sci. *19*:457 (1981).
6. S. A. Wise, W. J. Bonnett, and W. E. May, in *Polynuclear Aromatic Hydrocarbons: Chemistry and Biological Effects*, A. Bjørseth and A. J. Dennis (Eds)., Battelle Press, Columbus, Ohio, 1980, p. 791.
7. K. Ogan and E. Katz, J. Chromatogr. *188*:115 (1980).
8. E. Katz and K. Ogan, J. Liq. Chromatogr. *3*:1151 (1980).
9. A. L. Colmsjö and J. C. MacDonald, Chromatographia *13*:350 (1980).
10. P. Amos, J. Chromatogr. *204:469 (1981)*.
11. A. P. Goldberg, Anal. Chem. 54:342 (1982).
12. J. G. Atwood and J. Goldstein, J. Chromatogr. Sci. *18*:650 (1980).
13. S. A. Wise and W. E. May, Anal. Chem. 55:1479 (1983).

14. S. A. Wise, L. C. Sander, and W. E. May, J. Liq. Chromatogr., 6: 2709 (1984).

15. K. Ogan, E. Katz, and W. Slavin, Anal. Chem. 51:1315 (1979).

15a. K. Ogan and E. Katz, Perkin-Elmer Corp., unpublished data, 1979.

16. M. Popl, V. Dolansky, and J. Mostecky, J. Chromatogr. 117:117 (1976).

17. J. Chmielowiec and H. Sawatzky, J. Chromatogr. Sci. 17:245 (1979).

18. L. R. Snyder, J. Chromatogr. 179:167 (1979).

19. L. C. Sander and S. A. Wise, Anal. Chem. 56:504 (1984).

20. L. C. Sander and S. A. Wise, J. Chromatogr. 316:163 (1984).

21. S. A. Wise and L. C. Sander, J. High Resolut. Chromatogr. Chromatogr. Commun. 8:248 (1985).

22. H. Issaq, G. M. Janini, B. Poehland, R. Shipe, and G. M Muschik, Chromatographia 14:655 (1981).

23. M. N. Hasan and P. C. Jurs, Anal. Chem. 55:263 (1983).

24. R. J. Hurtubise, T. W. Allen, and H. F. Silver, J. Chromatogr. 235: 517 (1982).

25. J. F. Schabron, R. J. Hurtubise, and H. F. Silver, Anal. Chem. 49: 2253 (1977).

26. A. Matsunaga, Anal. Chem. 55:1375 (1983).

27. K. G. Liphard, Chromatographia 13:603 (1980).

28. W. Holstein, Chromatographia 14:468 (1981).

29. H. Deymann and W. Holstein, Erdoel Kohle Erdgas Petrochem. 34:353 (1981).

30. P. L. Grizzle and J. S. Thomson, Anal. Chem. 54:1071 (1982).

31. W. Ecknig, Bui Trung, R. Radeglia, and U. Gross, Chromatographia 16:178 (1982).

32. J. Chmielowiec, J. Chromatogr. Sci. 19:296 (1981).

33. A. L. Colmsjö and C. Östman, in Polynuclear Aromatic Hydrocarbons: Physical and Biological Chemistry, M. Cooke, A. J. Dennis, and G. L. Fisher (Eds.), Battelle Press, Columbus, Ohio, 1982, p. 201.

34. S. A. Wise, R. M. Campbell, W. E. May, M. L. Lee, and R. N. Castle, in Polynuclear Aromatic Hydrocarbons: Formation, Metabolism and Measurement, M. Cooke and A. J. Dennis (Eds.), Battelle Press, Columbus, Ohio 1983, p. 1247.

35. F. J. Yang, in Ultrahigh Resolution Chromatography (S. Ahuja, Ed.), ACS Symp. Ser. 250:91 (1984).

36. F. J. Yang, J. Chromatogr. 236:265 (1982).

37. T. Takeuchi and D. Ishii, J. Chromatogr. 213:25 (1981).

38. J. Hirata and K. Jinno, J. High Resolut. Chromatogr. Chromatogr. Commun. 6:196 (1983).

39. M. Verzele and C. Dewaele, J. High Resolut. Chromatogr. Chromatogr. Commun. 5:245 (1982).

40. J. L. DiCesare, M. W. Dong, and L. S. Ettre, Chromatographia 14: 257 (1981).

41. M. W. Dong and J. L. DiCesare, J. Chromatogr. Sci. 20:517 (1982).

42. M. W. Dong, K. Ogan, J. L. DiCesare, in Polynuclear Aromatic Hydrocarbons: Physical and Biological Chemistry, M. Cooke, A. J. Dennis, and G. L. Fisher (Eds.), Battelle Press, Columbus, Ohio, 1982, p. 237.

43. S. A. Wise, S. L. Bowie, S. N. Chesler, W. F. Cuthrell, W. E. May, and R. E. Rebbert, in Polynuclear Aromatic Hydrocarbons: Physical and Biological Chemistry, M. Cooke, A. J. Dennis, and G. L. Fisher (Eds.), Battelle Press, Columbus, Ohio, 1982, p. 919.

44. W. E. May and S. A. Wise, Anal. Chem. *56*:225 (1984).
45. H. Hellmann, Fresenius Z. Anal. Chem. *302*:115 (1980).
46. National Bureau of Standards Certificate of Analysis, SRM 1649 Urban Dust/Organics, 1982.
47. I. Stöber and R. Reupert, Vom Wasser *56*:115 (1981).
48. H. Matsushita, T. Shiozaki, Y. Kato, and S. Gato, Bunseki Kagaku *30*:362 (1981).
49. H. Obana, S. Hori, and T. Kashimoto, Bull. Environ. Contam. Toxicol. *26*:613 (1981).
50. H. Obana, S. Hori, T. Kashimoto, and N. Kunita, Bull. Environ. Contam. Toxicol. *27*:23 (1981).
51. F. L. Joe, Jr., J. Salemme, and T. Fazio, J. Assoc. Off. Anal. Chem. *65*:1395 (1982).
52. Y. Kodama, K. Arashidani, and M. Yoshikawa, J. Chromatogr. *261*: 103 (1983).
53. M. P. Fogarty, D. C. Shelly, and I. M. Warner, J. High Resolut. Chromatogr. Chromatogr. Commun. *4*:561 (1981).
54. D. C. Shelly, M. P. Fogarty, and I. M. Warner, J. High Resolut. Chromatogr. Chromatogr. Commun. *4*:616 (1981).
55. D. R. Choudhury and B. Bush, Anal. Chem. *53*:1351 (1981).
56. J. W. Readman, L. Brown, and M. M. Rhead, Analyst *106*:122 (1981).
57. H. Schaper, J. Electroanal. Chem. *129*:335 (1981).
58. D. C. Locke, B. S. Dhingra, and A. D. Baker, Anal. Chem. *54*:447 (1982).
59. S. Y. Su, E. P. C. Lai, and J. D. Winefordner, Anal. Lett. *15*:439 (1982).
60. D. W. Armstrong, W. L. Hinze, K. H. Bui, and H. N. Singh, Anal. Lett. *14*:1659 (1981).
61. S. A. Wise, C. F. Allen, S. N. Chesler, H. S. Hertz, L. R. Hilpert, W. E. May, R. E. Rebbert, and C. R. Vogt, NBSIR-82-2595, National Bureau of Standards, Washington, D.C., 1982.
62. S. A. Wise, S. N. Chesler, L. R. Hilpert, W. E. May, R. E. Rebbert, and C. R Vogt, in *Polynuclear Aromatic Hydrocarbons: Mechanisms, Methods, and Metabolism* (M. Cooke and A. J. Dennis, Eds.), Battelle Press, Columbus, Ohio, 1984, p. 1413.
63. H. S. Hertz, J. M. Brown, S. N. Chesler, F. R. Guenther, L. R. Hilpert, W. E. May, R. M. Parris, and S. A. Wise, Anal. Chem. *52*: 1650 (1980).
64. W. E. May, J. Brown-Thomas, L. R. Hilpert, and S. A. Wise, in *Chemical Analysis and Biological Fate of Polynuclear Aromatic Hydrocarbons*, M. Cooke and A. J. Dennis (Eds.), Battelle Press, Columbus, Ohio, 1981, p. 1.
65. P. L. Konash, S. A. Wise, and W. E. May, J. Liq. Chromatogr. *4*: 1339 (1981).
66. W. F. Kline, S. A. Wise, and W. E. May, J. Liq. Chromatogr. *8*:223 (1985).
67. W. J. Sonnefeld, W. H. Zoller, W. E. May, and S. A. Wise, Anal. Chem. *54*:723 (1982).
68. W. J. Sonnefeld, W. H. Zoller, W. E. May, and S. A. Wise, in *Polynuclear Aromatic Hydrocarbons: Physical and Biological Chemistry*, M. Cooke, A. J. Dennis, and G. L. Fisher (Eds.), Battelle Press, Columbus, Ohio, 1982, p. 755.
69. B. A. Tomkins, H. Kubota, W. H. Griest, J. E. Caton, B. R. Clark, and M. R. Guerin, Anal. Chem. *52*:1331 (1980).

70. J. Chmielowiec, J. E. Beshai, and A. E. George, Fuel 59:838 (1980).

71. T. Katoh, S. Yokoyama, and Y. Sanada, Fuel 59:845 (1980).

72. K. Ogan and E. Katz, Anal. Chem. 54:169 (1982).

73. J. M. Colin, G. Vion, M. Lamotte, and J. Joussot-Dubien, J. Chroma-
     togr. 204:135 (1981).

74. M. Ewald, M. Lamotte, P. Garrigues, J. Rima, A. Veyres, R. Lapouyade,
     and G. Bourgeois, in Advances in Organic Geochemistry, 1981, Wiley,
     Chichester, West Sussex, England, 1983, p. 705.

75. P. Garrigues, R. Vazelhes, J.-M. Schmitter, and M. Ewald, in Poly-
     nuclear Aromatic Hydrocarbons: Formation, Metabolism, and Measure-
     ment, M. Cooke and A. J. Dennis (Eds.), Battelle Press, Columbus,
     Ohio, 1983, p. 545.

76. A. Colmsjö and U. Stenberg, Anal. Chem. 51:145 (1979).

77. A. L. Colmsjö, Y. U. Zebühr, and C. E. Ostman, Anal. Chem. 54:
     1673 (1982).

78. W. E. May, S. N. Chesler, H. S. Hertz, S. A. Wise, Int. J. Environ.
     Anal. Chem. 12:259 (1982).

79. Y. Yang, A. P. D'Silva, V. A. Fassel, and M. Iles, Anal. Chem. 52:
     1350 (1980).

80. Y. Yang, A. P. D'Silva, and V. A. Fassel, Anal. Chem. 53:2107
     (1981).

81. V. B. Conrad, W. J. Carter, E. L. Wehry, and G. Mamantov, Anal.
     Chem. 55:1340 (1983).

82. Method 610: Polycyclic Aromatic Hydrocarbons, Fed. Regist. 44(233):
     65914 (Dec. 3, 1979).

83. R. A. Velapoldi, P. A. White, W. E. May, and K. R. Eberhardt, Anal.
     Chem. 55:1896 (1983).

84. T. Takeuchi and D. Ishii, J. Chromatogr. 216:153 (1981).

85. T. Takeuchi and D. Ishii, J. Chromatogr. 218:199 (1981).

86. T. Takeuchi and D. Ishii, J. Chromatogr. 244:23 (1982).

87. J. Grimalt and J. Albaiges, J. High Resolut. Chromatogr. Chromatogr.
     Commun. 5:255 (1982).

88. J. L. DiCesare and M. W. Dong, Chromatogr. Newslett. 10(1):12
     (1982).

89. R. Alexander, K. M. Cumbers, and R. I. Kagi, Int. J. Environ. Anal.
     Chem. 12:161 (1982).

90. D. T. Wang and O. Meresz, in Polynuclear Aromatic Hydrocarbons:
     Physical and Biological Chemistry, M. Cooke, A. J. Dennis, and G. L.
     Fisher (Eds.), Battelle Press, Columbus, Ohio, 1982, p. 885.

91. J. J. Black, T. F. Hart, Jr., and P. J. Black, Environ. Sci. Technol.
     16:247 (1982).

92. T. Romanowski, W. Funcke, J. Koenig, and E. Balfanz, Anal. Chem.
     54:1285 (1982).

93. S. P. Levine and L. M. Skewes, J. Chromatogr. 235:532 (1982).

94. K. Anderson, J. -O. Levin, and C. -A. Nilsson, Chemosphere 12:197
     (1983).

95. J. Grzybowski, A. Radecki, and G. Rewkowska, Environ. Sci. Technol.
     17:44 (1983).

96. G. Becher and A. Bjørseth, Cancer Lett. 17:301 (1983).

97. D. Karlesky, D. C. Shelly, and I. M. Warner, J. Liq. Chromatogr.,
     6:471 (1983).

98. D. A. Haugen, V. C. Stamoudis, M. J. Peak, and A. S. Boparai, in
     Polynuclear Aromatic Hydrocarbons: Formation, Metabolism, and Measure-

ment, M. Cooke and A. J. Dennis (Eds.), Battelle Press, Columbus, Ohio, 1983, p. 607.

99. G. R. Sirota, J. F. Uthe, A. Sreedharan, R. Matheson, C. J. Musial, and K. Hamilton, in *Polynuclear Aromatic Hydrocarbons: Formation, Metabolism, and Measurement,* M. Cooke and A. J. Dennis (Eds.), Battelle Press, Columbus, Ohio, 1983, p. 1123.

100. W. A. Dark, J. Liq. Chromatogr. *6*:325 (1983).

101. A. Bylina, L. Gluzinski, K. Lesnak, B. Radwanski, and P. A. Panczek, Chromatographia *17*:132 (1983).

102. D. Ishii and T. Takeuchi, J. Chromatogr. *255*:349 (1983).

103. Sj. van der Wal and F. J. Yang, J. High Resolut. Chromatogr. Chromatogr. Commun. *6*:216 (1983).

104. T. V. Alfredson, J. Chromatogr. *218*:715 (1981).

6

Recent Advances in the Analysis of Polycyclic Aromatic Compounds by Gas Chromatography

KEITH D. BARTLE / Department of Physical Chemistry, University of Leeds, Leeds, United Kingdom

I.   Introduction   193

II. Columns for Gas Chromatography   194

    A. Technology   194
    B. Stationary phases for GC of PAC   196
    C. Liquid-crystal and mesogenic stationary phases   198

III. Injection Methods   204

IV. Detectors   205

V.   Retention Indices and Their Use in PAC Identification   214

VI. Mechanism of Retention in the GC of PAC   215

VII. Applications of GC in PAC Analysis   219

    A. Source-specific PAH   219
    B. Nitrogen-containing PAC   220
    C. Nitro PAH   223
    D. Sulfur- and oxygen-containing PAC   228
    E. Applications to different sample types   228

VIII. Conclusions   228

    References   229

I. INTRODUCTION

The separation of mixtures of polycyclic aromatic hydrocarbons (PAH) by gas chromatography (GC) has evolved [1] over the past 20 years from packed columns to the first reported analysis in 1964 on a relatively inefficient 35 m glass open tubular column coated with SE-30. Improvements in resolution, as judged from the isomer pairs benz[a]anthracene/chrysene and benzo[e]- and benzo[a]pyrenes, were noted three years later with an SE-52-coated glass column with temperature programming. Near baseline resolution of these pairs was obtained on a stainless steel column (OV-17 stationary phase), but was unfortunately accompanied by peak tailing and loss of trace constituents. Approximately 10 years ago there were marked improvements in the resolution of PAH when it was found that removal of Lewis acids from glass and reaction of the surface silanol groups with derivatizing reagents gave more efficient and deactivated columns. Four recent publications have extensively reviewed the literature of PAH analysis by GC up to 1980 [1-4].

However, the current revolution in open-tubular-column GC brought about by the development of fused-silica columns, which have performance similar or superior to that of glass columns but are much more robust, has brought the technique within the scope of every analytical laboratory specializing in the analysis of polycyclic aromatic compounds (PAC). Moreover, better understanding of the chemistry of the fused-silica surface, coupled with developments in the cross-linking and synthesis of stationary phases, are rapidly extending the method. The use of existing detectors is also being extended, while new detectors are being applied to PAC. In this chapter, recent advances (mainly since January 1981) in the analysis of PAH and their derivatives and analogs by open-tubular-column GC are reviewed.

Particular emphasis is placed on the analysis of higher molecular weight (MW) PAH, in which there is much interest currently [5], and of the nitrogen-containing derivatives: heterocyclic (PANH), amino derivatives (APAH), and nitro compounds (NPAH). The classification of PAC in Ref. 3 is adopted.

II. COLUMNS FOR GAS CHROMATOGRAPHY

A. Technology

The column is at the center of the analytical gas chromatograph; the quality of the separation achieved by the whole system can only be that of the column. Until comparatively recently, the preparation of efficient open-tubular-columns for GC was incompletely understood in spite of a voluminous literature. There is now, however, a much better comprehension of the chemistry of the column surface and its interaction with the stationary phase, and of the requirements for a stable phase film [6] suitable for PAC analysis. Moreover, the introduction of flexible fused-silica columns in 1979 by Dandeneau [7] began a new era in open-tubular-column GC.

The crucial observation was that of Lee et al. [1,8]; the metallic oxides in a glass surface can act as Lewis acid adsorption sites for polar and aromatic molecules. Successful procedures for the removal of these metal ions were proposed. Of course, the pure fused-silica surface is free of these sites. The second important structural feature of glass, and to a lesser extent fused silica, is that hydroxyl groups are attached to surface silicon atoms [1,6,8]. Especially for glass, these hydroxyl groups comprise active sites which must be deactivated, generally by silylation; high temperatures are necessary for complete reaction [9], preferably with methylcyclosiloxanes [e.g., octamethyltetrasiloxane (D4)], which is suitable for the deactivation of fused silica as well as glass [9,10].

For the construction of an efficient open-tubular column, a uniform and homogeneous film of stationary phase must be applied to the inner surface which should not rearrange as the temperature is increased. Clean glass and fused-silica surfaces are high-energy surfaces (i.e., are wettable by most organic liquids), but after the deactivation step only stationary phases of the appropriate surface tension will spread since the critical surface tension of the surface is reduced [9]. For this reason, coating with nonpolar or weakly polar phases is better understood; these are the general phases of interest in PAC analysis. A typical (now routine) chromatogram of a PAC mixture on a fused-silica SE-52 column is shown in Fig. 1.

The wettability of a column surface is not influenced by temperature even up to 250°C since changes in the surface tension of the phase are matched by changes in the critical surface tension of the surface [9]. However, of

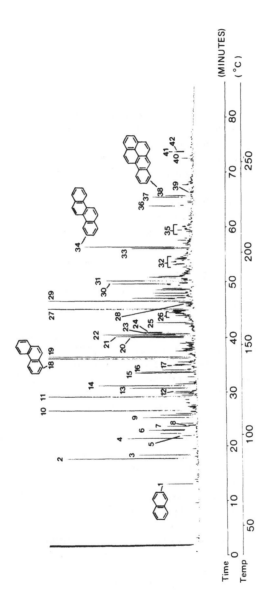

Figure 1. Chromatogram of the PAH fraction of a coal gasification condensate product. 20 m SE-52 fused-silica column, H_2 carrier gas, splitless injection. (Reproduced with permission from Ref. 55. Copyright John Wiley & Sons.)

vital importance is the viscosity of the coating film and its changes with temperature [11]. The rate of film disruption, for which the driving forces are vibration, gravity, and so on, is inversely proportional to the viscosity of the film [12]. The nonpolar polysiloxane phases commonly used in GC of PAC have viscosities that are virtually independent of temperature, whereas phases with phenyl and other polar groups incorporated in their structures have viscosities that fall more rapidly with increasing temperatures [11]. Film disruption at high temperatures is much more likely for the latter. The in situ cross-linking of coated stationary phases discussed below has now overcome this problem.

B. Stationary Phases for GC of PAC

Until the advent of cross-linking, "conventional" gum phases were most commonly used with methylpolysiloxanes (SE-30, OV-1, OV-101), 5% phenyl methylpolysiloxane (SE-52), and 5% phenyl 1% vinyl methylpolysiloxanes (SE-54) predominant [1]. These phases allowed chromatograms extending up to coronene (MW 300) to be routinely obtained on columns greater than 15 m long, although Grob was able [13] to chromatograph rubrene (MW 532) on a short (5.5 m) OV-101 column at 260°C. The heavy aromatic constituents of carbon black (MW up to 376) were analyzed by Hirata et al. [14] on a well-deactivated and silanized glass column coated with SE-52 and operated at 350°C, while Romanowski et al. carried out GC of an air particulates fraction containing compounds in the MW range 300-402 on a fused-silica column coated with a thin film of SE-54 [15].

Among newer GC phases, OV-73 and OV-1701 both give less column bleed than SE-52 and OV-17, respectively, because of their lower content of low molecular weight material shown to be present by size-exclusion chromatography [16]. Low molecular weight material may also be removed from OV-17 by extraction with supercritical carbon dioxide [17]. The high-temperature-phase polyoxyarylsulfonylarylene used in analysis of PAC [18] has a polarity similar to that of Carbowax [19] and can be used up to 390°C, but its usefulness is limited by its high viscosity below 200°C.

The increase in thermostability necessary to extend appreciably the working temperature range of polysiloxanes has recently been made possible by in situ cross-linking to form "bonded" or nonextractable stationary phases [6,8,20]. Such materials have the added advantage of allowing columns to be washed free of nonvolatile deposits, and also result in little stripping of the phase by inject solvent. Changes in viscosity with temperature are also minimized—particularly important for polar polysiloxane phases, where the presence of phenyl and other bulky groups distorts the regular helical conformation of the polysiloxane molecule [11,21].

Both Madani [22,23] and Blomberg [24-26] accomplished the in situ synthesis of polysiloxanes by thermally condensing hydroxyl and alkoxyl groups to split out water, alcohols, or ethers. Very stable methyl and phenylpolysiloxanes were formed since Si-O-Si bonds were formed; columns prepared in this way have been used up to 320°C routinely in PAC analysis, with the elution of compounds with MW 326 [27] and 376 [14]. Nonextractable cyanosilicone rubbers cross-linked by the Blomberg procedure have been found suitable for the analysis of PAH and azaarenes [26].

Unfortunately, the columns prepared by this cross-linking method are often less efficient and more active than those coated with conventional phases [6,8,28]. Free-radical cross-linking of polysiloxane stationary phases is

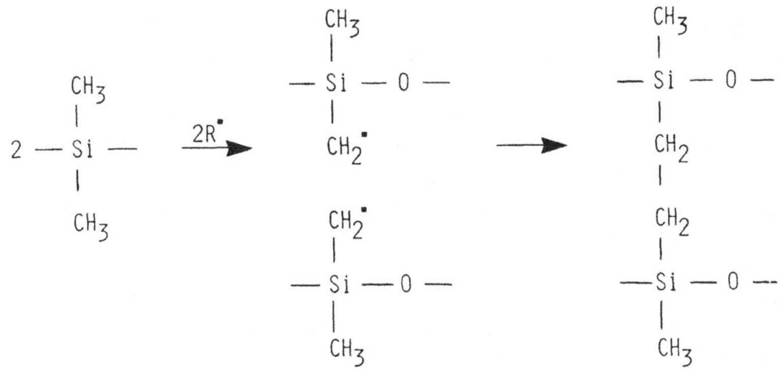

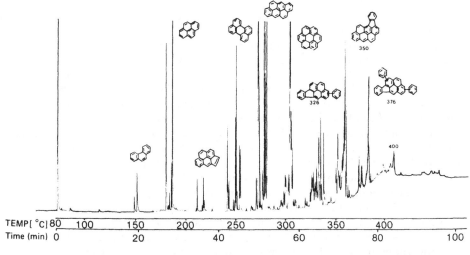

Figure 2. Free-radical cross-linking of a methylpolysiloxane stationary phase.

preferred [20,28-30], with formation of insoluble rubbers as a result of the carbon-carbon bonds formed between methyl groups attached to silicon atoms, (i.e., Si—C—C—Si) (Fig. 2). Vinyl groups are also very susceptible to cross-linking to methyl groups [28] so that nonextractable phases are readily prepared from SE-54. A fused-silica column coated with such a phase (DB-5) was used in analyses of the high molecular weight (up to 456) PAH of coal tar [31].

Of possible free-radical initiators, azo compounds rather than peroxides have minimal effect on phase polarity and column activity, although peroxides do give the greatest degree of polymerization [28,32]. Phenylpolysiloxanes have high intrinsic thermal stability, but unfortunately phenyl groups inhibit cross-linking. Peaden et al. synthesized [21] a range of phenylmethylpolysiloxanes with up to 70% phenyl groups and 4% vinyl groups; the latter phase was cross-linked using dichlorobenzoyl peroxide to produce a remarkably thermostable column which could be programmed up to 400°C and on which PAC with MW > 400 from a carbon black extract were separated (Fig. 3) [21].

Figure 3. Chromatogram of a carbon black extract. 12 m 70% phenyl methylphenylpolysiloxane fused-silica column; H_2 carrier gas. (Reproduced with permission from Ref. 21. Copyright Friedr. Vieweg and Sohn.)

Methyl-p-tolylpolysiloxane polymers containing 50% [33,34] and 70% [34] tolyl groups are easier to cross-link (with azo-t-butane) than are the corresponding phenyl-containing material.

The interactions between solutes and phenylsiloxane stationary phases and the selectivity [35] of the latter arise from the polarizability of the phenyl groups [36]. Lee et al. [37] have therefore suggested a new approach to selective phases for PAC by preparing a cross-linked biphenylmethylpolysiloxane phase and comparing its properties with those of methyl, phenyl, and cyanopropyl polysiloxanes for the separation of isomeric PAC. There were few advantages for nonpolar isomers (e.g., methylphenanthrenes), but the induced polarity of the biphenyl groups by polar solutes gives markedly improved separations for, for example, aminophenanthrene isomers (Fig. 4).

C. Liquid-Crystal and Mesogenic Stationary Phases

Remarkable selectivity has been achieved with liquid crystals as stationary phases in the GC of PAH on packed columns [38,39]. Rodlike molecules tend to pack in the solid crystal lattice with long axes parallel, and on melting, liquid crystals pass through a number of mesophases which retain the parallel molecular orientation. Solutes with rodlike molecules interact more strongly and are hence more soluble in the smectic (layers with long axes parallel which can slip over one another) and nematic (spatially averaged parallel orientation of long axes) phases. A number of important PAH isomer separations are thus facilitated [40] on N,N'-bis(p-n-alkoxybenzylidene)-$\alpha\alpha'$-bi-p-toluidine liquid crystals (BABT; Fig. 5 and Table 1) which can be correlated (see also Sec. VI) with molecular shape (e.g., phenanthrene from anthracene, triphenylene from chrysene, and benzo[a]pyrene from benzo[e]pyrene). New liquid-crystal stationary phases with naphthalene nuclei in the molecule [41] have also been found effective in packed columns for the separation of anthracene from other PAC [42].

These separations are dramatic for simple mixtures of standards, but the limited efficiency of packed columns even for coatings on glass microbeads [43] means that the complex mixtures of environmental PAH are not fully resolved and BABT columns are generally applied only to specific PAH in real mixtures (e.g., in rat brain tissues [44]). Zielinski et al. investigated [45] the separation of PAH on liquid-crystal phases coated on stainless steel open-tubular columns, but columns efficiencies were generally low. Somewhat better results were obtained for BMBT and BPhBT coated by in situ synthesis on barium carbonate or carbonized glass [46-48], but plate numbers were still inadequate. Laub et al. showed [49,50], however, that blends of up to 20% BBBT with SE-52 coated on leached and silanized glass retained the selectivity with only a small loss in column efficiency [50].

The applicability of a variety of liquid-crystal BABT stationary phases in mixtures with OV-73 and SE-52 coated on glass capillaries to the profiling of PAH from emission sources, especially the combustion of different fuels and coal liquefaction processes, was investigated by Bartle et al. [51]. They found that although BMBT, BPrBT, and BBBT are preferred because of their wide nematic range (Table 1), they are insufficiently thermally stable for routine use. The nematic range of BPhBT starts at too high a temperature for columns to be applicable to mixtures with two- and three-ring PAC. The nematic range of BHxBT is narrow, but rather similar retention behavior is observed in the smectic II range, and the columns are sufficiently thermally stable for use over long periods up to temperatures near the limit of the

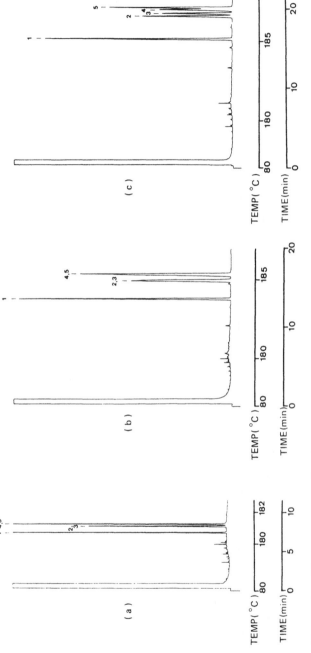

Figure 4. Chromatogram of aminophenanthrenes 12 m × 0.31 mm i.d. fused-silica columns coated with a 0.25 μm film thickness of (a) SE-54; (b) 50% phenyl; (c) 25% biphenyl polysiloxanes. H_2 carrier gas. Peak identifications: 1, 4-aminophenanthrene; 2, 1-aminophenanthrene; 3, 9-aminophenanthrene; 4, 3-aminophenanthrene; 5, 2-aminophenanthrene. (Reproduced with permission from Ref. 37. Copyright Friedr. Vieweg and Sohn.)

Table 1. Transition Temperatures (°C) of Liquid Crystals Coated on Capillary Columns

BRBT[a]	Solid–nematic	Solid–smectic	Smectic I–smectic II	Smectic–nematic	Nematic–isotropic
BMBT	181				337
BPrBT		169		176	311
BBBT		159		188	303
BHxBT		127	203	229	274
BPhBT	257				403

[a]R: M, methoxy; Pr, n-propoxy; B, n-butoxy; Hx, n-hexyloxy; Ph, phenoxy.
Source: Ref. 40.

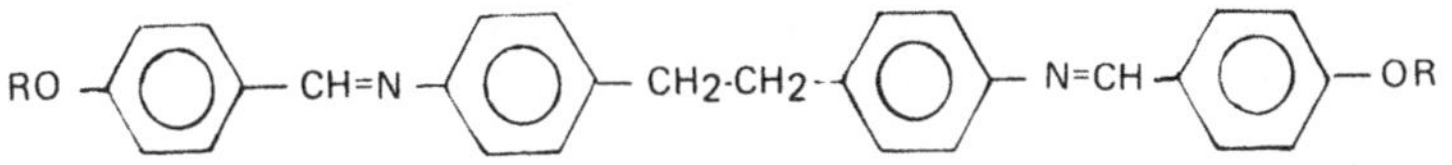

RO —⟨O⟩— CH=N —⟨O⟩— CH2-CH2 —⟨O⟩— N=CH —⟨O⟩— OR

Figure 5. Structure of N,N'bis(p-n-alkoxybenzylidene)$\alpha\alpha'$-bi-p-toluidine liquid crystals.

nematic range (274°C). On 15-m-long columns coated with a mixture of 20% BHxBT in OV-73, a temperature program in the nematic range eluted tetra- cyclic PAH by 245°C and pentacyclics at 260°C. For elution of six-ring aromatics, a short (7-m-long) column was necessary.

The application of BHxBT/OV-73 to the separation of PAH used as in- dicators for emission sources (see below) and to EPA "consent decree" and World Health Organization-nominated PAH was studied [51]. The improved resolution of anthracene and phenanthrene, benzo[a]- and benzo[e]pyrenes, of dibenz[a,c]- and dibenz[a,h]anthracenes and among the C_{18} tetra and pentacyclic PAH is advantageous. The resolution of chrysene and triphenyl- ene from each other and from other components is especially useful (Fig. 6).

In the studies above [50,51] it was found that increasing the liquid- crystal content of the mixture improved isomer resolution, but there was a concomitant fall in column efficiency. The optimum BABT loading for a series of separations may be determined by the window diagram method [50,52,53].

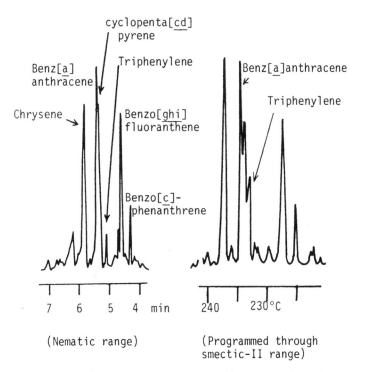

Figure 6. C_{18} region of chromatogram of coal combustion PAH on 15 m glass open-tubular column coated with 20% BxBT/OV-73 mixture. (Reproduced with permission from Ref. 51. Copyright Butterworth Publishers.)

The capacity factors of solutes (k') can be related to the weight fraction of, for example, BBBT in the stationary-phase mixture via

$$k'_M = k'_{BBBT} W_{BBBT} + k'_{SE-52} W_{SE-52} \qquad (1)$$

where k'_M, k'_{BBBT}, and k'_{SE-52} are the capacity factors of solutes on the mixed phase, BBBT, and SE-52 columns, respectively, and W_{BBBT} and W_{SE-52} are the weight fractions of each phase in the mixed phase. Graphs of relative retention α (ratio of capacity factors) against phase composition for all the solute pairs form a window diagram (e.g., for four-ring PASH in Fig. 7). Since only a minimum value of α is required, regions above line intersections can be neglected. From Fig. 7 the optimum resolution of the 12 four-ring PASH should be achieved for 16% BBBT/SE-52, with α for the least well separated pair, 1.011. In fact, 11 peaks were obtained for the 12 compounds on a 23 m × 0.32 mm column coated with 16% BBBT.SE-52. The required efficiency (N_{req}) for baseline resolution of all four-ring S-PAC isomers is calculated [from eq. (2) [53]]:

$$N_{req} = 36 \left(\frac{\alpha}{\alpha - 1} \right)^2 \left(\frac{k' + 1}{k'} \right)^2 \qquad (2)$$

as 3.68×10^5 plates for k' > 10. The practical application of mixed BABT/SE-52 phases to the resolution of PASH in HPLC fractions of a coal liquid has been demonstrated [52]. The essential limitations of liquid-crystal phases even in blends is that they cannot be applied to solutes with high vapor pressure because the nematic transition temperature is too high. Poly(mesogen methyl)siloxanes (PMMS), however, are gumlike phases which show high

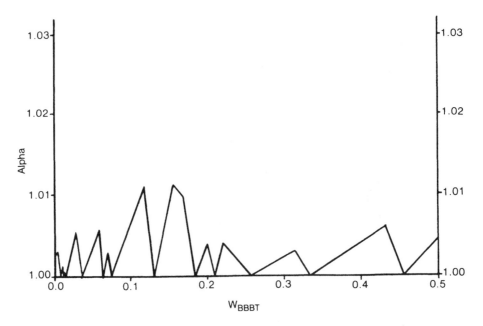

Figure 7. Window diagram (graph of relative retention, α, against phase composition) for four-ring polycyclic aromatic thiophen isomers. (Reproduced with permission from Ref. 52. Copyright Preston Publications, Inc.)

column efficiencies and stabilities, but retain high selectivity for PAC isomers and have a wide nematic temperature range (70-300°C) [54]. The resolution afforded by the PMMS phase for the 12 four-ring PASH isomers is superior to that of the BBBT/SE-52 mixed phase. The new phase offers dramatic improvements in separation for PAH, especially among the tetra- and pentacyclic compounds; the baseline resolution of benzo[b]-, benzo[j]-, and benzo[k]-fluoranthenes, which coelute on SE-52, is notable. Improved selectivity was also obtained for the six methylchrysene isomers on PMMS, although 3- and 4-methylchrysenes are not resolved. A window diagram of relative retention against phase composition indicated that a 50:50 blend of PMMS and SE-52 would provide a 4σ separation for N_{req} = 45,000 plates at k' > 10. The predicted result was achieved for 19-m-long columns joined in tandem in the sequence SE-52:PMMS (Fig. 8).

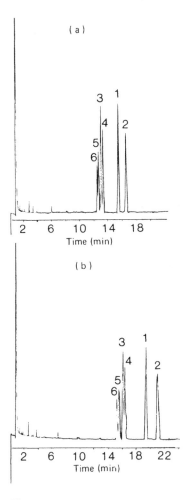

Figure 8. Chromatograms of the six methylchrysene isomers on tandem columns: (a) SE-52 (19 m):PMMS (19 m) fused-silica columns, 230°C isothermal, H_2 carrier gas; (b) PMMS (19 m):SE-52 (19 m) fused-silica columns, 230°C isothermal, H_2 carrier gas. Peak numbers refer to position of methyl substitution. (Reproduced with permission from Ref. 54. Copyright American Chemical Society.)

The excellent chromatographic properties of PMMS are conferred by its chemical structure, a combination of a polysiloxane backbone (imparting homogeneous coating characteristics with mesogenic side chains (imparting selectivity). The thermal stability of PMMS is comparable with that of SE-52, so that operation up to 270°C is possible with elution of low-volatile solutes such as indeno[1,2,3-cd]pyrene and dibenzo[ghi]perylene. Retention times are longer, but the selectivity means that shorter columns can be used. Figure 9 illustrates a chromatogram for a real mixture of PAH on PMMS.

III. INJECTION METHODS

The increasing necessity for quantitation in the analysis of PAC by capillary GC has led to a number of investigations of factors affecting the precision of these measurements. Injection technique is foremost among these [4]; the necessity to minimize the spreading of the injected band if substantial loss of column resolution is to be avoided led, initially, to split injection.

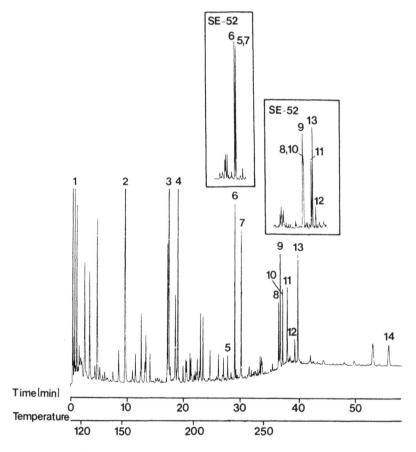

Figure 9. Chromatogram of coal tar on mesogenic polysiloxane stationary phase (20 m fused-silica column). Insets are portions of chromatograms obtained on SE-52 stationary phase. Selected peak assignments: 5, triphenylene, 6, benz[a]anthracene; 7, chrysene; 8, benzo[j]fluoranthene; 9, benzo-[b]fluoranthene; 10, benzo[k]fluoranthene; 11, benzo[e]pyrene; 12, perylene; 13, benzo[a]pyrene. (Reproduced with permission from Ref. 55. Copyright John Wiley & Sons.)

However, the loss of sensitivity in trace analysis, coupled with the difficulties in obtaining linear and reproducible split injection for samples containing compounds with a wide range of volatility, led to splitless and eventually, cold on-column injection. The latter completely avoids fractionation during sample vaporization and is more reproducible [55] since a liquid sample is injected directly into the inlet of the column.

Springer et al. compared [56] split and splitless injection for PAH, and found that if an internal standard were used, the response was linear over two orders of magnitude (2-200 ng), but there was discrimination against higher molecular weight compounds so that a standard with a molecular weight similar to that of the analyte was necessary. The effect was reduced in splitless injection [57], but there was still substantial variability from this source, which increased with difference in retention times of analyte and internal standard [56] (Table 2).

Standard deviations (SDs) of PAH peak areas for the same injected amount may be reduced if the injection time is increased [58], but only in cold on-column injection is reproducible injection observed. For example, the following ranges of SDs for 24 PAH relative to benzo[e]pyrene internal standard were observed [59]: split, 8-45%, splitless, 2.6-25%; cold on-column, 1.4-4%.

IV. DETECTORS

Although flame ionization, often coupled with electron impact mass spectrometry, is the most frequently applied detector in the GC of PAC, a range of more sensitive and selective detectors may also be employed. For example, the electron capture detector (ECD) has sensitivity in the picogram-femtogram range [55], and exhibits marked selectivity to PAC, showing increased response relative to that of nonaromatic hydrocarbons [2]. The ratio of ECD to flame ionization detector (FID) response varies with PAH structure [2] and has been used in the identification of air particulate PAH [60]. The ECD response to a given PAC is determined by the electron capture coefficient, K, which is related to the electron affinity and hence the adiabatic ionization potential, IP [61,62]:

$$K = kAT^{3/2} \exp\left[\frac{1000}{T}(70.17 - 8.539 IP)\right] \tag{3}$$

Table 2. Coefficient of Variation in Peak Area Relative to Internal Standard (Splitless Injection)

	n-Decane IS	n-Eicosane IS
Naphthalene	2.0	11.3
1-Methylnaphthalene	3.1	10.1
Acenaphthene	6.0	7.4
Phenanthrene	13.8	3.5
Pyrene	15.4	1.1

Source: Ref. 56.

where k and A are constants and T is absolute temperature. The ECD is especially applicable to nitro derivatives since these compounds have high electron affinity and hence large values of K [63]. The interference from carbazoles in chromatograms with FID and thermionic detectors is eliminated since these compounds have values of K approximately 10^3 times lower than those of nitro PAH. A disadvantage of the ECD method generally is the variation of response (e.g., by a factor of 60 among nitro PAH), so that individual calibration is necessary.

ECD response to PAC is greatly enhanced by the addition of small quantities (0.1-0.2%) of oxygen to the carrier gas [64,65] (Fig. 10). The mechanism is though [65] to involve the more rapid capture of electrons via the coupled reactions.

$$e^- + O_2 \rightleftharpoons O_2^- \tag{4}$$

$$O_2^- + A \longrightarrow products \tag{5}$$

rather than the direct

$$e^- + A \longrightarrow products \tag{6}$$

The relative rates of reactions (5) and (6) determine the response enhancement, which may vary between 3 and 400 (Table 3) and is correlated with molecular structure. By use of an atmospheric pressure ionization mass spectrometer, the foregoing mechanism has been confirmed; the product of reactions (5) and (6) is the negative parent ion M^- [66]. O_2 doping is clearly an aid to compound identification, and because of the improved detection limit, is also useful in trace analysis. The procedure has been extended to polycyclic aromatic amines (trifluoracetates and pentafluoropropionates) and phenols (methylated) [66]. Selective enhancement of ECD response can also be obtained by doping with N_2O [67]. The mechanism is thought to involve reaction of N_2O^- with the analyte at a rate faster than that of electron attachment. The signal of benzo[e]pyrene is enhanced 20-fold relative to that of benzo[a]-pyrene.

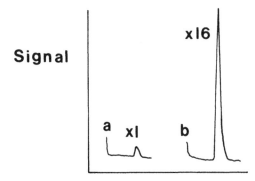

Figure 10. Response of electron-capture detector to 40 ng of benzo[e]pyrene: (a) without added oxygen; and (b) with 0.10% oxygen added to the carrier gas. Column temperature 200°C, detector temperature 250°C. (Reproduced with permission from Ref. 65. Copyright American Chemical Society.)

Table 3. Electron-Capture Detector Responses and Response Enhancements Caused by 0.20% Oxygen in the Carrier Gas at 250°C

no.	Compound	R_{EC}[a]	RE[b]
1	Phenanthrene	1.0	4.6
2	Anthracene	40	62
3	Triphenylene	1.5	3.0
4	Chrysene	13	140
5	Pyrene	20	185
6	Benz[a]anthracene	130	165
7	Naphthacene	4000	7.4
8	Benzo[e]pyrene	150	400
9	Benzo[a]pyrene	2300	29
10	Perylene	1000	15
11	Dibenz[a,c]anthracene	500	190
12	Dibenz[a,h]anthracene	640	67
13	Carbazole	110	200
14	Dibenzofuran	1.6	175
15	Dibenzothiophene	0.9	15
16	Acridine	1100	33
17	Xanthene	14	130
18	4-Azaphenanthrene	12	50
19	1-Azaphenanthrene	10	70
20	1-Chloroanthracene	2900	4.7
21	2-Chloroanthracene	3500	8.9
22	9-Chloroanthracene	2600	50
23	9-Chlorophenanthrene	4300	1.7
24	2-Methylanthracene	80	85
25	9-Methylanthracene	45	130
26	2-Methylbenz[a]anthracene	70	140
27	7-Methylbenz[a]anthracene	500	100
28	12-Methylbenz[a]anthracene	190	150
29	6,8-Dimethylbenz[a]anthracene	230	140
30	7,12-Dimethylbenz[a]anthracene	8500	8.9

[a]Normal ECD molar responses determined from peak areas all normalized with respect to the case of phenanthrene.
[b]Response enhancement induced by 0.20% oxygen, determined by ratio of peak heights obtained with and without oxygen in the carrier gas.
Source: Ref. 65.

The principle of operation of the photoionization detector (PID) [55]—
ionization of eluted compounds by ultraviolet (UV) light and collection of the
ions—makes it particularly applicable to PAC. Only radiation with energy
greater than the IP gives a signal, and the PID response is determined predominantly by the IP, which is generally lower for π-electron-rich polycyclic
molecules. The large linear dynamic range (seven orders of magnitude),
its sensitivity (10^2 times that of FID), and its concentration-dependent response all make it especially suitable for use with capillary columns [55].

The selectivity of the PID for PAH over alkanes is illustrated in Fig. 11;
here the PID/FID response ratios for the aromatic compounds are about 50,
compared with about 10 for the aliphatics [68]. Analyses of coal tars have
been made with the aid of these ratios [69]. If lamps with lower energy are
employed, the selectivity toward PAH is magnified (Table 4) [70]. Resonance-
enhanced two-photon ionization has advantages over conventional one-photon
ionization in spectral selectivity, improved ionization efficiency, and decreased
detector volume. Detection limits of the order of 10 pg have been determined
with this detector [71].

The combination of ECD and PID detectors has recently been found possible [72,73]. The response ratios for these two detectors allow nitro-PAH
to be easily distinguished from the parent PAH (e.g., the ECD/PID response
for naphthalene is 2.8×10^{-5} but is 3.8 for 2-nitronaphthalene) [73].

The mass spectrometer (MS) is the most versatile detector in open-tubular-column GC because it allows combination of efficient separation with a
high-performance identification technique [55]. The MS can be used as a
low-picogram range general detector in the total-ion-current mode; a selective
detector in the selected ion monitoring or multiple-ion detector modes; and for
positive identification to support retention data in the scanning mode. The
recent substantial improvements in interfaces for the coupling of the capillary column to the MS ion source, in particular open-split coupling [55],
have led to the widespread application of MS detection in the GC of PAH [3].
(see also Sec. VII).

Electron-impact and chemical ionization are most widely used, although
in both cases the spectra of PAH isomers are virtually identical. A variety
of methods have been proposed to allow isomers to be differentiated [74].
For example, the use of a mixed-charge-exchange-chemical ionization reagent
gas (Ar-CH_4) results in differences in the ratio of the abundance of $(M + 1)^+$
ion to the abundance of the M^+ ion brought about a variation for isomers of

Table 4. Photoionization Detector Responses for Standard Components

	IP (eV)	PID response[a] 10.2-eV lamp	PID response[a] 9.5-eV lamp	Ratio 9.5/10.2
Benzene	9.25	1.0	1.0	1.0
Toluene	8.82	8.4	11.6	1.4
Naphthalene	8.12	12.1	89.3	7.4
Anthracene	7.23	3.8	34.2	9.0

[a]Relative to response of benzene.
Source: Ref. 70.

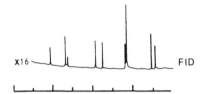

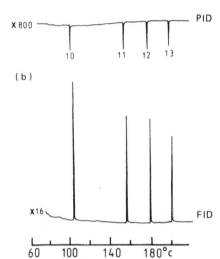

Figure 11. Photoionization and flame ionization detector responses of (a) aromatic hydrocarbons (5 ng), and (b) alkanes (10 ng). Peak identifications: 1, naphthalene, 2, methylnaphthalene; 3, 1-methylnaphthalene; 4, dimethylnaphthalene, 5, acenaphthene; 6, phenanthrene; 7, anthracene; 8, triphenylmethane; 9, terphenyl; 10, dodecane; 11, hexadecane; 12, octadecane; 13, eicosane. (Reproduced with permission from Ref. 68. Copyright Gordon and Breach, Science Publishers.)

IP and hence structure [75]; simultaneous monitoring of selected ions during chromatography of the isomeric compounds allows identification. Table 5 compares abundance ratios for several anthracenes and phenanthrenes with those from peaks in the chromatogram of the PAH fraction of air particulates, and allows identification of these compounds as methylphenanthrenes rather than methylanthracenes.

Ion generation in GC/MS of PAH by multiphoton photoionization with a laser has been demonstrated [76]. High sensitivity detection (limits as low as 200 fg) and linearity over a range of 5×10^4 were observed; the resulting

Table 5. Abundance Ratios for Several Anthracenes,
Phenanthrenes, and GC Peaks from Air Particulates

	$(M + 1)^+/M^+$ ratio[a]
Compound	
Anthracene	0.82
2-Methylanthracene	0.77
9-Methylanthracene	0.79
Phenanthrene	1.57
1-Methylphenanthrene	1.20
9-Methylphenanthrene	1.20
GC peaks	1.26
	1.20
	1.24
	1.24

[a]Corrected for ^{13}C natural abundance.
Source: Ref. 75.

mass spectra were dominated by the parent ion, with only low relative abundances of fragment ions. Selectivity can be obtained for isomers with different IP; KrF excimer laser radiation ionizes both chrysenes and triphenylenes (IP 7.8 and 8.1 eV, respectively), but XeCl excimer laser radiation ionizes only chrysenes at moderate light intensity [76].

Electron-impact mass spectra of isomeric primary, secondary, and tertiary amines, and azaarenes are very similar. The determination of the type of nitrogen species is extremely important in view of the variation in biological activity which these display. Derivatives (e.g., trimethylsilyl) are formed by primary and secondary amines [77] which can be distinguished from tertiary amines, but not from each other. A solution is presented by MS with ammonia chemical ionization [78], which brings about ionization of amines by transfer of H^+ from NH_4^+:

$$R_{3-n}NH_n + NH_4^+ \longrightarrow [R_{3-n}NH_{n+1}]^+ + NH_3 \tag{7}$$

where n is 0, 1, or 2 (for a tertiary, secondary, or primary amine). If ND_3 is used as reagent gas, not only is the amine ionized by the transfer of a deuterium atom, but there is also exchange of the hydrogen atoms on the nitrogen with deuterium:

$$R_{3-n}NH_n + ND_4^+ \longrightarrow [R_{3-n}ND_{n+1}]^+ + NH_nD_{3-n} \tag{8}$$

Typical behavior is shown in Table 6: the deuterated ion exhibits a change in mass from the NH_3 CI mass spectrum of one, two, or three mass units depending on whether the compound is a tertiary, secondary, or primary amine, respectively. The method can also be applied to compounds containing

Table 6. NH_3/ND_3 CI Mass Spectra of Nitrogen-Containing Compounds

Compound	Mol. wt.	m/z NH_3 CI	m/z ND_3 CI	Degree of substitution
Amines				
Aniline	93	94	97	1°
2-Aminobiphenyl	169	170	173	1°
1-Aminonaphthalene	143	144	147	1°
1-Aminoanthracene	193	194	197	1°
1-Aminopyrene	217	218	221	1°
N-Ethylaniline	121	122	124	2°
Diphenylamine	169	170	172	2°
N-Methylaminoanthra-quinone	237	238	240	2°
N,N-Dimethylaniline	121	122	123	3°
Triphenylamine	245	246	247	3°
Azaarenes				
Quinoline	129	130	131	3°
Acridine	179	180	181	3°
Carbazole	167	168	170	3°
Compounds containing two nitrogens				
5-Aminoquinoline	144	145	148	1° + 3°
1-Aminoacridine	194	195	198	1° + 3°
Phenazine	180	181	182	3° + 3°
O-Tolidine	212	213	218	1° + 1°
1,4-Diaminoanthra-quinone	238	239	244	1° + 1°

Source: Ref. 78.

two nitrogen atoms. The main limitation of the NH_3/ND_3 CI method is the failure to distinguish tertiary aromatic amines from azaarenes; recourse to the detail of the fragment-ion intensities can be of help here, however, The procedure has been applied with considerable success to coal-derived oils [78]. A further advantage is that only compounds with proton affinities greater than that of NH_4^+ are ionized, and OH- and SH-substituted compounds do not interfere [79].

A very sensitive and selective detection method in GC of nitro-PAH is electron-capture negative-ion CIMS [80]. With methane as reagent gas, electrons with thermal energies are generated in the ion source, and nitro-PAH form stable negative ions by electron capture with a mechanism similar to that of the ECD but increased sensitivity [80,81]. The mass spectra are characterized by an intense M^- peak with $(M - 16)^-$ and $(M - 30)^-$ also observed for mono- and dinitro-PAH [80]; sensitive (1 pg detection limit) selective-ion monitoring is possible, and ions from nitro compounds in low abundance in EI spectra are readily detected. Figure 12 shows chromatograms from a mixture of PAH (1-25 ng) with nitro-PAH (200-400 pg) under EI and electron-capture negative-ion (ECNI) CI conditions. Applications to an extract of carbon black [80] and to air particulate matter [63,82] have been reported.

Selective spectroscopic detection other than MS is afforded by UV and infrared (IR) detection. Early GC/UV combinations (e.g., [83]) were relatively insensitive, operating at the microgram level, but more recently, Novotny et al. succeeded in coupling a variable-wavelength UV detector via a 50-μl volume flow cell [84,85]. Detection limits of the order of nanograms were demonstrated for PAH; non-UV-absorbing aliphatic compounds are, of course, not detected.

Fourier transform (FT) IR spectroscopy "on the fly" can be carried out on GC effluents by transferring the gas stream to a reflecting light pipe which is the sample cell. Interferograms are accumulated at the rate of 1 per second and the chromatogram is reconstructed by a number of methods [55] for the total absorbance or for various spectral windows (e.g., of the carbonyl absorption region) and graphed against time to produce a chemical-function group chromatogram. Although the inherent sensitivity and resolution are low, the procedure was used [86] to distinguish alkyl-9-fluorenones from the isomeric benzo[c]cinnolines in the GC of a diesel exhaust particulate fraction. Matrix-isolation FT IR spectroscopy allows high-resolution spectra to be recorded, which can be used for fingerprinting and to distinguish between isomers in complex samples and allows greater sensitivity because of the much greater number of scans possible [87]. In a system proposed by Wehry et al. [88,89], compounds eluted from a support-coated capillary column are deposited in a nitrogen matrix on faces of a 12-sided movable gold-plated disk mounted within a cryostat. FR IR spectra can be recorded later at 1 cm^{-1} resolution for 500-1200 scans. Analysis of a mixture of naphthalene derivatives allowed isomeric dihydro, methyl, and dimethyl compounds to be distinguished, and applications to coal liquids were also reported.

The most selective GC detector for nitro-PAH is that based on chemiluminescence [90]. The operating principle is illustrated in Fig. 13. The carrier gas and eluted compounds are swept through a catalytic pyrolyzer held at 900°C to yield first the NO_2^- radical, and then the NO$^{\cdot}$ radical. All other materials are frozen out in a trap at -130°C, before the NO$^{\cdot}$ is reacted with ozone in a low-pressure chamber to yield excited singlet-state NO_2. The

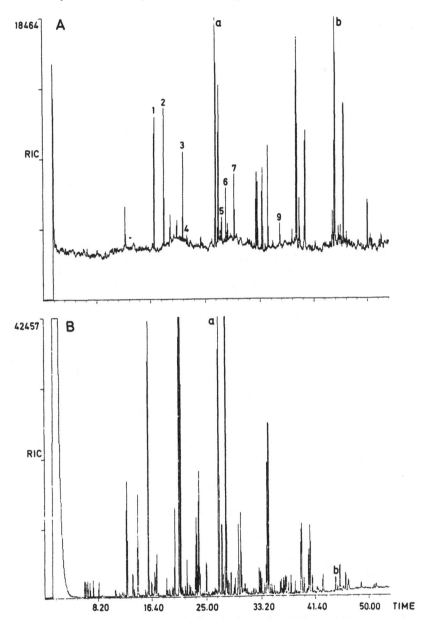

Figure 12. Reconstructed ion chromatograms of a mixture of PAH (1-25 ng) spiked with nitro-PAH (200-400 pg): (A) ECNICI condition; (B) EI conditions. Column, 30 m DB-5 fused silica. Selected peak identifications: 1, 2-methyl-1-nitronaphthalene; 3, 3-nitrobiphenyl; 6, 9-nitroanthracene; 7, 2-nitrofluorenone; 9, 1-nitropyrene. (Reproduced from Ref. 80. Copyright American Chemical Society.)

$$Ar - NO_2 \quad \xrightarrow[900°C]{Pyrolysis} \quad Ar^{\bullet} + NO_2^{\bullet}$$

$$NO_2^{\bullet} \quad \xrightarrow[900°C]{Pyrolysis} \quad NO^{\bullet} + \tfrac{1}{2}O_2$$

$$NO^{\bullet} + O_3 \quad \longrightarrow \quad NO_2^{*} + O_2$$

$$NO_2^{*} \quad \longrightarrow \quad NO_2 + h\nu \ (600 \text{ nm})$$

Figure 13. Operating principle of the chemiluminescence detector for nitro-PAH.

decay to the ground state of this species is monitored from the emission between 600 and 2800 nm, the intensity of which is proportional to the concentration of the original nitro compound. Combination with both packed and capillary columns has been reported with detection limits between 40 and 80 pg for mono-nitro PAH. Complete discrimination against PAH and derivatives containing sulfur, nitrogen, nitrile, amino, and carbonyl functionalities was observed.

V. RETENTION INDICES AND THEIR USE IN PAC IDENTIFICATION

It is generally accepted that correspondence between the retention data of an unknown and a reference compound on two columns of different polarity is sufficient for positive identification, although spectroscopic evidence lends further weight. The use of systems of reproducible retention parameters [91] has therefore been proposed. Variations in chromatographic conditions and film thickness makes the use of a retention index, I, based on a homologous series of reference standards most useful [91]. For PAH, the Kovats system is less reliable than that of Lee et al. [92], which is based on the internal standards naphthalene, phenanthrene, chrysene, and picene rather than n-alkanes:

$$I_x = 100Z + 100 \left(\frac{T_{R_X} - T_{R_Z}}{T_{R_{Z+1}} - T_{R_Z}} \right) \tag{9}$$

where T_{R_X} is the elution temperature of compound x in linear temperature-programmed GC and Z and Z + 1 are the number of rings in the bracketing standards.

The original system has been critically evaluated by Vassilaros et al. [93], and the effects of variations in programming rate, column internal diameter, and initial temperature have been discussed. Values of I for a total of 310 PAC have been listed (78 PAH, 115 PASH, and 117 PANH and APAH), with standard deviations (generally less than 0.10 unit) and 95% confidence limits [93]. The description of relative elution order and position is possibly more valuable than absolute values of I. For three- and four-ring PAC, values of I are independent of the initial column temperature below 60°C. A 2-min initial time at the injection temperature is recommended [93].

The Lee system [92,93] has the advantage that the first three reference standards are generally present in real PAC mixtures. Picene is often not present, and rather than add this compound a converging algorithm is used to calculate its T_R from those of chrysene and benzo[e]pyrene.

This system has been applied to the analysis of a variety of complex PAC mixtures (e.g., from coal extracts [94] and pyrolysis products [95], diesel particulates [96,97], synthetic fuels [98,99], tissues and sediment [100-103], air particulates [51,104] and combustion effluents [51,105]). An illustrative example of the value of the Lee system is its use by Krishnan and Hites [105] to confirm that the compound alleged to be acephenanthrylene throughout the PAH literature was the same as their (synthesized) authentic sample. Table 7 compares the values of the retention index for acephenanthrylene and its neighbors in the chromatograms of a variety of real PAH mixtures. The various values are consistent with one another, and given the approximations involved in retention-time measurements from chromatograms in the literature, within the expected limits of experimental error.

Novotny et al. proposed [106] a similar retention index system for PANH, based on the standards quinoline (200.00), acridine (300.00), benz[a]acridine (400.00), and dibenz[a,j]acridine (500.00), which has been extended to other stationary phases [107] (e.g., OV-73, OV-61, and SP-2340).

For unknown compounds for which reference standards are not available, correlations between retention properties and molecular structure may help in identification. The connectivity index, χ, a readily computable topological molecular parameter [108], has been claimed [109,110] as a significant identification aid for unknown PAH in mixtures since linear relationships have been shown between χ and various retention indices on a number of phases. The very high (>0.990) correlation coefficients [109,110] for these relations disguise, however, quite erroneous sequences of elution of numerous compound pairs [111]. For example, the relative retentions on all of SE-52, OV-101, and OV-17 of anthracene and phenanthrene, of benzo[a]- and benzo[b]fluorenes, and of benzo[a]- and benzo[e]pyrenes are all wrongly predicted on the basis of χ, so that the empirical value of this procedure is questionable [111]. The average molecular polarizability has also been claimed to correlate with retention index [112,113]. As shown below, such correlations are merely fortuitous consequences of the thermodynamics of GC.

VI. MECHANISM OF RETENTION IN THE GC OF PAC

The sequence of elution of PAH is independent of the polarity of the stationary phase, with only minor selectivities for some isomer groups on a more polar phase [18]. A consideration [111] of the theory of GC [114] suggests that the retention properties of PAH on conventional phases are strictly related only to the logarithm of the vapor pressure, p, of the PAH at column temperature. This conclusion is a consequence of the relation between log retention volume, V_g, and I, which arises if elution temperature is replaced by log V_g in Eq. (9). Hence

$$I \propto \log V_g = \log \frac{273R}{\gamma PM_p} \tag{10}$$

Table 7. Retention Indices for Fluoranthene, Acephenanthrylene, and Pyrene in Different Studies

| | Retention index on SE-52 | | |
	Fluoranthene	Acephenanthrylene	Pyrene
Mixture of authentic compounds [105]	344.7	347.9	351.3
Combustion products [122] of:			
Coal	344.0	347.6	350.9
Wood	344.2	348.2	352.0
Kerosene	344.0	347.7	
Carbon black [105]	344.7	348.6	352.1
Standard [92]	344.01		351.22

where Mp is the molecular weight of the stationary phase, and the activity coefficient γ is constant for the PAH series. Less rigorous correlations with PAH boiling point, T_b [18,115], and molecular weight, M, follow from the empirical relations between log p and T_b (via the Clausius-Clapeyron equation and the Trouton rule) and between T_b and M. In fact, χ correlates well with M (r = 0.995) [111], as does average molecular polarizability, for PAH, so that all these parameters are related in a cyclical manner (Fig. 14) and χ is not linked with retention in a causative way.

The connectivity has, however, been incorporated in more complex regression equations; the retention indices of 231 PAC were accounted for by a four-variable equation with a multiple correlation coefficient of 0.990 [116]. This work also encompassed PANH, for which Guiochon also considered [107] relations between I and molecular properties; a general correlation with T_b was noted, but steric hindrance and location and degree of alkyl substitution also influenced retention behavior. Bartle et al. [111] also found different proportionality constants between I and T_b for alkyl and nonplanar PAH on the one hand, and planar PAH on the other.

Shape factors have been suggested [115] as contributing to retention on conventional phases; for example, among isomers, larger retention is apparently associated with "more extended" molecules. This would correspond of course, to different values of γ in Eq. (10). On the other hand, for PAH in SE-52, OV-101, and OV-17, elution strictly follows the sequence of T_b or p, where these are precisely known (e.g., anthracene and phenanthrene, and methyl and ethylphenanthrenes) [111].

The mechanism of liquid-crystal and mesogenic stationary-phase selectivity can be understood [38,51,111], however, in terms of molecular ordering

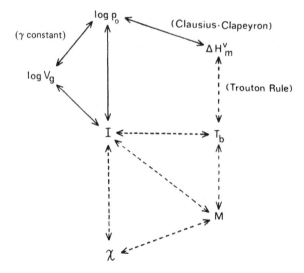

Figure 14. Interrelation of GC retention parameters and molecular properties. Direct, causative relations are shown with solid lines, indirect relations with dashed lines. (Reproduced with permission from Ref. 111. Copyright Friedr. Vieweg and Sohn.)

Table 8. Relative Retentions and Length-to-Breadth Ratios for Methyl-chrysene Isomers

Isomer	Relative retention on:		
	BBBT[a,b]	BPhBT[c,d]	L/B
6-Methylchrysene	1.37	1.124	1.48
5-Methylchrysene	1.44	1.164	1.48
3-Methylchrysene	1.55	1.235	1.63
4-Methylchrysene	1.62	1.253	1.51
1-Methylchrysene	1.92	1.515	1.71
2-Methylchrysene	2.17	1.720	1.85

[a] Relative to benz[a]anthracene = 1.00.
[b] From Ref. 117.
[c] Relative to chrysene = 1.000.
[d] From Ref. 48.

as a result of molecular geometry. Retention of isomers on such phases is, in general, in sequence of the length-to-breadth ratio L/B [51,111,117,118]. For example, the retention of methylchrysenes on PMMS [54] (Fig. 8), BBBT [117] and BPhBT [48] follows the values of L/B except for the relative sequence of the 3- and 4-methyl isomers (Table 8).

More generally, a graph of I against T_b, for example for BMBT (Fig. 15), is linear for the naphthalene, phenanthrene, and chrysene anchor points (L/B 1.24, 1.44, and 1.68, respectively), but deviations from the graph are shown by more extended molecules (e.g., anthracene, L/B = 1.58), or compact molecules (e.g., pyrene, L/B = 1.12).

These values of L/B reflect systematic changes in γ from the differently shaped PAH. γ is made up [119] of translational, rotational, and conformational contributions:

$$\gamma = \gamma_{trans}\gamma_{rot}\gamma_{conf} \tag{11}$$

and can be greater or less than 1 depending on the magnitude of each term. The treatment is simplified [38] by taking γ_{conf} as unity for rigid molecules such as PAH, where there is no restriction of internal rotation in solution [51]. In the nematic and certain smectic liquid-crystal stationary phases (and in mixtures of these with gum phases, where molecular ordering evidently persists), more polarizable and rodlike isomers (i.e., higher L/B) have smaller γ_{trans}; solubility is energetically favored because there are now more solvent-solute interactions. This enthalpy gain overcomes [120] the entropy loss (corresponding to $\gamma_{rot} > 1$) which arises from the greater degree of solute ordering in the liquid crystal for the higher-L/B isomer, which gives up more rotational freedom. The product γ is therefore less than unity for an isomer with higher L/B, and solubility is favored in the ordered phase [51].

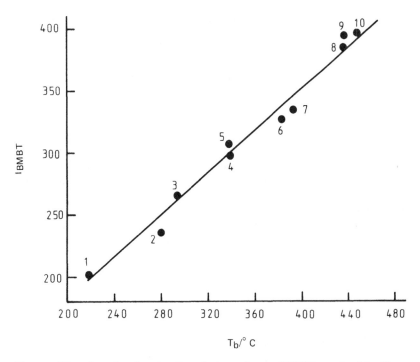

Figure 15. Graph of retention index, I, on BMBT against boiling point, T_b, for 10 planar PAH: 1, naphthalene; 2, 1-methylnaphthalene; 3, fluorene; 4, phenanthrene; 5, anthracene; 6, fluoranthene; 7, pyrene; 8, triphenylene; 9, benz[a]anthracene; 10, chrysene. (Reproduced with permission from Ref. 111. Copyright Friedr. Vieweg and Sohn.)

VII. APPLICATIONS OF GC IN PAC ANALYSIS

A. Source-Specific PAH

A number of PAH have been used as indicators for the origin of emissions [51, 121-127]. For example, coronene, benzo[ghi]perylene, and especially cyclopenta[cd]pyrene have been identified [124,125] as being typical of automobile exhausts, and the latter compound is also present at high concentration in the combustion products of kerosene [122,128]. Benzo[b]-naphtho[2,1-d]thiophen originates [124,125] almost solely from the burning of high-rank coal, while picene and its derivatives are thought [126] to distinguish brown coal combustion. Retene (1-methyl-7-isopropylphenanthrene) was until recently thought to arise in sediments from the natural degradation of abietic acid [129], but has now been shown [127] to be a unique molecular marker in air from the combustion of coniferous wood. High values of the concentration ratios benzo[a]pyrene/benzo[ghi]perylene [123], greater relative concentrations of alkylated PAH [122], but lower chrysene/triphenylene ratios [51] are all also thought to be typical of coal combustion products.

Resolution of many of these markers is fairly readily achieved on gum phases such as SE-52 or OV-101. As discussed in Sec. V, however, picene may overlap benzo[b]chrysene; benzo[b]naphtho[2,1-d]thiophen and retene are incompletely resolved [92] on SE-52 from, respectively, benzo[ghi]fluor-

anthene and benzo[b]fluorene, so that mesogenic phases may be employed
to advantage. Either a mesogenic or a highly polar phase (e.g., Poly S-179
[18]) is necessary to resolve triphenylene from chrysene; the analysis is
rapid on a liquid-crystal column, needing only 6 min on a 15 m 20% BHxBT
column at 235°C [51]. This column is, however, unsuitable for the determination of cyclopenta[cd]pyrene, which overlaps the benz[a]anthracene peak
[51] (Fig. 6).

Ratios of compounds with widely differing volatilities are probably less
reliable than ratios of isomers, for which collection efficiencies and evaporation losses are likely to be similar. Discrimination during injection onto
the column should also be minimal. It is vital that cold on-column injection
be used in determining indicator ratios not involving isomers. The following
pyrene/benzo[a]pyrene ratios were measured [51] for a coking plant emission
with different injection methods: split (100:1), 5.6; splitless, 4.2; cold on-
column, 3.4. Only the last method avoids sample fractionation during injection with loss of high molecular weight compounds.

B. Nitrogen-Containing PAC

Nitrogen-containing PAC (N-PAC) are present in petroleum, shale oils, and
coal liquids, and their analysis is of interest for a variety of reasons [130].
Thus N-PAC are effective catalyst poisons during upgrading; they are responsible for nitrogen oxide emissions during combustion, and many N-PAC
are mutagenic and/or carcinogenic. The different possible chemical classes
of N-PAC and the possibilities for isomerism (Table 9) make open-tubular-
column GC obligatory in their separation, but the strongly basic nature of
many of these compounds makes them difficult to chromatograph [131].

For satisfactory GC, surface deactivation of the column is necessary,
preferably by use of a polar stationary phase such as Carbowax 20M or UCON
50-HB-2000 to neutralize the absorptive effects of the column wall [131];
further deactivation may be achieved by adding agents such as tetraphenyl-
boron sodium [106] to the coating solution. The polar phases are selective
and offer many advantages for the analysis of N-PAC, but cannot be used
at temperatures much above 200°C, so that the analysis is restricted to one-
or two-ring compounds (e.g., from marijuana smoke [132], coal tar [133],
and solvent-refined coal recycle oil) [106]. For N-PAC with three rings or
more, gum phases must be used, so far the nonpolar silicones, which offer
little selectivity. Nonetheless, useful separations of N-PAC with two, three,
and four rings have been achieved [99,130,134,135] on SE-52-coated fused-
silica columns, which may be deactivated with D4 before coating. Recent
progress with the cross-linking of moderately polar phases should lead to
more selectivity. The GC of an APAH fraction from a coal liquid on a cross-
linked 90% cyanopropyl, 10% tolyl polysiloxane phase has already been demonstrated [34]. The selectivity of the polarizable biphenyl-group containing
polysiloxane phase for aminophenanthrenes [37] was discussed in Sec. II. B.

An N-PAC fraction may be separated from fossil fuels and synfuels [130,
134] and from the PAC concentrate from environmental samples [137] by
chromatography on alumina. GC of this fraction with simultaneous nitrogen-
selective detection (NPD) and flame ionization detection differentiates between
the nitrogen-containing components and other polar compounds not containing
nitrogen (Fig. 16). The selectivity and sensitivity of the NPD are well known
[55]—the selectivity for N-PAC over PAH is 10^4:1, while a fused-silica effluent splitter causes only a 5-10% loss in efficiency when the two detectors
are used [136].

Table 9. Possible Classes of Nitrogen-Containing PAC

	Name	Chemical class	Acronym	Number of isomers
	Acridine	Tertiary polycyclic aromatic nitrogen heterocycle (azaarenes)	3°-PANH	8
	Carbazole	Secondary PANH	2°-PANH	6
	2-Aminoanthracene	Amino polycyclic aromatic hydrocarbon	APAH	8
	2-Cyanoanthracene	Cyano polycyclic aromatic hydrocarbon	CPAH	8
	2-Nitroanthracene	Nitro polycyclic aromatic hydrocarbon	NPAH	8

Sources: Refs. 3 and 130.

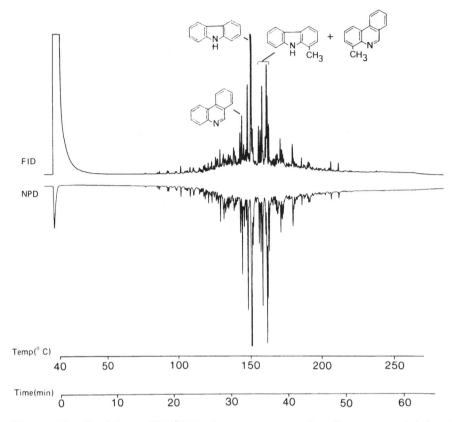

Figure 16. Dual-trace FID/NPD chromatograms of a nitrogen-containing PAC fraction from an SRC II coal liquid. 15 m SE-52 fused-silica column with fused-silica effluent splitter. (Reproduced with permission from Ref. 136. Copyright Dr. Alfred Huethig Publishers.)

Further separation of the N-PAC fraction on silicic acid yields 2°-PANH, enriched APAH, and 3°-PANH fractions [134], although 3°-PANH with the nitrogen lone pair shielded by alkyl groups or a ring overlap the APAH fraction. The use of ammonia CI MS for distinguishing isomeric N-PAC (e.g., methylbenzoquinolines and aminophenanthrenes) according to chemical class [78] was described in Sec. IV. An alternative is derivatization of the enriched APAH fraction to form either the corresponding trifluoroacetyl (TFA) or pentafluoropropyl (PFP) amides of the APAH followed by GC with ECD detection or GC/MS [134]; 3°-PANH do not react in this procedure and give no derivatives; the selectivity for derivatized over underivatized amines is > 1000:1, so that excellent profiles of APAH can be obtained [134]. Formation of the TFA or PFP derivatives of APAH is also especially useful in GC/ MS of the enriched APAH fraction since the molecular weight is increased by 96 or 146 mass units, respectively; selective ion monitoring is therefore especially effective. The fragmentation patterns are also informative since major ions at $(M - 97)^+$ (TFA derivatives) and $(M - 147)^+$ (PFP) are also found [134].

APAH are the most potent mutagens among the N-PAC, and indeed the principal microbial mutagens in coal liquids [137]; considerable interest therefore attaches to the isolation of "clean" APAH fractions [135] to aid interpretation of biological activity and to allow more detailed characterization. Complete separation of APAH-PFP derivatives from 3°-PANH was obtained by gel permeation chromatography, followed by catalytic hydrolysis of the derivatives back to amino compounds. GC/MS was used to determine that hydrolysis was complete and that the 3°-PANH fraction contained no APAH-PFP derivatives [135].

C. Nitro PAH

In 1978, Pitts et al. [138] demonstrated that PAH react with nitrogen oxides to form nitro PAH (NPAH), some of which have been shown to be potent direct-acting mutagens [139]. A large portion of the biological activity of engine (particularly diesel) exhaust emissions has been attributed to NPAH [140], which have also been identified in a variety of other environmental samples [63,80,141-149]. The large number of possible NPAH isomers has demanded analytical methods of high resolution and specificity; fused-silica open-tubular-column gas chromatography with nitrogen selective detection, and GC/MS are therefore preeminent.

Nonpolar gum phases —SE-52, SE-54, or DB-5 (cross-linked)—have been found suitable for these compounds, but greater isomer resolution may again be expected on the new cross-linked polar and polarizable gum phases (Sec. II.B.). Retention indices for a large number of synthetic NPAH have been measured [143]. Analyses are made for enriched HPLC fractions because of the low concentrations of each NPAH in the environmental samples (e.g., <0.03 ng/m^3, compared with 0.6 ng/m^3 benzo[a]-pyrene in urban air [141,142]), but the presence of other compounds of similar polarity makes use of a nitrogen-specific detector obligatory. Simultaneous detection by FID and NPD is most often used; relative NPD/FID responses have been used to allow isomers (e.g., dinitropyrenes) to be distinguished [145]. Detection limits for NPAH of 50 pg and linearity of response up to 50 ng have been noted [143]. The interference of N-PAC such as 9-methylcarbazole [146] in the analysis of NPAH fractions may be overcome by the use of ECD detection [63], discussed in Sec. IV. Of course, the chemiluminescence detector [90], also discussed previously, is the ultimate specific detector for NPAH. Electron-impact MS with selective-ion monitoring [144,146,147] is particularly useful for detection since NPAH give abundant M^+, $(M - NO)^+$, and $(M - NO_2)^+$ ions [146]. The use of negative-ion CI MS in the GC/MS of NPAH [80,148] was noted in Sec. IV. Care must be exercised in GC/MS, however, since NPAH (such as 1-nitro-pyrene) may decompose before detection [149]. Similarly, high injection temperatures are best avoided, especially if hydrogen is used as carrier gas [145]; cold on-column injection is preferred [143,145].

Early work used retention data with nitrogen selective detection followed by GC/MS to determine and quantify the most abundant APAH in environmental samples [140-145] (e.g., 9-nitroanthracene, 1-nitropyrene, 2-nitro-pyrene, 3-nitrofluoranthene, 7-nitrobenz[a]anthracene, and 6-nitrobenzo[a]-pyrene). A much larger number of NPAH are present in diesel particulates, and Paputa-Peck et al. [146] have positively identified 15 NPAH and two nitro N-PAC [using criteria of matching retention times (GC/NPD) and both low- and high-resolution mass spectra (GC/MS of standards]; 15 further NPAH present in too low a concentration for GC/MS were identified from retention data of standards; and a further 45 for which isomer structures were not available were identified by GC/MS.

Table 10. Further Applications of GC in the Analysis of PAC

Analyte	Column length (m)	Stationary phase	Special detector(s)	Reference
Fossil fuels				
Coal liquid	PAH, N-PAC	60	SE-52	MS(El and Cl)	155
Coal conversion oil	APAH, 3°-PANH	50	OV-101		156
Fossil fuels	APAH	3.7 (packed)	Dexsil 400		157
Coal liquids	PAC	45	SE-54		158
Coal tar and pitch	PASH	Various (SCOT)	OV-17 SP-2250	FPD, MS	159
Anthracene oil	Mixed heterocycles	Various (SCOT)	OV-17 SP-2250	FPD, MS	160
Petroleum	Carbazoles	25	OV-73	NPD	161
Coal liquids	N-PAC	50	SP-2250	NPD, MS	162
Coal tar	3°-PANH	15	Superox-4	MS	163
Crude oil	PAH	50	CP sil 5	MS	164
Petroleum	N-PAC	70	OV-73		165
Fossil fuels	Hydroaromatics	30	DB-5	MS	166
Coal tar products	N-PAC	Various (SCOT)	SP-2250	NPD, MS	167
Coal tar	N-PAC	40 (SCOT)	SP-2250	NPD, MS	168

Sample	Compound	Length (m)	Phase	Detector	Ref.
Coal gasifier products	PAC		Silicone	MS(EI and CI)	169
Hydrotreated coal liquids	PAC	25		MS	170
Lignite gasification products	PAC	50	OV-101	MS	171
Coal liquids	PAC	25	DB-5		172
Asphaltene pyrolysis products	PAC	50	SE-54	MS	173
Air particulate matter	PAC	25	OV-17		174
	PAH	30	SE-52		175
	PAC	40	SE-30		176
	PAC	30	OV-1		177
	NPAH, oxygenated PAH	30	SE-54	MS	178
	PAC	9	SE-54		179
	NPAH	30	DB-5	ECD, NPD, MS	180
Combustion emissions					
Wood and straw burning	Oxygenated PAH	30	DB-5	MS	181
Fluidized-bed combustion products	PAC	15	DB-5	MS	182
Wood burning	PAC	60	DB-5	MS	183
Domestic oil burning	PAC	15	DB-5	MS	184
Fly ash	PAC	30	DB-5		185
Coal liquid burning	PAC		SE-52	MS	186

Table 10. Continued

Analyte	Column length (m)	Stationary phase	Special detector(s)	Reference
Engine emissions				
Diesel particulates — PAH	40	SE-52	MS	187
Used engine oil — PAC	50	CP sil 5	MS	188
Aircraft turbine particulates — PAH	40	OV-1	MS	189
Diesel particulates — NPAH	50	SE-54	MS	190
Diesel emissions — PAC	30	DB-5	MS	191
Diesel particulates — PAH	50	OV-101	MS	192
Diesel emissions (vegetable oil fuel) — PAH	50	OV-101	MS	193
Diesel emissions (aromatic fuels) — PAH	50	OV-101	MS	194
Biota and environmental samples				
Wastewater — PAH	25	SE-30		195

Environmental samples					
Marine sediments and biota	PAC	20	OV-1		196
	PAC				197
Water	3°-PANH	5 (packed)	PEG-HT		198
Fish exposed to fuel oil	PAH metabolites	30	DB-5	MS	199
Lake sediments	PAC	25	SE-30	MS	200
River sediments, fish, and water	PAH, PASH, N-PAC	20	SE-52	MS	201
Miscellaneous					
Lubricating oil	PAC	50	CP sil 5	MS	202
Vegetable oils	PAC	30	OV-17/SE-30		203
Cellulose pyrolysis products	Alkylated tricyclics	25	Carbowax 20M		204
Tobacco smoke	High MW PAH		Dexsil 400	MS	205

D. Sulfur- and Oxygen-Containing PAC

Polycyclic aromatic sulfur heterocycles (PASH) are well-known constituents of petroleum and synfuels, and hence are also present in the environment [101] with a high degree of persistence [100,103]; PASH are of considerable importance in view of their potential health effects [150].

Analysis of complex samples for PASH are generally carried out by GC on gum phases (e.g., SE-52). Other stationary phases may be used to resolve specific PASH isomers: 50% Superox 20M/SE-52 mixed phase for three-ring isomers; a mesogenic polysiloxane for the methylated three-ring PASH and the four-ring PASH isomers; and a 50 % BBBT/SE-52 mixed phase for the peri-condensed five-ring PASH isomers with molecular weight 258 [52,54]. MS or dual-trace FID/flame photometric (FPD) detection are usually employed. The FPD is highly selective for PASH, but care must be taken in its use because of the nonlinear response and the quenching which may occur in the presence of coeluting compounds [151].

The large number of interfering compounds means that prefractionation is usually necessary. The PASH of environmental samples are found in the neutral PAC fraction obtained by chromatography on alumina [103]; further cleanup by gel permeation chromatography is necessary for biogenic samples [54]. Preliminary screening for PASH with FID/FPD and the use of retention indices (Sec. V) is followed by confirmatory GC/MS [52,54,98, 103,152]. Sulfur heterocycle fractions may be isolated from the neutral PAC fraction by a chemical procedure [98]; the PASH are oxidized with H_2O_2 to their corresponding sulfones, which are separated from the PAC on silica gel, and reduced with $LiAlH_4$ to the original PASH. Evaluation [152] of the scheme has revealed discrimination against certain structural types, but the method has proved extremely useful in the analysis of fuels.

Many oxygenated PAC are associated with the particulate matter of diesel engine exhausts and ambient air. They may be concentrated in the moderately molar fraction eluted from silica gel with benzene [153] or from silicic acid with toluene [96]. Analysis of the fractions by GC/MS on nonpolar columns (e.g., SP-2100 or OV-73) showed the presence of numerous ketones, quinones, anhydrides, and aldehydes [96,153]. Phenalen-1-one, identified in part by GC/MS as a combustion product of fossil fuels [154], is a particularly potent mutagen.

E. Applications to Different Sample Types

The developments in GC techniques have been discussed in the previous sections with illustrations from applications to a range of sample types. Table 10 lists other applications in the period under review.

VIII. CONCLUSIONS

The recent revolution in GC brough about by the routine availability of durable fused-silica open-tubular columns with highly thermostable cross-linked stationary-phase films has been reflected in the analysis of PAC. Not only are very high molecular weight PAH now analyzed by this method, but the nitrogen, sulfur, and oxygen derivatives, which often have greater biological activity, have been brought into the province of GC, especially when the highly selective detectors now available are applied. No other separation method offers the high resolution, speed, possibilities for positive identification, and small-sample-size requirement of GC with open-tubular columns. Their use is consequently now rising, apparently exponentially, in the analysis of PAC.

ACKNOWLEDGMENTS

It is a pleasure to express my thanks to my friends Milton L. Lee (Brigham Young University), Douglas W. Later (Battelle Pacific Northwest Laboratory), and Bernard Frere (Leeds University) for their help and advice during the preparation of this chapter.

REFERENCES

1. M. L. Lee and B. W. Wright, J. Chromatogr. 184:235 (1980).
2. M. L. Lee, M. V. Novotny, and K. D. Bartle, Analytical Chemistry of Polycyclic Aromatic Compounds, Academic Press, New York, 1981, Chap. 7.
3. K. D. Bartle, M. L. Lee, and S. A. Wise, Chem. Soc. Rev. 10:113 (1981).
4. B. S. Olufsen and A. Bjørseth, in Handbook of Polycyclic Aromatic Hydrocarbons, A. Bjørseth (Ed.), Marcel Dekker, New York, 1983, Chap. 6.
5. K. D. Bartle and M. Zander, Erdoel Kohle Erdgas Petrochem. 36:15 (1983).
6. M. L. Lee, F. J. Yang, and K. D. Bartle, Open Tubular Column Gas Chromatography—Theory and Practice, Wiley, New York, 1984, Chap. 3.
7. R. Dandeneau and E. H. Zerenner, J. High Resolut. Chromatogr. Chromatogr. Commun. 2:351 (1979).
8. B. W. Wright, B. E. Richter, and M. L. Lee, in Recent Advances in Capillary Chromatography, M. Novotny (Ed.), 1985, in press.
9. K. D. Bartle, B. W. Wright, and M. L. Lee, Chromatographia 14:387 (1981).
10. M. L. Lee, B. W. Wright, and K. D. Bartle, in Proceedings of the Fourth International Conference on Capillary Chromatography, R. E. Kaiser (Ed.), Hüthig, Heidelberg, West Germany, 1981, p. 505.
11. B. W. Wright, P. A. Peaden, and M. L. Lee, J. High Resolut. Chromatogr. Chromatogr. Commun. 5:413 (1982).
12. K. D. Bartle, unpublished calculations, 1983.
13. K. Grob, Chromatographia 7:94 (1974).
14. Y. Hirata, M. Novotny, P. A. Peaden, and M. L. Lee, Anal. Chim. Acta 127:55 (1981).
15. T. Romanowski, W. Funcke, J. König, and E. Balfanz, J. High Resolut. Chromatogr. Chromatogr. Commun. 4:209 (1981).
16. K. D. Bartle and M. J. Mulligan, unpublished measurements, 1982.
17. E. F. Barry, P. Ferioli, and J. A. Hubball, J. High Resolut. Chromatogr. Chromatogr. Commun. 6:173 (1983).
18. H. Borwitzky and G. Schomburg, J. Chromatogr. 170:99 (1979).
19. F. Sellier, G. Tersac, and G. Guiochon, J. Chromatogr. 219:203 (1981).
20. K. Grob and G. Grob, J. Chromatogr. 213:211 (1981).

21. P. A. Peaden, B. W. Wright, and M. L. Lee, Chromatographia *15*:335 (1982).

22. C. Madani, E. M. Chambaz, M. Rigaud, P. Chebroux, J. C. Breton, and F. Berthou, Chromatographia *10*:466 (1977).

23. C. Madani and E. M. Chambaz, Chromatographia *11*:725 (1978).

24. L. Blomberg, J. Buitjen, J. Gawdzik, and T. Wännman, Chromatographia *11*:521 (1978).

25. L. Blomberg and T. Wännman, J. Chromatogr. *168*:81 (1979).

26. L. Blomberg, K. Markides, and T. Wännman, J. Chromatogr. *203*:217 (1981).

27. U. Stenberg, T. Alsberg, L. Blomberg, and T. Wännman, in *Polynuclear Aromatic Hydrocarbons*, P. W. Jones and P. Leber (Eds.), Ann Arbor Science, Ann Arbor, Mich., 1979, pp. 313-326.

28. B. W. Wright, P. A. Peaden, M. L. Lee, and T. J. Stark, J. Chromatogr. *248*:17 (1982).

29. P. Sandra, G. Redant, E. Schacht, and M. Verzele, J. High Resolut. Chromatogr. Chromatogr. Commun. *4*:411 (1981).

30. L. Blomberg, J. Buitjen, K. Markides, and T. Wännman, J. Chromatogr. *239*:51 (1982).

31. T. Romanowski, W. Funcke, I. Grossmann, J. König, and E. Balfanz, Anal. Chem. *55*:1030 (1983).

32. B. E. Richter, J. C. Kuei, N. J. Park, S. J. Crowley, J. S. Bradshaw, and M. L. Lee, J. High Resolut. Chromatogr. Chromatogr. Commun. *6*:371 (1983).

33. J. Buijten, L. Blomberg, K. Markides, and T. Wännman, Chromatographia *16*:183 (1982).

34. B. E. Richter, J. C. Kuei, J. I. Shelton, L. W. Castle, J. S. Bradshaw, and M. L. Lee, J. Chromatogr. *229*:21 (1983).

35. L. Blomberg, J. High Resolut. Chromatogr. Chromatogr. Commun. *5*:413 (1982).

36. T. J. Stark, P. A. Larson, and R. D. Dandeneau, J. Chromatogr., *279*:31 (1983).

37. J. C. Kuei, J. I. Shelton, L. W. Castle, R. C. Kong, B. E. Richter, J. S. Bradshaw, and M. L. Lee, J. High Resolut. Chromatogr. Chromatogr. Commun., *7*:186 (1984).

38. G. M. Janini, Adv. Chromatogr. *17*:231 (1979).

39. Z. Witkiewicz, J. Chromatogr. *251*:311 (1982).

40. W. L. Zielinski and G. M. Janini, J. Chromatogr. *186*:237 (1979).

41. Z. Witkiewicz, J. Szulc, A. Ziolek, R. Dubrowski, and J. Dziaduszek, J. Chromatogr. *246*:37 (1982).

42. K. Kubica and Z. Witkiewicz, J. Chromatogr. *241*:33 (1982).

43. M. Parisinou and K. D. Bartle, unpublished measurements (1979).

44. R. Modica, M. Fiume, and I. Bartoszk, J. Chromatogr. *247*:352 (1982).

45. W. L. Zielinski, R. A. Scanlan, and M. M. Miller, J. Chromatogr. *209*:87 (1981).

46. F. Janssen and T. Kalidin, J. Chromatogr. *235*:323 (1982).

47. F. Janssen, Chromatographia *15*:33 (1982).

48. F. Janssen, Int. J. Environ. Anal. Chem. *13*:37 (1982).

49. R. J. Laub, W. L. Roberts, and C. A. Smith, J. High Resolut. Chromatogr. Chromatogr. Commun. *3*:355 (1980).

50. R. J. Laub, W. L. Roberts, and C. A. Smith, in *Polynuclear Aromatic Hydrocarbons: Chemistry and Biological Effects*, M. Cooke and A. J. Dennis (Eds.), Battelle Press, Columbus, Ohio, 1981, p. 287.

51. K. D. Bartle, A. El-Nasri, and B. Frere, in *Identification and Analysis of Organic Pollutants in Am.*, L. H. Keith (Ed.), Butterworth , Boston, 1984, p. 183.

52. R. C. Kong, M. L. Lee, Y. Tominaga, R. Pratap, M. Iwao, R. N. Castle, and S. A. Wise, J. Chromatogr. Sci. *20*:502 (1982).

53. J. H. Purnell, J. Chem. Soc. 1268 (1960).

54. R. C. Kong, M. L. Lee, Y. Tominaga, R. Pratap, M. Iwao, and R. N. Castle, Anal. Chem. *54*:1802 (1982).

55. M. L. Lee, F. J. Yang, and K. D. Bartle, *Open Tubular Column Gas Chromatography—Theory and Practice*, Wiley, New York, 1984, Chap. 4.

56. D. L. Springer, D. W. Phelps, and R. E. Schirmer, J. High Resolut. Chromatogr. Chromatogr. Commun. *4*:638 (1981).

57. A. Bjørseth, Anal. Chim. Acta *94*:21 (1977).

58. L. Kolarovic and H. Traitler, J. High Resolut. Chromatogr. Chromatogr. Commun. *4*:523 (1981).

59. C. K. Huynh, Trav. Chim. Aliment. Hyg. *71*:532 (1980).

60. A. Bjørseth and G. Eklund, J. High Resolut. Chromatogr. Chromagogr. Commun. *2*:22 (1979).

61. L. Wojnarovits and G. Földich, J. Chromatogr. *206*:511 (1981).

62. L. Wojnarovits and G. Földich, J. Chromatogr. *234*:451 (1982).

63. M. Oehme, S. Manø, and H. S. Kay, J. High Resolut. Chromatogr. Chromatogr. Commun. *5*:417 (1982).

64. E. P. Grimsrud, D. A. Miller, R. G. Stebbins, and S. H. Kim, J. Chromatogr. *197*:51 (1980).

65. D. A. Miller, K. Skogerboe, and E. P. Grimsrud, Anal. Chem. *53*:464 (1981).

66. J. A. Campbell, E. P. Grimsrud, and L. R. Hageman, Anal. Chem. *55*:1335 (1983).

67. M. A. Wizner, S. Singhawangcha, R. M. Barkley, and R. E. Sievers, J. Chromatogr. *239*:45 (1982).

68. C. R. Vogt, S. Kapila, and S. E. Manahan, Int. J. Environ. Anal. Chem. *12*:27 (1981).

69. S. Kapila and C. R. Vogt, J. High Resolut. Chromatogr. Chromatogr. Commun. *4*:233 (1981).

70. J. N. Driscoll, J. Chromatogr. Sci. *20*:91 (1982).

71. C. M. Klimcak and J. E. Wessel, Anal. Chem. *52*:1233 (1980).

72. S. Kapila, D. J. Bornhop, S. E. Manahan, and G. L. Nickell, J. Chromatogr. *259*:205 (1983).

73. I. S. Krull, M. S. Swartz, R. Hilliard, and K. H. Xie, J. Chromatogr. *260*:347 (1983).

74. M. L. Lee, M. V. Novotny, and K. D. Bartle, *Analytical Chemistry of Polycyclic Aromatic Compounds*, Academic Press, New York, 1981, Chap. 8.

75. M. L. Lee, D. L. Vassilaros, W. S. Pipkin, and W. L. Sorensen, *Trace Organic Analysis*, NBS Special Publication 519, U.S. Government Printing Office, Washington, D.C., 1979.

76. G. Rhodes, R. B. Opsal, J. T. Meek, and J. P. Reilly, Anal. Chem. *55*:280 (1983).

77. K. Blau and G. S. King, *Derivatives for Chromatography*, Heyden, London, 1978, p. 39.

78. M. V. Buchanan, Anal. Chem. *54*:570 (1982).

79. M. V. Buchanan, C. H. Ho, B. R. Clark, and M. R. Guerin, in *Polycyclic Aromatic Hydrocarbons*, Vol. 5, M. Cooke and A. J. Dennis (Eds.), Battelle Press, Columbus, Ohio, 1981, p. 133.

80. T. Ramdahl and K. Urdal, Anal. Chem. *54*:2256 (1982).

81. D. F. Hunt and F. W. Crow, Anal. Chem. *50*:1781 (1978).

82. T. Ramdahl, G. Becher, and A. Bjørseth, Environ. Sci. Technol. *16*:861 (1982).

83. J. Merritt, F. Comendant, S. T. Abrams, and V. N. Smith, Anal. Chem. *35*:1461 (1963).

84. M. Novotny, F. J. Schwende, M. J. Hartigan, and J. E. Purcell, Anal. Chem. *52*:736 (1980).

85. F. J. Schwende, M. Novotny, and J. E. Purcell, Chromatogr. Newslett. *8*:1 (1980).

86. M. D. Erickson, D. L. Newton, E. D. Pellizzari, and K. B. Tomer, J. Chromatogr. Sci. *17*:449 (1979).

87. E. L. Wehry and G. Mamantov, Anal. Chem. *51*:643A (1979).

88. E. L. Wehry, G. Mamantov, D. M. Hembree, and J. R. Maple, in *Polynuclear Aromatic Hydrocarbons: Chemistry and Biological Effects*, A. Bjørseth and A. J. Dennis (Eds.), Battelle Press, Columbus, Ohio, 1980, p. 1005.

89. D. M. Hembree, A. A. Garrison, R. A. Crocombe, R. A. Yokley, E. L. Wehry, and G. Mamantov, Anal. Chem. *53*:1738 (1981).

90. W. C. Wu, in *Polynuclear Aromatic Hydrocarbons: Formation, Metabolism and Measurement*, M. Cooke and A. J. Dennis (Eds.), Battelle Press, Columbus, Ohio, 1983, p. 1267.

91. M. L. Lee, F. J. Yang, and K. D. Bartle, *Open Tubular Column Gas Chromatography—Theory and Practice*, Wiley, New York, 1984, Chap. 5.

92. M. L. Lee, D. L. Vassilaros, C. M. White, and M. Novotny, Anal. Chem. *51*:768 (1979).

93. D. L. Vassilaros, R. C. Kong, D. W. Later, and M. L. Lee, J. Chromatogr. *252*:1 (1982).

94. G. Alexander and I. Hazai, J. Chromatogr. *217*:19 (1981).

95. Z. Stompel, K. D. Bartle, and B. Frere, Fuel *61*:817 (1982).

96. M. -L. Yu and R. A. Hites, Anal. Chem. *53*:951 (1981).

97. J. A. Yergey, T. M. Risby, and S. S. Lestz, Anal. Chem. *54*:354 (1982).

98. C. Willey, M. Iwao, R. N. Castle, and M. L. Lee, Anal. Chem. *53*:400 (1981).

99. D. W. Later, M. L. Lee, K. D. Bartle, R. C. Kong, and D. L. Vassilaros, Anal. Chem. *53*:1612 (1981).

100. D. L. Vassilaros, P. W. Stoker, G. M. Booth, and M. L. Lee, Anal. Chem. *54*:106 (1982).

101. D. L. Vassilaros, D. A. Eastmond, W. R. West, M. L. Lee, and G. M. Booth, in *Proceedings of the 6th International Symposium on Polynuclear Aromatic Hydrocarbons (Physical and Biological Chemistry)*, Columbus, Ohio, Oct, 1981, p. 845.

102. G. V. Graas, J. W. de Leeuw, P. A. Schenk, and J. Haverkamp, Geochim. Cosmochim. Acta 45:2465 (1981).

103. M. L. Lee, D. L. Vassilaros, and D. W. Later, Int. J. Environ. Anal. Chem. *11*:251 (1982).

104. K. D. Bartle, C. Gibson, and N. Taylor, unpublished measurements, 1983.

105. S. Krishnan and R. A. Hites, Anal. Chem. *53*:342 (1981).

106. M. Novotny, R. Kump, F. Merli, and L. J. Todd, Anal. Chem. *52*:401 (1980).

107. J. M. Schmitter, I. Ignatidis, and G. Guiochon, J. Chromatogr. *248*:203 (1982).

108. L. B. Kier and L. H. Hall, *Connectivity in Chemistry and Drug Research*, Academic Press, New York, 1976.

109. R. Kaliszan and H. Lamparczyk, J. Chromatogr. Sci. *16*:246 (1978).

110. R. C. Lao, S. Win Lee, and R. S. Thomas, in *Proceedings of the 5th International Symposium on Polynuclear Aromatic Hydrocarbons*, Columbus, Ohio, Oct. 1980, p. 407.

111. K. D. Bartle, M. L. Lee, and S. A. Wise, Chromatographia *14*:69 (1981).

112. R. Lamparczyk, D. Wilczynska, and A. Radecki, Chromatographia *14*:707 (1981).

113. R. Lamparczyk, D. Wilczynska, and A. Radecki, J. High Resolut. Chromatogr. Chromatogr. Commun. *6*:390 (1983).

114. H. Purnell, *Gas Chromatography*, Wiley, New York, 1962.

115. H. Beernaert, J. Chromatogr. *173*:109 (1979).

116. E. K. Whalen-Pedersen and P. C. Jurs, Anal. Chem. *53*:2184 (1981).

117. S. A. Wise, W. J. Bonnett, F. R. Guenther, and W. E. May, J. Chromatogr. Sci. *19*:457 (1981).

118. A. Radecki, H. Lamparczyk, and R. Kaliszan, Chromatographia *12*:595 (1979).

119. J. M. Schnur and D. E. Martine, Mol. Cryst. Liq. Cryst. *26*:213 (1974).

120. G. M. Janini and M. T. Ubeid, J. Chromatogr. *236*:329 (1982).

121. R. J. Jeltes, J. Chromatogr. Sci. *12*:399 (1974).

122. M. L. Lee, G. P. Prado, J. B. Howard, and R. A. Hites, Biomed. Mass Spectrom. *4*:182 (1977).

123. J. M. Daisey, M. A. Leyko, and T. J. Kneip, in *Polynuclear Aromatic Hydrocarbons*, P. W. Jones and P. Leber (Eds.), Ann Arbor Science, Ann Arbor, Mich., 1979, p. 201.

124. G. Grimmer, in *Handbook of Polycyclic Aromatic Compounds*, A. Bjørseth (Ed.), Marcel Dekker, New York, 1983, Chap. 4.

125. G. Grimmer, K. W. Naujack, and D. Schnieder, Z. Anal. Chem. *311*:475 (1982).

126. G. Grimmer, J. Jacob, K. W. Naujack, and G. Dettbarn, Anal. Chem. *55*:892 (1983).

127. T. Ramdahl, Nature *306*:580 (1983).

128. K. D. Bartle, unpublished measurements, 1982.

129. S. G. Wakeham, C. Schaffner, and W. Giger, Geochim. Cosmochim. Acta *44*:415 (1980).

130. D. W. Later, M. L. Lee, R. A. Pelroy, and B. W. Wilson, in *Proceedings of the 6th International Symposium on Polynuclear Aromatic Hydrocarbons (Physical and Biological Chemistry)*, Columbus, Ohio, 1981, p. 427.

131. M. L. Lee, F. J. Yang, and K. D. Bartle, *Open Tubular Column Gas Chromatography—Theory and Practice*, Wiley-Interscience, New York, 1984, Ch. 6.

132. F. Merli, D. Wiesler, M. P. Maskarinec, M. Novotny, D. L. Vassilaros, and M. L. Lee, Anal. Chem. *53*:193 (1981).

133. M. Novotny, J. W. Strand, S. L. Smith, D. Wiesler, and F. J. Schwende, Fuel *60*:213 (1981).

134. D. W. Later, M. L. Lee, and B. W. Wilson, Anal. Chem. *54*:117 (1982).

135. D. W. Later, T. G. Andros, and M. L. Lee, Anal. Chem. *55*:2126 (1983).

136. D. W. Later, B. W. Wright, and M. L. Lee, J. High Resolut. Chroma-togr. Chromatogr. Commun. *4*:406 (1981).

137. D. W. Later, R. A. Pelroy, D. D. Mahlum, B. W. Wright, M. L. Lee, W. C. Welmer, and B. W. Wilson, in *Proceedings of the 7th International Symposium on Polynuclear Aromatic Hydrocarbons,* Columbus, Ohio, 1982.

138. J. N. Pitts, Jr., K. A. Cauwenberghe, D. Grosjean, J. T. Schmid, D. R. Fitz, W. L. Belser, Jr., G. B. Knudson, and P. M. Hynds, Science *202*:515 (1978).

139. T. C. Pederson and J. S. Siak, J. Appl. Toxicol. *1*:54 (1981).

140. D. Schuetzle, F. S. C. Lee, T. J. Prater, and S. B. Tejada, Int. J. Environ. Anal. Chem. *9*:93 (1981).

141. T. Nielsen, B. Seitz, A. M. Hansen, K. Keiding, and B. Westerberg, in *Proceedings of the 7th International Symposium on Polynuclear Aromatic Hydrocarbons,* Columbus, Ohio, 1982, p. 961.

142. T. L. Gibson, Atmos. Environ. *16*:2037 (1982).

143. P. A. D'Agostino, D. R. Narine, B. E. McCarry, and M. A. Quilliam, in *Proceedings of the 7th International Symposium on Polynuclear Aromatic Hydrocarbons,* Columbus, Ohio, 1982, p. 365.

144. T. Nielsen, Anal. Chem. *55*:286 (1983).

145. T. Ramdahl, K. Kveseth, and G. Becher, J. High Resolut. Chromatogr. Chromatogr. Commun. *5*:22 (1982).

146. M. C. Paputa-Peck, R. S. Marano, D. Schuetzle, T. L. Riley, C. V. Hampton, T. J. Prater, L. M. Skewes, and T. E. Jensen, Anal. Chem. *55*:1946 (1983).

147. D. Schuetzle, T. L. Riley, T. J. Prater, M. Harvey, and D. F. Hunt, Anal. Chem. *54*:265 (1982).

148. D. L. Newton, M. D. Erickson, K. B. Tomer, E. D. Pellizzari, and P. Gentry, Environ. Sci. Technol. *16*:206 (1982).

149. J. A. Sweetman, F. W. Karasek, and D. Schuetzle, J. Chromatogr. *247*:245 (1982).

150. T. McFall, G. M. Booth, M. L. Lee, Y. Tominaga, R. Pratap, M. Tedjamulia, and R. N. Castle, Mutat. Res. *135*:97 (1984).

151. B. Wenzel and R. L. Aiken, J. Chromatogr. Sci. *17*:503 (1979).

152. R. C. Kong, M. L. Lee, M. Iwao, Y. Tominaga, R. Pratap, R. D. Thompson, and R. N. Castle, Fuel *63*:707 (1984).

153. J. König, E. Balfanz, W. Funcke, and T. Romanowski, Anal. Chem. *55*:599 (1983).

154. J. A. Leary, A. L. Lafleur, H. L. Liber, and K. Biemann, Anal. Chem. *55*:758 (1983).

155. B. W. Wilson, M. R. Peterson, R. A. Pelroy, and J. T. Cresto, Fuel *60*:289 (1981).

156. D. A. Haugen, M. J. Peak, K. M. Suhrbler, and V. C. Stanoudis, Anal. Chem. *54*:32 (1982).

157. B. A. Tomkins and C. -L. Ho, Anal. Chem. *54*:91 (1982).

158. J. T. Swansiger, M. T. Best, D. A. Danner, and T. L. Yourgless, Anal. Chem. *54*:2576 (1982).

159. P. Burchill, A. A. Herod, and E. Pritchard, J. Chromatogr. *242*:51 (1982).

160. P. Burchill, A. A. Herod, and E. Pritchard, J. Chromatogr. *242*:65 (1982).

161. M. Dorban, J. Schmitter, P. Arpino, and G. Guiochon, J. Chromatogr. *246*:255 (1982).

162. P. Burchill, A. A. Herod, and E. Pritchard, J. Chromatogr. 246:271 (1982).

163. M. Novotny, D. Wiesler, and F. Merli, Chromatographia 15:374 (1982).

164. G. Grimmer, J. Jacob, and K. W. Naujack, Fresenius' Z. Anal. Chem. 314:29 (1983).

165. J. M. Schmitter, I. Ignatidis, P. Arpino, and G. Guiochon, Anal. Chem. 55:1685 (1983).

166. T. J. Wozniak and R. A. Hites, Anal. Chem. 55:1791 (1983).

167. P. Burchill, A. A. Herod, and E. Pritchard, Fuel 62:11,20 (1983).

168. P. Burchill, A. A. Herod, J. P. McMahon, and E. Pritchard, J. Chromatogr. 265:223 (1983).

169. A. A. Bradley, J. K. Ferrell, and R. M. Felder, in Polynuclear Aromatic Hydrocarbons: Formation, Metabolism, and Measurement, M. Cooke and A. J. Dennis (Eds.), Battelle Press, Columbus, Ohio, 1983, p. 147.

170. M. V. Buchanan, J. Flanagan, I. B. Rubin, and M. R. Guerin, in Polynuclear Aromatic Hydrocarbons: Formation, Metabolism, and Measurement, M. Cooke and A. J. Dennis (Eds.), Battelle Press, Columbus, Ohio, 1983, p. 211.

171. D. A. Haugen, V. C. Stanoulis, M. J. Peak, and A. S. Bopara, in Polynuclear Aromatic Hydrocarbons: Formation, Metabolism, and Measurement, M. Cooke and A. J. Dennis (Eds.), Battelle Press, Columbus, Ohio, 1983, p. 607.

172. B. W. Wilson, R. A. Pelroy, D. D. Mahlum, M. E. Frazier, D. W. Later, and C. W. Wright, Fuel 63:46 (1984).

173. S. J. W. Grigson, W. Kemp, P. R. Ludgate, and W. Steedman, Fuel 62:604 (1983).

174. G. Grimmer, K. W. Naujack, and D. Schneider, Int. J. Environ. Anal. Chem. 10:265 (1981).

175. S. A. Wise, S. L. Bowie, S. N. Chesler, W. F. Cuthrell, W. E. May, and R. E. Roberts, in Proceedings of the 6th International Symposium on Polynuclear Aromatic Hydrocarbons (Physical and Biological Chemistry), Columbus, Ohio, 1981, p. 919.

176. D. M. Choudhury and R. A. Hites, Anal. Chem. 53:1351 (1981).

177. T. Spitzer, J. Chromatogr. 237:273 (1982).

178. J. N. Pitts, D. M. Lokensgard, W. Harper, T. S. Fisher, V. Mejia, J. T. Schuler, G. M. Scorzell, and Y. Katzenstein, Mutat. Res. 103:241 (1982).

179. T. Romanowski, W. Funcke, J. König, and E. Balfanz, Anal. Chem. 54:1285 (1982).

180. R. L. Tanner and R. Fajer, Int. J. Environ. Anal. Chem. 14:231 (1983).

181. T. Ramdahl and G. Becher, Anal Chim. Acta 144:83 (1982).

182. K. S. Chia, P. M. Walsh, J. M. Beer, and K. Biemann, in Polynuclear Aromatic Hydrocarbons: Formation, Metabolism, and Measurement, M. Cooke and A. J. Dennis (Eds.), Battelle Press, Columbus, Ohio, 1983, p. 319.

183. R. C. Lao, R. S. Thomas, M. Lanoy, and S. Win Lee, in Polynuclear Aromatic Hydrocarbons: Formation, Metabolism, and Measurement, M. Cooke and A. J. Dennis (Eds.), Battelle Press, Columbus, Ohio, 1983, p. 745.

184. J. A. Leary, A. L. Lafleur, J. P. Longwell, W. A. Peters, E. L. Krenzel, and K. Biemann, in *Polynuclear Aromatic Hydrocarbons: Formation, Metabolism, and Measurement*, M. Cooke and A. J. Dennis (Eds.), Battelle Press, Columbus, Ohio, 1983, p. 799.

185. B. A. Tomkins, R. R. Reagen, M. P. Maskarinec, S. H. Harmon, W. M. Greist, and J. A. Caton, in *Polynuclear Aromatic Hydrocarbons: Formation, Metabolism, and Measurement*, M. Cooke and A. J. Dennis (Eds.), Battelle Press, Columbus, Ohio, 1983, p. 1173.

186. M. Toqan, J. M. Beer, J. B. Howard, W. Farmayan, and W. Rovesti, in *Polynuclear Aromatic Hydrocarbons: Formation, Metabolism, and Measurement*, M. Cooke and A. J. Dennis (Eds.), Battelle Press, Columbus, Ohio, 1983, p. 1205.

187. B. A. Peterson, C. C. Chuang, T. L. Hayes, and D. A. Trayser, SAE Technical Paper 820774, Society of Automotive Engineers, Detroit, Mich., 1982.

188. G. Grimmer, J. Jacob, K. W. Naujack, and G. Dettbarn, Fresenius' Z. Anal. Chem. *309*:13 (1982).

189. T. Spitzer and W. Danniker, J. High Resolut. Chromatogr. Chromatogr. Commun. *5*:98 (1982).

190. M. Y. Tong, J. A. Sweetman, and F. W. Karasek, J. Chromatogr. *264*:231 (1983).

191. T. E. Jensen and R. A. Hites, Anal. Chem. *55*:594 (1983).

192. A. G. Howard and G. A. Mills, Int. J. Environ. Anal. Chem. *14*:43 (1983).

193. G. A. Mills and A. G. Howard, J. Inst. Energy 131 (1983).

194. G. A. Mills, A. G. Howard, and J. S. Howarth, J. Inst. Energy 273 (1984).

195. R. Kadar, K. Nagy, and D. Fremstad, Talanta *27*:227 (1980).

196. L. Szepesy, K. Lakszner, L. Ackermann, L. Pod, and P. Literathy, J. Chromatogr. *206*:611 (1981).

197. L. S. Ramos and P. G. Prohaska, J. Chromatogr. *211*:284 (1981).

198. R. Shinohara, A. Kido, and Y. Okamoto, J. Chromatogr. *256*:81 (1982).

199. M. M. Krahn and D. C. Malins, J. Chromatogr. *248*:99 (1982).

200. Y. L. Tan and M. Heit, in *Polynuclear Aromatic Hydrocarbons: Formation, Metabolism, and Measurement*, M. Cooke and A. J. Dennis (Eds.), Battelle Press, Columbus, Ohio, 1983, p. 567.

201. M. R. West, P. A. Smith, P. W. Stoker, G. M. Booth, T. Smith-Oliver, and B. E. Bullwill, in *Proceedings of the 8th International Symposium on Polynuclear Aromatic Hydrocarbons*, Columbus, Ohio, 1983.

202. G. Grimmer, J. Jacob, and K. W. Naujack, Fresenius' Z. Anal. Chem. *306*:347 (1981).

203. L. Kolarovic and H. Traitler, J. Chromatogr. *237*:263 (1982).

204. J. D. Adams, E. L. LaVoie, and D. Hoffman, J. Chem. Sci. *20*:274 (1982).

205. R. Burrows, M. Cooke, and D. G. Gillespie, J. Chromatogr. *260*:168 (1983).

7

Determination of Occupational Exposure to PAH by Analysis of Body Fluids

GEORG BECHER / Toxicological Department, National Institute of Public Health, Oslo, Norway

ALF BJØRSETH / Petroleum Research Center, Norsk Hydro, Bergen, Norway

I.   Introduction   237

II. Adsorption, Distribution, and Excretion of PAH   238

III. Metabolism of PAH   239

IV. Effects of Particulate PAH   241

V.   Biological Monitoring of Occupational Exposure to PAH   243

    A. PAH in human urine   243
    B. Analysis of blood samples   246

VI. Conclusion   249

    References   250

I. INTRODUCTION

Polycyclic aromatic hydrocarbons (PAH) are formed by combustion reactions or high-temperature processes involving carbonaceous materials such as coal, coal tar and pitch, asphalt, and oil [1]. Since such processes are frequent in many industries, PAH have been detected in a large number of workplace environments, such as [2-7]:

Coke plants
Iron and steel industries
Aluminum smelters
Ferroalloy industry
Petroleum refineries
Coal gasification and liquefaction plants
Oil shale industry

The potential carcinogenic effect of PAH is the basic reason for concern about the levels of these compounds in the working environment. Some PAH are shown to cause cancer in animals and are suspected to be human carcinogens as well [1,8]. Furthermore, in some cases an association has been demonstrated between occupational exposure to coal tar pitch volatiles (CTVP) containing PAH and the frequency of lung cancer and some other types of cancer [9-11]. As a result of this, CTPV are included in the official list of threshold limit values. To control the levels, analyses of PAH in the working environment are routinely performed in many large industries, especially those listed above. However, it has been shown that the method used at present, which is based on gravimetric determination of benzene or cyclohexane soluble matter in the air particulates, has several deficiencies and does not reflect the analytical state of the art. Critical reviews of the method have been published as well as alternative methods for measuring airborne PAH in the working environment [2,3,12-14].

Frequently, the concentration of PAH in workplace atmospheres exceeds more than 1000 times the concentration found in urban atmospheres, suggesting that workers exposed to such atmospheres might be at an increased risk of cancer of the respiratory tract. Somewhat surprisingly, the high exposure is not reflected to a great extent in the epidemiological evidence presently available from many industries. In particular, this is the case for the aluminum industry [15,16]. This might be due to the following:

1. Only a small fraction of airborne PAH is absorbed in the body and is biologically active (e.g., because of the mode in which PAH is adsorbed to particles).
2. Man may be much less susceptible than experimental animals to PAH-mediated cancer, and thus the occupational hazard ascribed to PAH may be considerably overestimated.
3. Inhibitory effects caused by the presence of antagonistic compounds (or lack of synergistic compounds) may result in a reduced cancer frequency compared to other types of exposure.

Thus important questions may be raised as to the significance of PAH air monitoring values as a measure of the relative safety of the working environment. Questions along these lines were raised as early as 1959 by Kreyberg [17], who stated: "If these two facts are correlated: (i.e.) the enormous amount of 3,4-benzpyrene in the air and the very moderate excess of lung cancer in the gasworks, a serious fallacy is evidently involved"; and "This represents a serious warning against any conclusions as to causative relationships between any substance and lung cancer based upon the mere finding of substances deposited on paper filtering the air." To get a better understanding of the occupational hazard connected with PAH exposure, it seems necessary to monitor real exposure of PAH by analyzing body fluids.

II. ADSORPTION, DISTRIBUTION, AND EXCRETION OF PAH

Extensive studies have been published on adsorption, distribution, metabolism, and excretion of PAH in vitro as well as in animal systems. In this section only a few relevant papers are reviewed to give some insight into the problem area. Most studies have been performed using benzo[a]pyrene (BaP), a well-known animal carcinogen, as an indicator substance for PAH exposure.

Adsorption of PAH through the lungs has particular relevance to occupational exposure. The major studies related to health effects of inhalation exposure involve local lesions in the respiratory tract, and only a few investigators have measured the levels of real PAH exposure. However, regardless of its route of administration, PAH, once adsorbed, becomes localized in a wide variety of body tissues. The distribution of radioactivity derived from [14C]BaP in the rat and mouse was determined following subcutaneous, intravenous, and intratracheal administration [18]. The pattern of distribution was found to be similar in all cases, except for high local pulmonary concentrations following intratracheal instillation (Table 1). Concentrations of BaP-derived radioactivity in the liver reach a maximum within only 10 min after injection and represented 12% of the total dose. Radioactivity in the liver was

Table 1. Distribution and Elimination of Radioactivity After Intratracheal Instillation of [^{14}C]BaP into Rats[a]

| Site | Percent of administered dose after: | |
	1 hr	24 hr
Feces	0	28.0
Urine	0	1.2
Stomach	0	0
Intestine	37.0	11.1
Kidney	0	2.4
Liver	1.3	4.3
Lung	43.2	38.6
Other organs	0.08	0.43
Total percent recovered	81.58	86.03

[a]Dose: 25 μg of [^{14}C]BaP in 0.3 ml of H_2O.
Source: Ref. 18.

reduced to 1 to 3% of the administered dose within 24 hr. Similarly, maximum blood levels of BaP following intravenous injection were reached very quickly, and radioactivity became barely detectable after 10 min. Minimal tissue localization of BaP and/or its metabolites occurred in the spleen, kidney, lung, and stomach; maximum radioactivity derived from labeled BaP was recovered in the bile and feces. Levels of radioactivity in fat, skin, and muscle were not determined, nor was the amount of unchanged BaP measured in any tissue. Bock and Dao [19] later showed that relative to other tissues, unmetabolized BaP was localized extensively in the mammary gland and general body fat after a single feeding of the carcinogen (10-30 mg). This accumulation of BaP was greater than that resulting from 3-methylcholanthrene (MCA), 7,12-dimethyl-benz[a]anthracene (DMBA), or phenanthrene. In all cases, the level of carcinogens detected in the tissue was directly related to the dose administered, and was dependent on the use of a lipid vehicle.

As early as 1936, it was recognized that various PAH were excreted primarily through the hepatobiliary system and the feces [20]. With the advent of radiotracer methodologies, it became possible to study more quantitatively the elimination of PAH. When [^{14}C]BaP was given intratracheally to rats, or by subcutaneous and intravenous injection, maximum excretion of radioactivity always occurred through feces [19,21]. The amount of PAH excreted in urine is about one-fourth to one-fifth of the amount excreted through feces [18].

III. METABOLISM OF PAH

It is well known that PAH are metabolized by the microsomal mixed-function oxidase system (MFO), often termed aryl hydrocarbon hydroxylase (AHH), which is most abundant in the liver. This enzyme system has been studied extensively and is the subject of several reviews [22,23]. While it is known that

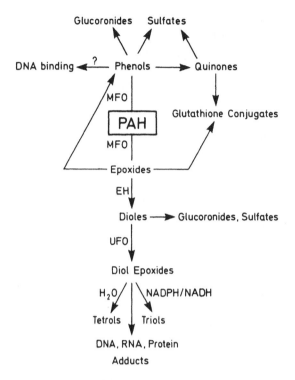

Figure 1. Mechanism of enzymatic activation of PAH: MFO, microsomal mixed-function oxidase; EH, epoxide hydrolase.

this enzyme complex is involved with detoxification of xenobiotics in conjunction with various P_{450}-type cytochromes, it is apparent that this system is also directly involved with the metabolism of polycyclic hydrocarbons to their active species (Fig. 1).

The first step in this metabolic activation, catalyzed by the cytochrome P_{450} system, gives rise to epoxide and phenolic groups in different positions on the polycyclic ring system [24]. A second microsomal enzyme, epoxide hydrolase (EH), converts epoxides into vicinal glycols. Since some epoxides are more active carcinogens than the parent hydrocarbon, this enzyme would probably affect both carcinogenesis and detoxification.

Information on epoxide hydrolase has been summarized recently [25,26], and its importance in the formation of the three known dihydrodiols of BaP has been demonstrated [27,28]. Dihydrodiols may be further oxidized by the MFO system to dihydrodiol epoxides [29]. There is now considerable evidence that a particular structural class of diol epoxides, the bay-region dihydrodiol epoxides, operate as the ultimate carcinogenic form of PAH, which react easily with cellular macromolecules, in particular DNA [30,31].

Another reaction involved in the metabolism of PAH is enzymatic conjugation of the oxygenated intermediates to glucuronic acid, sulfate, and glutathione [32-34]. These water-soluble conjugates are readily removed from the organism through bile, feces, and urine and have generally been viewed as detoxification products.

Identification of metabolites has been performed for some PAH, BaP being the compound studied most extensively. The metabolite profile of BaP has

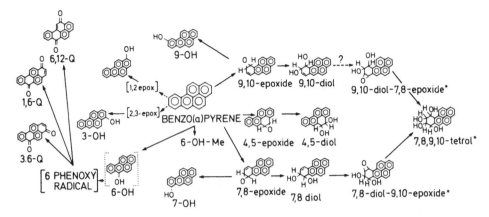

Figure 2. Metabolites of benzo[a]pyrene. An asterisk designates that the stereochemistry is not indicated.

been largely worked out using high-performance liquid chromatography, as shown in Fig. 2. BaP metabolites found in microsomal incubation are 3-hydroxy-BaP, 6-hydroxy-BaP, 7-hydroxy-BaP, and 1-hydroxy-BaP. The BaP-4,5-epoxides have been isolated and identified as precursors of the BaP-4,5-diol. Other studies indicate that epoxides are precursors of 7,8-diol and 9,10-diol as well. There have been no intermediates isolated as phenol precursors, although recent evidence using deuterium labelling suggests that at least a portion of 3-OH-BaP is derived from an intermediate 2,3-epoxide [35]. In addition to the hydroxylated metabolites, 1,6-, 3,4-, and 6,12-BaP quinone have been identified [36]. These are produced enzymatically by microsomes and nonenzymatically from air oxidation of phenols [37].

Further oxidation of 7,8-dihydroxy-7,8-dihydrobenzo[a]pyrene by the MFO system leads to the formation of the highly reactive and probably ultimate carcinogenic metabolites, the isomeric 7,8-dihydroxy-9,10-oxy-7,8,9,10-tetrahydrobenzo[a]pyrene, in which the epoxy ring is adjacent to the "bay" of the hydrocarbon. These diol epoxides react rapidly by electrophilic attack with cellular macromolecules. Adduct formation has been observed with DNA, RNA, and proteins [31,38].

It is known that the various forms of cytochrome P_{450} may be preferentially induced by different chemicals, such as phenobarbital, β-naphthoflavone, and various PAH [39]. Thus induction may alter the balance of the metabolic pathways of PAH. This has been demonstrated in an excellent way by Jacob and Grimmer [40] using high-resolution gas chromatography/mass spectrometry (GC/MS). They could show that different PAH induce the enzymatic oxidation of a given PAH to quite a different extent and that certain inducers stimulate the metabolism of various PAH to a different extent. Some weak carcinogenic or noncarcinogenic PAH are potent inducers, leading to increased amounts of activated metabolites of carcinogenic PAH [41].

IV. EFFECTS OF PARTICULATE PAH

Because of low vapor pressures, most carcinogenic PAH in working environments are adsorbed onto particles. Thus it is reasonable to assume that the nature of the particle may influence the effect of adsorbed PAH. A number of

Table 2. Biological Availability of Carcinogens from
Carbon Black

| | Result | |
Test method	Whole carbon black	Benzene extract of carbon black
Skin painting	Negative	Positive
Inhalation	Negative	Positive
Feeding	Negative	Positive
Injection	Negative	Positive

Source: Ref. 47.

studies have been published on the ability of particulates to enhance or reduce the cellular uptake of PAH. Synergistic effects between particles and
PAH have been observed in studies with experimental animals. In the absence
of particulate matter, intratracheal instillation of BaP results in only a low in-
cidence of lung cancer, even at high dosages [42]. However, a wide variety
of particles, while coinstilled with PAH, results in a high tumor yield. These
include hematite [43], asbestos [44], carbon particles [45], india ink [46],
and aluminum and titanium oxides [45]. Thus particulates in general, irrespective of chemical composition, appear to act as cocarcinogens. So far, human epidemiological evidence points to asbestos as being the most significant
cocarcinogen.

On the other hand, data resulting from animal studies show that the PAH
adsorbed on carbon black is not available for cellular uptake. The results of
these animal tests are summarized in Table 2. In this extensive series of tests
[47], whole carbon black and the benzene extract of carbon black were tested
in laboratory animals. These tests included skin painting, inhalation, feeding,
and subcutaneous injection. The tests for the whole carbon black and the benzene extract of the identical carbon black were performed under identical conditions. In each case the animal tests of the whole carbon black gave negative
results, while the benzene extract of the carbon black yielded positive results.
Carcinogenic materials can be removed from carbon black by benzene extraction, but the carcinogenic materials adsorbed on the carbon black apparently
are not a carcinogenic hazard.

When inducing lung cancer, there seem to be several essential processes:

1. The inhaled carcinogen must be retained by the respiratory tract. Ad-
    sorption of PAH on particulates, which are themselves efficiently retained,
    can increase the total exposure level to PAH.
2. The carcinogens must enter the cells that are subsequently transformed.
    Transport can be facilitated by particles that release adsorbed carcino-
    gens rapidly in comparison with the clearance rates of these particles
    from the lungs, by particles that are rapidly phagocytosed, or by those
    which penetrate cell membranes as a result of their shape or surface
    properties.

Particles that retain the carcinogenic material may possibly decrease the
carcinogenic activity. In this regard, Creasia and et al. [48] showed that when

BaP is adsorbed on large carbon particles (15-30 μm) and instilled into the lungs, 50% of both the BaP and the carrier particles were cleared from the lungs in 4-5 days. Little carcinogenic material was released from the carbon particles in this case, and therefore contact with the respiratory epithelium (and carcinogenicity) was low. With smaller carbon particles (0.5-1.0 μm), however, 50% particle clearance was not achieved until 7 days after administration. In this case, 15% of the adsorbed BaP was eluted from the particles and left free to react with the respiratory tissues. In the complete absence of carrier particles, BaP was cleared from the lungs at 20 times the rate for adsorbed BaP. This observation may explain the difficulty in producing experimental pulmonary tumors with BaP without the use of carrier particles. Other investigators [49] confirmed that carbon particle size affects BaP retention in the lungs, but also demonstrated that BaP retention was not affected by particle size when BaP was adsorbed on ferric oxide or aluminum oxide.

V. BIOLOGICAL MONITORING OF OCCUPATIONAL EXPOSURE TO PAH

There is extensive data available on the biological effect of PAH in general, and BaP in particular, as well as epidemiological studies and chemical characterization of workplace atmospheres. Despite all these data, there seems to be no clear understanding of the problems connected with occupational hazards of PAH exposure.

There are several prerequisites for particulate PAH to show any biological effect:

1. The particles must be deposited in the respiratory tract.
2. PAH must be eluted from the particles.
3. PAH must be taken up in cells, lymph, or blood.
4. PAH must be metabolized to reactive intermediates.

Therefore, determination of PAH concentrations in workplace atmospheres may not give a good measure of the effective dose of PAH received by the individual worker. However, in many cases in which air sampling is of little or no value, biological monitoring has been used successfully to obtain information on the uptake of hazardous chemicals in the body of exposed workers and to evaluate the significance of their airborne levels [50,51].

The ideal biological monitoring method would permit direct measurement of a toxic agent in the target organ, but the problem of "invasive" sampling on a routine basis makes this approach impractical. Therefore, body fluids, mainly urine and blood, are generally used for biological monitoring.

A. PAH in Human Urine

Data on urine levels of various industrial contaminants have frequently been used to supplement information on concentration of pollutants in air to which workers are exposed [51]. However, data on the analysis of PAH in urine samples are rather limited.

Low concentrations of fluorescent material, presumable unmetabolized PAH, have been observed in urine samples from tobacco smokers and non-smokers (Table 3) [52,53]. Malý [52] used acid hydrolysis of the urine samples, petrol ether extraction, and paper chromatography for the semiquantitative determination of dibenzo[a,l]pyrene. Repetto and Martinez [53] separated BaP from methylene chloride extracts of urine samples by preparative column chromatography on silica and quantitated BaP by spectrofluorimetry.

Table 3. Concentration of PAH in Human Urine Samples

PAH	Subject	Number of samples	Concentration (mean in μg/liter)	References
Dibenzo[a,l] pyrene	Active smoker	1	1.1	52
	Passive smoker	1	0.3	
Benzo[a] pyrene	Active smoker			53
	morning	8	0.17	
	evening	7	0.34	
	Passive smoker			
	morning	1	0.15	
	evening	1	0.23	
	Nonexposed			
	morning	1	0.01	
	evening	1	0.06	
	Topside coke oven workers			54
	After 6 hrs of work	19	4.67	
	After 18 hrs of rest	13	2.1	
	After 48 hrs of rest	8	1.6	
	Persons from urban/industrial area	451	0.690	55
	Persons from rural area	35	0.445	
Benz[a] anthracene	Persons from urban/industrial area	437	0.701	55
	Persons from rural area	34	0.489	
Sum of 11 prominent PAH[a]	Aluminum workers			58
	Smokers	7	68.2[b]	
	Nonsmokers	4	79.3[b]	
	Controls			
	Smokers	6	45.5[b]	
	Nonsmokers	4	13.2[b]	
Sum of 10 prominent PAH[a]	Aluminum workers			59,60
	Smokers	9	47.5[b]	
	Nonsmokers	6	24.2[b]	
	Controls			
	Smokers	4	43.4[b]	
	Nonsmokers	5	6.3[b]	

[a]Sum of PAH and PAH metabolites determined by reversed-metabolism method.
[b]Results converted from μg of PAH per mmol of creatinine using an average excretion of 12 mmol of creatinine per liter urine.

Szyja [54] found high amounts of unmetabolized BaP in urine samples of top-side coke oven workers collected after 6 hr of work (Table 3). The urinary PAH levels were significantly lower after an 18 hr and a 48 hr rest. Michels and Einbrodt [55] determined BaP and benz[a]anthracene (BaA) in more than 480 urine samples of randomly selected persons from a highly industrialized area and a rural area as reference. The excretion of BaP and BaA was found to be significantly higher in the polluted area than in the reference area (Table 3). No differences, however, were observed between urines from smokers and nonsmokers.

In the studies desribed above only the unmetabolized part of the PAH excreted in urine has been determined. However, as shown above, PAH are metabolized to a great extent to polar, water-soluble metabolites both in vitro and in vivo. For example, the relative amount of unmetabolized BaP found in urine from mice following intraperitoneal injection of ^{14}C-labeled BaP was as low as 0.7% of the total amount excreted in urine [56]. Thus an alternative approach has been suggested by Keimig et al. [57]—to analyze urine for metabolites of specific PAH compounds that are consistently prominent in environmental samples. They have identified 1-hydroxypyrene as a major metabolite in the urine of pigs and propose this metabolite for monitoring PAH-exposed workers. However, no application of this method in occupational hygiene has been reported so far.

Recently, Becher and Bjørseth [58] have described a method to determine multiple PAH compounds in urine specimens based on the reduction of excreted, oxidized metabolites of PAH back to the parent hydrocarbons. The analytical procedure included extraction of PAH and PAH metabolites from urine using cartridges containing C_{18}-modified silica, reduction of metabolites to PAH by refluxing hydriodic acid ("reversed metabolism"), and subsequent analysis of 11 prominent PAH by high-performance liquid chromatography with fluorimetric detection. Table 4 shows the results for two parallel analyses, one including the reduction of metabolites, the other omitting the reduction step. The table reveals that the amount of PAH identified increased approximately fivefold when the reduction step was included. The method has been applied to urine samples from aluminum workers with high exposure to PAH and to occupationally nonexposed control groups. Figure 3 shows a typical HPLC/fluorescence chromatogram of PAH isolated from an exposed worker's urine. The results are included in Table 3. In the control group, urine extracts from smokers show a significantly higher level of PAH than those from nonsmokers. However, the high concentrations of PAH found in the working atmosphere of aluminum plants are not reflected to a corresponding extent in the excretion of PAH in workers' urine.

In a more recent study simultaneous monitoring of PAH exposure of aluminum workers was performed using personal sampling of airborne particulate PAH as well as urine analysis [59,60]. The average particulate PAH exposure was 126 $\mu g/m^3$. The PAH profile in urine seems to correspond to the profile found in the work atmosphere if one takes into account that the sampling efficiency of the filter used for sampling particulates decreases with increasing volatility of the PAH. The means of PAH concentrations in the urine samples from the four different categories of personnel investigated are given in Table 3 and shown schematically in Fig. 4. A significant increase in the excretion of PAH and PAH metabolites is observed for smokers in the control groups compared to the nonsmokers in this group. In accordance with previous results, the high occupational PAH exposure is not reflected to a great extent in the excretion of PAH in the urine.

Table 4. PAH in an Exposed Worker's Urine Without and with Reduction of Metabolites (μg/liters)

PAH	Without reduction (unmetabolized PAH)[a]	With reduction (unmetab. + metabol. PAH)
Fluorene	ND	4.2
Phenanthrene	6.85	9.8
Anthracene	ND	4.4
Fluoranthene	1.12	9.8
Pyrene	0.37	5.5
Benzo[a]fluorene	ND	3.1
Benz[a]anthracene	0.17	2.0
Chrysene	0.18	Obscured
Benzo[e]pyrene	ND	0.7
Benzo[a]pyrene	0.01	0.13
Dibenz[a,h]anthracene	ND	0.57
Total	8.7	40.2

[a]ND, not detected.
Source: Ref. 58.

These results are particularly interesting when comparing the analytical data with epidemiological data on occupational cancer in the aluminum industry. There is no obvious correlation between exposure as measured by air analysis and lung cancer frequency. Recent epidemiological data indicate that there is only a slight excess of lung cancer in aluminum workers, while the exposure is up to three orders of magnitudes higher than that of urban atmospheres. On the other hand, analysis of PAH metabolites in urine fit well with the epidemiological data, as the excreted PAH shows only a slight increase in exposed workers. This corresponds well with the low or negligible increase in lung cancer found in epidemiological studies. These results are supported by the study of sister chromatid exchanges (SCE) in blood lymphocytes of the same aluminum workers [60]. The highly exposed workers had no significant increase in the SCE frequencies, while the SCE frequencies of cigarette smokers were significantly higher than those of the nonsmokers in both groups.

B. Analysis of Blood Samples

Blood has in many cases been recognized as a useful matrix for biological monitoring, as blood levels often reflect well the dose (i.e., the amount of toxic agent in the target organ or tissue). For example, a very good correlation has been demonstrated between the inorganic lead exposure and the level of this metal in blood [61].

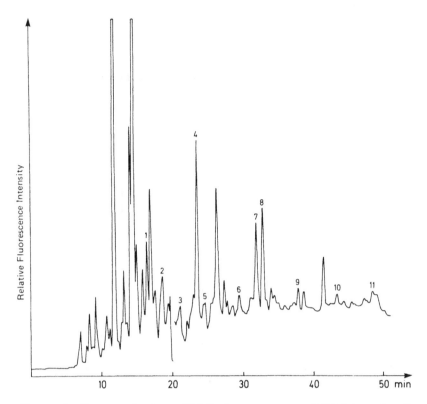

Figure 3. Reversed-phase HPLC chromatogram of PAH in urine from exposed workers after reduction of metabolites. Peak identities: 1, fluorene; 2, phenanthrene; 3, anthracene; 4, fluoranthene; 5, pyrene; 6, benzo[a]fluorene; 7, benz[a]anthracene; 8, chrysene; 9, benzo[e]pyrene; 10, benzo[a]pyrene; 11, dibenz[a,h]anthracene. (From Ref. 58.)

It has been shown that PAH form complexes with serum albumin by hydrophobic interaction [62]. This interaction seems to play an important role in the transport of PAH in the blood system. Recently, Hutcheon et al. [63] determined plasma BaP levels by radioimmunoassay. They observed a higher mean BaP level in subjects from an urban-industrialized area than from subjects from an outer suburban area.

Extensive interest has been focused on the formation of covalent PAH adducts with cellular macromolecules to evaluate the risks posed by exposure to mutagenic and carcinogenic chemicals [64-67]. The interaction of activated carcinogens with cellular targets (DNA and possibly RNA and proteins) is a critical early event in the action of a number of carcinogens, among them PAH, and is presumed to be directly involved in the carcinogenic process [66,67]. Therefore, it seems reasonable to assume that the quantitation of PAH-DNA adducts can be used as a particularly relevant measure of the biologically effective dose of the metabolically activated PAH. Furthermore, the carcinogenic potency of a number of PAH correlates with their ability to form covalent adducts with DNA [65]. Ultrasensitive immunoassays specific for BaP-DNA adducts have been developed [68,69] which allow the detection of one BaP-DNA adduct per 10^6 nucleotides. The methods could be sensitive

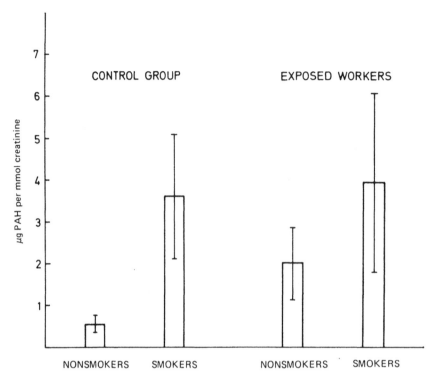

Figure 4. Mean ± SD of PAH levels in urine samples from aluminum workers and controls. (From Ref. 60.)

enough to detect adduct formation in vivo as a result of environmental exposure.

In a pilot project Perera et al. [69] observed low levels of BaP-DNA adducts in lung tissues from lung cancer patients. However, the number of subjects was too small to draw conclusions relating exposure history to the occurrence of PAH-DNA adducts. Monitoring the biological effective dose of PAH in exposed workers restricts the type of sample to blood as a source for DNA. Although some qualitative results from the adduct level of DNA from peripheral blood mononuclear cells are reported by Perera et al., it is not quite clear if the assay method used is sensitive enough to detect the low levels in blood cells.

Fluorimetry, with its intrinsically high sensitivity, has been used as an alternative to immunoassays for determining the binding of PAH to DNA [70]. Fluorescence studies of intact BaP-DNA adducts have utilized either photon counting [71] or low-temperature fluorimetry [72] or a combination of both of these methods [73] to detect as little as one molecule of bound BaP per 10^5 DNA nucleotides. Recently, Rahn et al. [74,75] described a fluorometric-HPLC assay based on the acid hydrolysis of DNA to liberate the BaP adducts in the form of the isomeric tetrols of BaP which are quantitated by normal fluorescence detection after HPLC separation.

Hitherto, no data have been published on the application of the fluorimetric methods to DNA samples from humans. However, extensive research in this field is presently going on in various research groups.

Measurements of the extent of adduct formation with hemoglobin in blood samples taken from exposed individuals have been proposed as markers of the biological effective dose [64,76]. This approach, called "hemoglobin dose monitor," is based on the fact that most activated carcinogens bind covalently not only to nucleic acid but also to proteins. In fact, the reactivity per unit weight of hemoglobin toward activated carcinogens is about 1.5-10 times higher than that of DNA. Thus hemoglobin serves as a convenient trapping agent for assessing the levels of activated carcinogens in vivo since it can be obtained in large amounts from red blood cells and is readily purified and analyzed. Furthermore, human erythrocytes have an average life span of about 125 days. Assuming that reaction with the activated carcinogen does not have a major influence on the lifetime of the erythrocytes, they will integrate the dose over a long time.

Data from experiments with mice show that a dose of 1 μg BaP/kg mouse gives 0.08-0.10 μg of BaP bound to 1 g of hemoglobin [77]. Taking into account the accumulations of adducts during the lifetime of the erythrocytes, this degree of binding should be sufficient to be measured by modern analytical methods.

VI. CONCLUSION

Potentially carcinogenic PAH have been detected in a great number of workplace atmospheres. Frequently, the concentration exceeds more than 1000 times the concentration found in urban atmospheres, suggesting that workers exposed to such an atmosphere might be at an increased risk for cancer of the respiratory tract. Somewhat surprisingly, this exposure is not reflected to a great extent in epidemiological evidence presently available on occupational cancer from several industries, such as the aluminum industry. Therefore, the relevance of PAH air monitoring data may be questioned as a measure of the occupational hazard connected with PAH exposure. It is suggested that the uptake of PAH in the body of exposed workers be monitored by analysis of body fluids.

There are two approaches to measuring the amount of PAH that has actually entered the body as a consequence of ingestion, inhalation, or skin adsorption (dose monitoring):

1. Detection and quantitation of PAH and PAH metabolites in the body fluids
2. Measuring the biological effective dose of PAH at the cellular level

A method has been developed to determine the PAH and PAH metabolites in human urine. The amount of PAH found in urine is a relative measure of the total amount of PAH adsorbed in the organism, metabolized, and excreted.

Highly sensitive analytical methods are available for detecting the levels of adducts of PAH and DNA or hemoglobin isolated from blood samples of exposed workers. These will give a measure of the biological effective dose of PAH at the cellular level under actual conditions of workers exposure (i.e., the amount of the metabolically activated PAH that has reacted with cellular targets).

The proposed methods have two main advantages over conventional exposure estimation, which merely determine the amount of PAH in the individual's immediate environment.

1. They give a better understanding of the degree of interindividual variation that exists in uptake, metabolism, and biological effect of PAH in humans.

2. They give a measure of the individual dose or biologically effective dose of PAH in the organism and thus allow a better prediction of worker's risks associated with exposure to PAH in different work environments.

REFERENCES

1. Committee on Biologic Effects of Atmospheric Pollutants, *Particulate Polycyclic Organic Matter*, National Academy of Sciences, Washington, D.C., 1972.
2. A. Bjørseth, O. Bjørseth, and P. E. Fjeldstad, Scand. J. Work Environ. Health 4:212 (1978).
3. A. Bjørseth, O. Bjørseth, and P. E. Fjeldstad, Scand. J. Work Environ. Health 7:223 (1981).
4. *NIOSH Criteria for a Recommended Standard—Occupational Exposure to Coal Tar Products*, DHEW (NIOSH) Publ. 78-107, U.S. Dept. of Health, Education and Welfare, Washington, D.C., 1977.
5. G. Lindstedt and J. Sollenberg, Scand. J. Work Environ. Health 8:1, (1982).
6. A. Bjørseth, O. Bjørseth, and P. E. Fjeldstad, Scand. J. Work. Environ. Health 4:224 (1978).
7. H. Blome, Staub—Reinhalt. Luft 41:225 (1981).
8. A. Dipple, in *Chemical Carcinogens*, C. E. Searle (Ed.), ACS Monograph 173, American Chemical Society, Washington, D.C., 1976, p. 245.
9. M. Kawai, H. Amanoto, and K. Harada, Arch. Environ. Health 14:859 (1967).
10. J. W. Lloyd, J. Occup. Med. 13:53 (1971).
11. R. Doll, M. P. Vessey, R. W. R. Beasley, A. R. Buckley, E. C. Fear, R. E. W. Fisher, E. J. Gammon, W. Gunn, G. O. Hughes, K. Lee, and B. Norman-Smith, Br. J. Ind. Med. 29:394 (1972).
12. H. J. Seim, W. W. Hannemann, L. R. Barsotti, and T. J. Walker, J. Am. Ind. Hyg. Assoc. 35:718 (1974).
13. K. A. Schulte, D. J. Larsen, R. W. Hornung, and J. V. Crable, *Report on Analytical Methods Used in Coke Oven Effluent Study. The Five Oven Study*, DHEW (NIOSH) Publ. 74-105, National Institute of Occupational Safety and Health, Cincinnati, Ohio, 1974.
14. H. Boden and R. Roussel, Light Met. (Warrendale, Pa), 1049 (1983).
15. Aa. Andersen, B. E. Dahlberg, K. Magnus, and A. Wannag, Int. J. Cancer 29:295 (1982).
16. M. E. Rockette and V. C. Arena, J. Occup. Med. 25:549 (1983).
17. L. Kreyberg, Br. J. Cancer 13:618 (1959).
18. P. Kotin, H. L. Falk, and R. Busser, J. Natl. Cancer Inst. 23:541 (1959).
19. F. G. Bock and T. L. Dao, Cancer Res. 21:1024 (1961).
20. P. R. Peacock, Br. J. Exp. Pathol. 17:164 (1936).
21. C. Heidelberger and S. M. Weiss, Cancer Res. 11:885 (1951).
22. Y. Ishimura, T. Iizuka, I. Morishima, and O. Hayaishi, in *Polycyclic Hydrocarbons and Cancer*, Vol. 1, H. V. Gelboin and P. O. P. Ts'o (Eds.), Academic Press, New York, 1978, p. 321.
23. R. W. Estabrook, J. Werringloer, J. Capdevila, and R. A. Prough, in *Polycyclic Hydrocarbons and Cancer*, Vol. 1, H. V. Gelboin and P. O. P. Ts'o (Eds.), Academic Press, New York, 1978, p. 285.
24. P. Sims and P. L. Grover, Adv. Cancer Res. 20:165 (1974).

25. T. M. Guenthner and F. Oesch, in *Polycyclic Hydrocarbons and Cancer*, Vol. 3, H. V. Gelboin and P. O. P. Ts'o (Eds.), Academic Press, New York, 1981, p. 183.

26. F. Oesch, P. Bentley, and H. R. Glatt, in *Biological Reactive Intermediates*, D. K. Jollow, J. J. Kocsis, R. Snyder, and H. Vainio (Eds.), Plenum Press, New York, 1977, p. 181.

27. J. K. Selkirk, R. G. Croy, and H. V. Gelboin, Science *184*:169 (1974).

28. J. K. Selkirk, R. G. Croy, P. P. Roller, and H. V. Gelboin, Cancer Res. *34*:3474 (1974).

29. P. Sims, P. L. Grover, A. Swaisland, K. Pal, and A. Hewer, Nature *252*:326 (1974).

30. H. W. S. King, M. R. Osborne, F. A. Beland, R. G. Harvey, and P. Brookes, Proc. Natl. Acad. Sci. USA *73*:2679 (1976).

31. D. H. Phillips and P. Sims, in *Chemical Carcinogens and DNA*, Vol. 2, P. L. Grover (Ed.), CRC Press, Boca Raton, Fla., 1979, p. 29.

32. N. Nemoto and H. V. Gelboin, Biochem. Pharmacol. *25*:1221 (1976).

33. N. Nemoto, S. Takayama, and H. V. Gelboin, Chem. Biol. Interact. *23*:19 (1978).

34. L. F. Chasseaud, Adv. Cancer Res. *29*:175 (1979).

35. S. K. Yang, P. P. Roller, P. P. Fu, R. G. Harvey, and H. V. Gelboin, Biochem. Biophys. Res. Commun. *77*:1176 (1977).

36. S. P. Lesko, W. Caspary, R. Lorentzen, and P. O. P. Ts'o, Biochemistry *14*:3978 (1975).

37. R. J. Lorentzen, W. J. Caspary, S. A. Lesko, and P. O. P. Ts'o, Biochemistry *14*:3970 (1975).

38. I. B. Weinstein, A. M. Jeffrey, S. Leffler, P. Pulkabek, H. Yamasaki, and D. Grunberger, in *Polycyclic Hydrocarbons and Cancer*, Vol. 2, H. V. Gelboin and P. O. P. Ts'o (Eds.), Academic Press, New York, 1978, p. 4.

39. J. DiGiovanni and T. J. Slaga, in *Polycyclic Hydrocarbons and Cancer*, Vol. 3, H. V. Gelboin and P. O. P. Ts'o (Eds.), Academic Press, New York, 1978, p. 259.

40. J. Jacob, in *Handbook of Polycyclic Aromatic Hydrocarbons*, A. Bjørseth (Ed.), Marcel Dekker, New York, 1983, p. 617.

41. J. Jacob, G. Grimmer, and A. Schmoldt, in *Polynuclear Aromatic Hydrocarbons: Chemistry and Biological Effects*, A. Bjørseth and A. J. Dennis (Eds.), Battelle Press, Columbus, Ohio, 1980, p. 807.

42. U. Saffiotti, F. Cefis, L. H. Kolb, and P. Shubik, J. Air Pollut. Control Assoc. *15*:23 (1965).

43. U. Saffiotti, F. Cefis, and L. H. Kolb, Cancer Res. *28*:104 (1968).

44. L. N. Pylev and L. M. Shabad, in *Biological Effects of Asbestos*, P. Bogovski, V. Timbrell, J. C. Gilson, and J. C. Wagner (Eds.), IARC Scientific Publ. 8, International Agency for Research on Cancer, Lyon, France, 1973, p. 99.

45. F. Stenbäck, J. Rowland, and A. Sellakumar, Oncology *33*:29 (1976).

46. L. N. Pylev, Bull. Exp. Biol. Med. USSR *52*:1316 (1961).

47. C. A. Nau, J. Neal, V. Stembridge, and R. N. Cooley, Arch. Environ. Health *4*:598 (1962), and previous papers in this series.

48. D. A. Creasia, J. K. Poggenburg, and P. Nettesheim, J. Toxicol. Environ. Health *1*:967 (1976).

49. M. C. Henry and D. G. Kaufman, J. Natl. Cancer Inst. *51*:1961 (1973).

50. A. L. Linch, *Biological Monitoring for Industrial Chemical Exposure Control*, CRC Press, Boca Raton, Fla., 1974.

51. R. Lauwerys, Scand. J. Work Environ. Health *1*:139 (1975).
52. E. Malý, Bull. Environ. Contamin. Toxicol. *6*:442 (1971).
53. M. Repetto and D. Martinez, J. Eur. Toxicol. 7:234 (1974).
54. J. Szyja, Z. Ges. Hyg. Grenzgeb. 7:440 (1977).
55. S. Michels and H. J. Einbrodt, Wiss. Umwelt 3:107 (1979).
56. G. Becher and G. Löfroth, unpublished results.
57. S. D. Keimig, K. W. Kirby, D. P. Morgan, J. E. Keiser, and T. D. Hubert, Xenobiotica *13*:415 (1983).
58. G. Becher and A. Bjørseth, Cancer Lett. *17*:301 (1983).
59. G. Becher and A. Bjørseth, in *Polynuclear Aromatic Hydrocarbons: Mechanisms, Method and Metabolism*, M. Cooke and A. J. Dennis (Eds.), Battelle Press, Columbus, Ohio, 1984, p. 145.
60. G. Becher, Aa. Haugen, and A. Bjørseth, Carcinogenesis 5:647 (1984).
61. R. A. Kehoe, J. Royal Inst. Publ. Health Hyg. *24*:101 (1961).
62. R. Franke, Biochim. Biophys. Acta *160*:378 (1968).
63. D. E. Hutcheon, J. Kautrowitz, R. N. van Gelder, and E. Flynn, Environ. Res. *32*:104 (1983).
64. L. Ehrenberg and S. Osterman-Golkar, Teratogen. Carcinogen. Mutagen. *1*:105 (1980).
65. W. K. Lutz, Mutat. Res. *65*:289 (1979).
66. F. Perera and I. B. Weinstein, J. Chron. Dis. *35*:581 (1982).
67. B. A. Bridges, Arch. Toxicol., *Suppl. 3*:271 (1980).
68. I.-C. Hsu, M. C. Poirier, S. H. Yuspa, D. Grunberger, I. B. Weinstein, R. H. Yolken, and C. C. Harris, Cancer Res. *41*:1090 (1981).
69. F. P. Perera, M. C. Poirier, S. H. Yuspa, J. Nakayama, A. Jaretzki, M. M. Curnen, D. M. Knowles, and I. B. Weinstein, Carcinogenesis, *3*:1405 (1982).
70. P. Vigny and M. Duquesne, in *Chemical Carcinogens and DNA*, Vol. 1, P. L. Grover (Ed.), CRC Press, Boca Raton, Fla., 1979, p. 85.
71. P. Daudel, M. Duquesne, P. Vigny, P. L. Grover, and P. Sims, FEBS Lett. *57*:250 (1975).
72. V. Ivanovic, N. E. Geactinov, and I. B. Weinstein, Biochem. Biophys. Res. Commun. *70*:1172 (1976).
73. R. O. Rahn, S. S. Chang, J. M. Holland, T. J. Stephens, and L. H. Smith, J. Biochem. Biophys. Methods 3:285 (1980).
74. R. O. Rahn, S. S. Chang, J. M. Holland, and L. R. Shugart, Biochem. Biophys. Res. Commun. *109*:262 (1982).
75. L. Shugart, J. M. Holland, and R. O. Rahn, Carcinogenesis 4:195 (1983).
76. C. J. Calleman, L. Ehrenberg, B. Jansson, S. Osterman-Golkar, D. Segerbäck, K. Svensson, and C. A. Wachtmeister, J. Environ. Pathol. Toxicol. *2*:427 (1978).
77. G. Löfroth and L. Ehrenberg, unpublished results.

8

Analysis of 6-Nitrobenzo[a]pyrene in Mammalian Cells and Microsomes by High-Pressure Liquid Chromatography

SAMUEL TONG and JAMES K. SELKIRK / The University of Tennessee–
Oak Ridge Graduate School of Biomedical Sciences, and Biology Division,
Oak Ridge National Laboratory, Oak Ridge, Tennessee

I. Introduction   253

II. Nitrogen-Containing PAH: Metabolism of 6-Nitrobenzo[a]pyrene   255

   References   261

I. INTRODUCTION

The induction of chemical carcinogenesis in both laboratory animals and man by environmental pollutants, such as polycyclic aromatic hydrocarbons, is a topic of major concern. This has become an area of extensive research and many workers are attempting to understand mechanisms underlying the action of carcinogens to produce malignant transformation. One of the most fundamental steps leading to such a process is the biological activation of a potential carcinogen to its reactive intermediate, since most chemical carcinogens are inert in their parent form. This metabolic event is catalyzed by an enzyme complex localized in the endoplasmic reticulum of the cell. These drug-metabolizing enzymes, also known as monooxygenases [1], are found mainly in the liver, although their presence in the extrahepatic tissues have also been detected [2-8]. Electrophilic metabolites produced as a result may accumulate and attack cellular components to destroy or disrupt normal regulatory functions in the cell.

The same microsomal enzyme system, however, is also responsible for detoxifying foreign compounds or xenobiotics when an organism is exposed to these chemicals. As potentially toxic or carcinogenic intermediates are enzymatically formed in the parent compound, they may undergo detoxification by conjugatory reactions catalyzed by cytoplasmic enzymes, such as glucuronyl transferases, which render the products more polar, thereby facilitating their excretion from the animal body. The metabolism of a compound is a complex process and may differ according to specific conditions encountered. Factors such as species, strain, sex, tissue, and even diet differences can all play determining roles in the eventual fate of a foreign compound in the animal body [9-12].

An event that is generally being regarded as a crucial step in chemical carcinogenesis is the interaction of reactive metabolites, which are electron deficient and have high affinity toward nucleophilic sites in cells, with biological macromolecules such as DNA, RNA, and proteins. Although not definitively proven, and epigenetic events are possible [13,14], the attempt to relate the DNA alkylating reactions of a chemical in its "activated" form with malignant transformation is currently under intense study by many laboratories [15-17].

To illustrate the metabolic events that lead to the interaction of a chemical with DNA, the metabolism of a typical carcinogenic polycyclic aromatic hydrocarbon, benzo[a]pyrene (BaP), is exemplified. BaP is an environmental pollutant and is ubiquitously distributed in our biosphere [18]. Its biological activity as both a mutagen and a carcinogen is also well known [19-22]. The identification of BaP metabolites has been greatly advanced since the application of the technique of high-pressure liquid chromatography (HPLC) [23]. Other methods, such as thin-layer chromatography and gas chromatography, in particular, are also capable to a certain extent of metabolite separation, although HPLC has in recent years been the generally preferred technique for a quicker and more efficient means of obtaining metabolic profiles of polycyclic hydrocarbon metabolism. By such a technique, samples may also be quantitatively recovered for analysis of their chemical structures and reused for other studies as required.

Figure 1 shows the overall metabolic chart of BaP. Most of the free oxygenated metabolites shown, as a result of the metabolism of BaP by eukaryotic systems, have been isolated intact. Metabolites shown in brackets, however, are those that have not been unequivocally determined or only hypothesized (as indicated by question marks). Among these pathways, the reaction by which BaP forms its major DNA binding species has now been identified. Apparently, the formation of BaP 7,8-diol, via an epoxide intermediate with catalysis by aryl hydrocarbon hydroxylase (AHH), a microsomal monooxygenase [24,25], and then epoxide hydrase [26,27], is an important step in the process (Fig. 2). Reactivation of this diol by AHH results in the formation of the more mutagenic and carcinogenic 7,8-diol 9,10-epoxide [28-30], which is relatively unstable and opens to form a triol-carbonium ion. Stereochemical studies have also indicated that the anti isomer instead of the syn isomer, of this diol epoxide is important in expressing both the mutagenic and carcino-

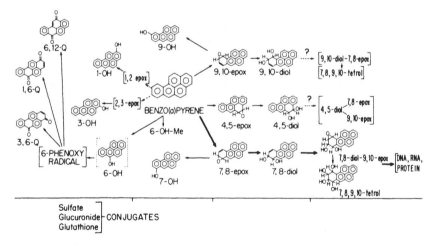

Figure 1. Metabolism of benzo[a]pyrene by eukaryotic systems. Free oxygenated metabolites may become conjugated with endogenous substrates, such as sulfates, glucuronic acid, and glutathione, to form polar metabolites that may be excreted from the animal body. Alternatively, they may accumulate and become covalently bound to cellular macromolecules.

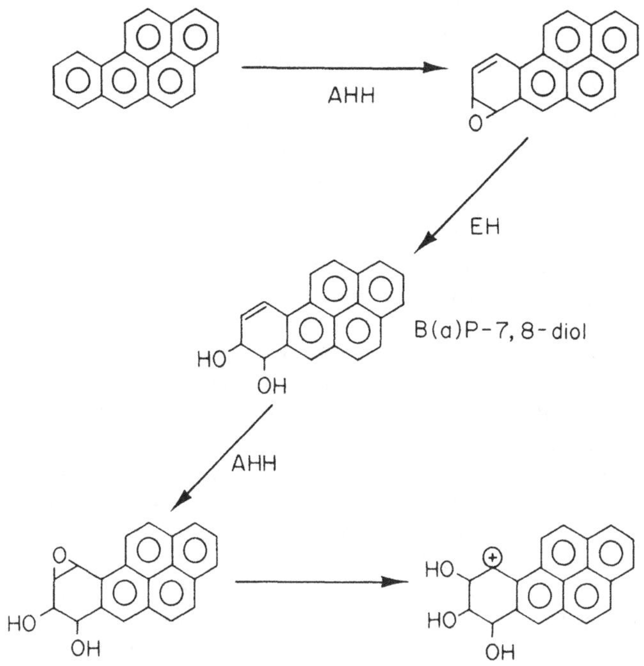

Figure 2. Metabolic pathway leading to the production of the ultimate car-cinogenic product of benzo[a]pyrene. AHH, aryl hydrocarbon hydroxylase; EH, epoxide hydrase.

genic properties of BaP [28,31]. The anti isomer is the major alkylating me-tabolite of BaP and readily interacts with the exocyclic N-2 of guanine as the predominant DNA binding adduct [32]. A cautionary note here is that the in-teraction of this metabolite with other nitrogen bases have also been demon-strated [33-36] and whether the major DNA adduct is responsible for expres-sing the carcinogenic properties of BaP is not totally certain at present.

II. NITROGEN-CONTAINING PAH: METABOLISM OF 6-NITRO-BENZO[a]PYRENE

In recent years, increasing attention has been focused on nitrogen-containing polycyclic hydrocarbons, which have been discovered in a variety of sources as environmental contaminants. Automobile exhaust [37,38], tobacco smoke condensate [39,40], shale oil fractions [41,42], urban airborne particulate matter [43,44], and xenographic toners [45] are some examples of the wide-spread presence of these chemicals. A number of these compounds have been found to be strongly mutagenic in the *Salmonella typhimurium* assay [46-49], which suggests their possible involvement as environmental carcinogens. It is also found recently that nitrated polycyclic hydrocarbons are able to induce unscheduled DNA synthesis in cultured human HeLa cells [50].

Among these compounds, the nitropyrenes have received considerable attention due to their powerful mutagenic activity, which has been related to the covalent interaction of the hydroxylamines with DNA, produced as a re-sult of the nitro reduction of the parent compounds [51]. The production of the hydroxyamino metabolites is catalyzed by nitroreductases present in the bacterial systems used in the mutagenicity assay, and therefore mammalian systems normally used to metabolically activate chemical carcinogens are not required for nitropyrenes to express their mutagenic properties [52]. In fact, it was found that the addition of hepatic post-mitochondrial fractions (S-9) actually decrease the mutagenic potency of a number of these chemicals [46]. It appears that microsomal enzymes may actually detoxify these compounds to mutagenically inactive quinones, diols, or phenolic products. Therefore, the mechanism of action of the nitropyrenes may differ from some other nitrogen-containing aromatic compounds where metabolic activation by an S-9 fraction is essentially required to enhance their mutagenic activities [53-55].

The universal occurrence and the positive mutagenic properties of nitrated compounds, together with our knowledge of BaP metabolism, has led to the investigation of the metabolic fate of 6-nitrobenzo[a]pyrene (6-NBaP), a nitro derivative of BaP, when incubated with a whole cell system, employing hamster embryonic fibroblasts, and a cell-free system as in hepatic microsomal preparation. The nitrogen-containing aromatic hydrocarbons are expected to undergo comparable chemical reactions as their unsubstituted counterparts since the Kekulé configuration, in which atoms within the ring are joined alternatively by single and double bonds, is maintained in the molecule. It is, however, not known whether the substitution of a nitro atom or addition of a nitro group to, for example, a metabolically important site of the BaP molecule would cause any significant changes in the biological effects of the compounds.

6-NBaP is an air pollutant [56] and its production can be simulated when BaP is exposed to an atmosphere containing traces of nitric acid and nitrogen dioxide [57]. It is an active mutagen in the *Salmonella typhimurium* test, particularly in the presence of a liver S-9 fraction, which contains monooxygenases to metabolically activate the compound [53-55]. When the mutagenicity of 6-NBaP was tested in *Salmonella* strains lacking in nitroreductases, activity of the compound did not differ from that in strains that contain the enzymes [53]. This implies that the production of the aryl hydroxylamine may not be important for this compound to manifest its mutagenic action. It must be mentioned that *Salmonella* tester strains are known to possess distinct nitroreductases with different specificity toward different compounds [52, 58], and the possibility that the particular strains tested in the experiment may possess activating potential for 6-NBaP cannot be totally excluded.

HPLC has permitted the effective separation of various metabolites produced by the incubation of tritium-labeled 6-NBaP with the two mammalian systems studied. Employing hepatic microsomal preparations from 3-methylcholanthrene-pretreated rats, which have been found to possess the highest activity toward the metabolism of 6-NBaP compared to untreated or phenobarbital-induced preparations, 6-NBaP was metabolized to various radioactively detectable products. The metabolic profile resulting from such an incubation is shown in Fig. 3. From a 30 min reaction, one major metabolite (peak 5) was produced, with several minor peaks also found along the HPLC tracing. Peak 5 represents a phenolic product(s) due to the occurrence of a "red shift" when its ultraviolet-visible spectra was obtained under alkaline conditions [59]. This bathochromic shift is characteristically observed as the phenolic moiety is ionized to the oxide ion in an alkaline solution.

The chemical nature of the metabolite at peak 3 was identified to be a dihydrodiol. This was demonstrated when addition of the inhibitor, 1,2-epoxy-3,3,3-trichloropropane, in the microsomal incubation abolished the formation

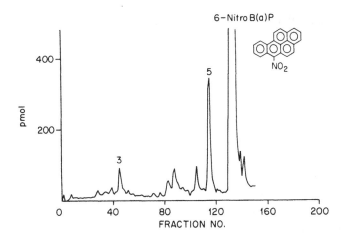

Figure 3. High-pressure liquid chromatographic separation of the microsomal metabolites of 6-nitrobenzo[a]pyrene. A Dupont 850 chromatograph employing a 4.6 mm x 1.5 cm Dupont C8 column with a flow rate at 0.9 ml/min (50°C) was used. For metabolite separation the following program was employed: 10 min, linear gradient, 40-55% methanol; 16 min, 55% methanol; 2 min, linear gradient, 55-62% methanol; 5 min, 62% methanol; 20 min, linear gradient, 62-80% methanol; hold at 80% methanol.

of this peak and surrounding peaks (fractions 20-60). This chemical is a specific inhibitor of epoxide hydrase, which converts epoxides into dihydrodiols. The metabolism of 6-NBaP by rat hepatic microsomes therefore yielded a major phenolic product with some formation of dihydrodiols.

The major phenolic product formed during the hepatic microsomal incubation of 6-NBaP has been identified by Fu et al. [53,60] to be 3-hydroxy-6-NBaP, although smaller amounts of 1-hydroxy-6NBaP were also found. The monohydroxylated metabolites are potentially important since they have been demonstrated to be more mutagenic than 6-NBaP, compared to when both have been activated by the liver S-9 fraction, in the *Salmonella typhimurium* assay [53]. Since metabolic activation is essential for 3- and 1-hydroxy 6-NBP to exert their mutagenic properties, additional steps in metabolism after the formation of the phenolic compounds are clearly necessary. However, the exact chemical structures of the proximate mutagens have yet to be determined. It is possible that further metabolism may have occurred at the bay region (i.e. the 7,8 position of the monohydroxylated compounds) to produce the respective diols which become the final reactive products.

Cell-free systems differ from intact or whole cell systems in that they lack conjugatory enzymes, which constitute an important detoxifying mechanism in the animal. Microsomes consist of fragments of cellular endoplasmic reticulum, but the complete machinery for conjugatory reactions (e.g., cofactors) is lost during the preparation of this fraction from homogenized tissues. Although it seems clear that monohydroxylated products of 6-NBaP have led to the observed mutagenicity of the compound, metabolic studies in an intact cell system with the parent compound would more closely reflect physiological events that may actually occur in the animal body and should give a more lucid picture of the biological fate of 6-NBaP.

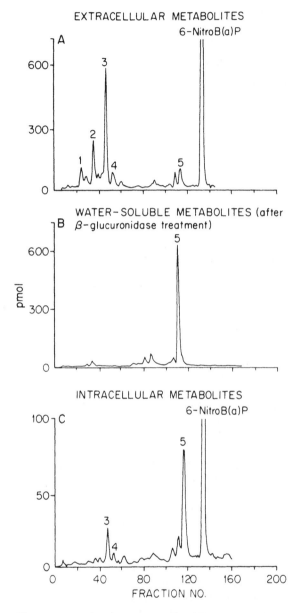

Figure 4. High-pressure liquid chromatographic separation of cellular metabolites produced by incubation of 6-nitrobenzo[a]pyrene with hamster embryonic fibroblasts: A, extracellular metabolites; B, water-soluble metabolites; C, intracellular metabolites. HPLC conditions were identical to that used for microsomal metabolism. Products were extracted with ethyl acetate, concentrated in methanol, and determined by HPLC.

Analysis of the HPLC profile produced from the incubation of 6-NBaP with hamster embryonic fibroblasts shows the accumulation of several dihydrodiols (peaks 1-4) as principal metabolites (Fig. 4A). The phenolic peak that was prominent in microsomal metabolism was present in only very small amounts in this case. These products, extracted with ethyl acetate from the culture medium after a 24-hr incubation, are termed organic solvent-soluble metabolites. What remains in the medium after this initial extraction consists of metabolites (water-soluble metabolites) that are associated with conjugatory substrates, such as glucuronic acid and glutathione. Treatment with the enzyme β-glucuronidase effectively dissociates metabolites that have been linked to glucuronic acid, and once released, the free metabolites can be extracted with ethyl acetate for HPLC separation. In the case of 6-NBaP, approximately 40% of the water-soluble metabolites was found to be linked to glucuronic acid. Analysis of the products released indicates that the major conjugatory product was the phenolic metabolite previously observed in microsomal metabolism (Fig. 4B). Intracellular metabolites, obtained by lysing cells with detergent [61], and consisting of less than 10% of the total radioactivity introduced in the medium, also showed the presence of the phenolic peak together with lesser amounts of the dihydrodiols (Fig. 4C).

To follow the formation and possibly disappearance of the various metabolites of 6-NBaP during its metabolism by hamster cells, a time course study (2-72 hr) was performed, allowing constant monitoring of cellular activities as extracellular products were formed. It was found that as the incubation time was prolonged, increasing amounts of the dihydrodiol product in peak 3 were formed (Fig. 5). In contrast, accumulation of the phenolic metabolite (peak 5) was not as substantial. This can be explained by the presence of cytoplasmic enzymes that effectively catalyze the conjugation of the phenol with endogenous substrates, such as glucuronic acid, thereby preventing accumulation of the product in its "free" form, which may be toxic or harmful to the cells. This also reflects that dihydrodiols are much poorer substrates for the conjugatory enzymes than the phenols, and "free" dihydrodiols may undergo further metabolism to produce reactive intermediates which may then interact with cellular macromolecules to produce harmful cellular effects. The exact chemical structures of the dihydrodiols are presently unknown, although work is under way for such determination using mass spectrometry and nu-

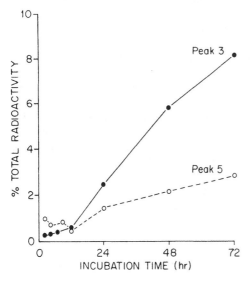

Figure 5. Time-course study of the formation of extracellular metabolites of 6-NBaP.

clear magnetic resonance studies. Since it is known that for BaP, formation of the 7,8-diol is an important pathway leading to its alkylating reaction with DNA, it seems necessary to know whether the presence of the nitro group in 6-NBaP would have any effect on the production of this or any other diol. A shift in metabolism of the diol to other sites away from the 7,8 position may significantly affect the biological activity of the compound. For instance, if the resulting diols is formed at a position that is less prone to be activated to interact with DNA, the mutagenicity and carcinogenicity of the compound may be substantially decreased. 6-NBaP has been demonstrated to interact with DNA, RNA, and nuclear protein from hamster cells [62] and such macromolecules isolated from various organs, such as the liver, lung, kidney, and spleen, after a single injection of the polycyclic hydrocarbon [63]. This interaction, especially with DNA, would suggest the possible carcinogenicity of 6-NBaP, particularly when the specific activities measured were similar or higher than that observed with BaP. At present, there is insufficient evidence to conclude as to the carcinogenicity of 6-NBaP, but in one report this compound was found to exhibit weak tumorigenic activity when tested on mouse skin [64]. This activity was, however, much lower than that exhibited by its parent compound, BaP, and indicates significant alteration in its biological activity when a nitro group is introduced in the molecule. Metabolism studies of 6-NBaP in the mouse skin may provide an explanation of this negative carcinogenic response, for it is not known to what extent the compound is metabolized or what affinity the products would have toward cellular macromolecules in this organ. If reactive products are produced, it is possible that they may effectively be removed as conjugatory products before they can disrupt normal cellular functions. These factors can all govern the carcinogenic activity of the compound.

In the study of cellular and microsomal metabolism of 6-NBaP, it is necessary to realize that metabolites produced in both systems are qualitatively similar, as determined by their identical individual ultraviolet-visible and fluorescence spectra and retention time through the HPLC column. This phenomenon has also been observed with BaP metabolism in that in all species and tissues, both resistant and susceptible to the chemical examined, qualitative similarities between these metabolic profiles existed consistently [65]. One suggestion is the possibility of a deficiency or impairment of conjugatory or other detoxifying processes existing in susceptible cells which would allow reactive intermediates to accumulate and produce adverse cellular effects. It is also possible that harmful metabolites are formed at such a rate in susceptible cells that normal cellular detoxifying mechanisms become incapable of preventing the accumulation of these intermediates, thus resulting in covalent macromolecular binding [66,67].

The differences between microsomal and cellular metabolism of 6-NBaP is borne out by the observation that due to the presence of conjugatory mechanisms in whole cell systems, phenols in their "free" form are present in only small quantities, whereas they are the major metabolites in microsomes. Although the phenolic products of 6-NBaP, 3-hydroxy- and 1-hydroxy-6NBaP, are mutagenic, it is not certain whether they would be found in substantial enough amounts in the animal body to be considered harmful. Since cell-free systems are preferably used in short-term mutagenicity testing, particularly in the widely used *Salmonella typhimurium* assay, caution is required when interpretating data from these tests. In the case of 6-NBaP, the positive mutagenic response may be misleading and the implication of this result to the biological activity of the compound in the animal is uncertain. The existence of a protective mechanism is obviously essential in the disposition of harmful materials, which are likely to produce cellular damage when allowed to accumulate. In the case of 6-NBaP, mutagenic products are effectively conjugated to polar metabolites for excretion from the animal body. Since both activation and detoxification of a carcinogen are manifested through the same enzyme system, the balance of such reactions is of critical importance in determining the potential carcinogenicity of a compound in a specific organ or animal species. The accumulation of dihydrodiols in the cellular system and not in microsomes is also an important aspect of 6-NBaP metabolism, due to the possible reactivity of these products undergoing further metabolism, which may result in their covalent binding to cellular macromolecules, such as the 7,8-diol epoxide in BaP metabolism. More studies involving whole animal experiments may be required to ascertain the biological action of this compound, and identification of the chemical structures of its metabolites, particularly the dihydrodiols, would also help us understand the mechanism of action of 6-NBaP and perhaps other nitrogen-containing polycyclic aromatic hydrocarbons that have become increasingly important in our environment as potential carcinogens.

ACKNOWLEDGMENT

Research sponsored by the Office of Health and Environmental Research, U.S. Department of Energy, under contract W-7405-eng-26 with the Union Carbide Corporation. S. Tong, Postdoctoral Investigator, was supported by subcontract 3322 from the Biology Division of Oak Ridge National Laboratory to The University of Tennessee.

REFERENCES

1. A. H. Conney, Pharmacol. Rev. 19:317-366 (1967).
2. D. Garfinkel, Comp. Biochem. Physiol. 8:367-369 (1963).
3. S. Orrenius, A. Ellin, S. V. Jakobsson, H. Thor, D. L. Cinti, J. B. Schenkman, and R. W. Estabrook, Drug Metab. Dispos. 1:350-356 (1973).
4. R. H. Wickramasinghe, Enzyme 19:348-376 (1975).
5. T. E. Gram, Drug. Metab. Rev. 2:1-32 (1973).
6. C. L. Miranda and R. S. Chhabra, Biochem. Pharmacol. 29:1161-1165 (1980).
7. J. K. Selkirk and F. J. Wiebel, Prog. Exp. Tumor Res. 24:61-72 (1979).
8. G. M. Cohen, A. C. Marchok, P. Nettesheim, V. E. Steele, F. Nelson, S. Huang, and J. K. Selkirk, Cancer Res. 39:1980-1984 (1979).
9. R. Kato, Pharmacol. Ther. 6:41-98 (1979).
10. E. S. Vesell, C. M. Lang, W. J. White, G. T. Passananti, R. N. Hill, T. L. Clemens, D. K. Lill, and W. D. Johnson, Fed. Proc. 35:1125-1132 (1976).
11. J. R. Fouts, Fed. Proc. 35:1162-1165 (1976).
12. D. W. Nebert and J. S. Felton, Fed. Proc. 35:1133-1141 (1976).
13. H. C. Pitot and C. Heidelberger, Cancer Res. 23:1694-1700 (1963).
14. J. H. Weisburger and G. M. Williams, in Environmental Mutagens and Carcinogenesis, T. Sugimura, S. Kondo, and H. Takebe (Eds.), University of Tokyo Press, Tokyo/Alan R. Liss, New York, 1982, pp. 283-294.

15. C. C. Irving, Methods Cancer Res. 7:189-244 (1973).
16. P. Brookes, in *Biological Reactive Intermediates: Formation, Toxicity and Activation*, D. J. Jollow, J. Kocsis, R. Snyder, and H. Vaino (Eds.), Plenum Press, New York, 1971, pp. 470-480.
17. W. K. Lutz, Mutat. Res. 65:289-356 (1979).
18. Committee on the Biological Effects of Atmospheric Pollutants, *Particulate Polycyclic Organic Matter*, National Academy of Science, Washington, D.C., 1972.
19. J. B. McCann and B. N. Ames, Proc. Natl. Acad. Sci. USA 73:950-954 (1976).
20. J. L. Hartwell and P. Shubik, *Survey of Compounds Which Have Been Tested for Carcinogenic Activity*, Public Health Service Publ. 149, U.S. Government Printing Office, Washington, D.C., 1951.
21. H. V. Gelboin, Physiol. Rev. 60:1107-1166 (1980).
22. A. W. Wood, W. Levin, A. Y. H. Lu, H. Yagi, O. Hernandez, P. M. Jerina, and A. H. Conney, J. Biol. Chem. 251:4882-4890 (1976).
23. J. K. Selkirk, R. G. Croy, and H. V. Gelboin, Science 184:169-171 (1974).
24. H. V. Gelboin, Adv. Cancer Res. 10:1-81 (1967).
25. D. W. Nebert and J. S. Felton, Fed. Proc. 35:1133-1141 (1976).
26. F. Oesch, Xenobiotica 3:305-340 (1972).
27. F. Oesch, P. Bentley, and H. R. Glatt, in *Biological Reactive Intermediates*, D. J. Jollow, J. J. Kocsis, R. Snyder, and H. Vaino (Eds.), Plenum Press, New York, 1977, pp. 187-206.
28. E. Huberman, L. Sachs, S. K. Yang, and H. Gelboin, Proc. Natl. Acad. Sci. USA 73:607-611 (1976).
29. R. F. Newbold and P. Brookes, Nature (London) 261:52-54 (1976).
30. W. Levin, A. W. Wood, H. Yagi, D. M. Jerina, and A. H. Conrey, Proc. Natl. Acad. Sci. USA 73:3867-3871 (1976).
31. T. J. Slaga, A. Viaje, W. M. Bracken, D. L. Berry, S. M. Fisher, D. R. Miller, and S. M. LeClerc, Cancer Lett. 3:23-30 (1977).
32. I. B. Weinstein, A. M. Jeffrey, K. W. Hennette, S. H. Blobstein, R. G. Harvey, C. Harris, H. Autrup, H. Kasai, and K. Nakanishi, Science 193:592-595 (1976).
33. A. M. Jeffrey, K. Grzeskowiak, I. B. Weinstein, K. Nakanishi, P. P. Roller, and R. G. Harvey, Science 206:1309-1311 (1979).
34. M. R. Osborne, R. G. Harvey, and P. Brookes, Chem. Biol. Interact. 20:123-130 (1978).
35. H. B. Gampar, A. S. C. Tung, K. Straub, J. C. Bartholomew, and M. Calvin, Science 197:671-674 (1977).
36. K. W. Jennette, A. M. Jeffrey, S. H. Blobstein, F. A. Beland, R. G. Harvey, and I. B. Weinstein, Biochemistry 16:932-938 (1977).
37. M. G. Nishioka, B. A. Petersen, and J. Lewtas, Comparison of nitroaromatic and direct-acting mutagenicity of diesel emissions, in *Polynuclear Aromatic Hydrocarbons: Physical and Biological Chemistry*, M. Cooke, A. J. Dennis, and G. L. Fisher (Eds.), Battelle Press, Columbus, Ohio, 1982, pp. 603-613.
38. D. Schuetzle, T. L. Riley, and T. J. Prater, Anal. Chem. 54:265-271 (1982).
39. B. L. Van Duuren, J. A. Bilbao, and C. A. Joseph, J. Natl. Cancer Inst. 25:53-61 (1960).
40. I. Schmeltz and D. Hoffman, Chem. Rev. 77:295-311 (1977).
41. D. E. Anders, F. G. Doolittle, and W. E. Robinson, Geochim. Cosmochim. Acta 39:1423-1430 (1975).

42. B. R. T. Simoneit, A. L. Burlingame, H. K. Schoes, and P. Haug, Chem. Geol. 7:123-141 (1971).

43. W. Cautreels and K. Van Cauwenberghe, Atmos. Environ. 10:447-457 (1976).

44. M. W. Dong, D. C. Locke, and D. Hoffmann, Environ. Sci. Technol. 11: 612-618 (1977).

45. G. Löfroth, G. Hefner, I. Alfheim, and M. Möller, Science 209:1037-1039 (1980).

46. R. Mermelstein, H. S. Rosenkranz, and E. McCoy, in Genotoxic Effects of Airborne Agents, Tice, Costa, and Schaich (Eds.), Plenum Press, New York, 1982, pp. 369-396.

47. C. W. Chiu, L. H. Lee, C. Y. Wang, and G. T. Bryan, Mutat. Res. 58:11-22 (1978).

48. S. Mizusaki, H. Okamoto, A. Akiyama, and Y. Fukyhara, Mutat. Res. 48:319-326 (1977).

49. T. C. Pedersen and J. S. Siak, Mutagenic activation and inactivation of nitro-PAH compounds by mammalian enzymes, in Polynuclear Aromatic Hydrocarbons: Physical and Biological Chemistry, M. Cooke, A. J. Dennis, and G. L. Fisher (Eds.), Battelle Press, Columbus, Ohio, 1982, pp. 623-639.

50. H. Campbell, G. C. Crumplin, J. V. Garner, R. C. Garner, C. N. Martin, and A. Rutter, Carcinogenesis 2:559-565 (1981).

51. H. S. Rosenkranz, E. C. McCoy, R. Mermelstein, and W. T. Speck, Mutat. Res. 91:103-105 (1981).

52. H. S. Rosenkranz, E. C. McCoy, D. R. Sanders, M. Butler, K. K. Demosthenes, and R. Mermelstein, Science 209:1039-1043 (1980).

53. P. P. Fu, M. W. Chou, S. K. Yang, F. A. Beland, F. F. Kadlubar, D. A. Casciano, R. H. Heflich, and F. E. Evans, Biochem. Biophys. Res. Commun. 105:1037-1043 (1982).

54. H. Tokiwa, R. Nakagawa, and Y. Ohnishi, Mutat. Res. 91:321-325 (1981).

55. J. N. Pitts, Jr., D. M. Lokensgard, W. Harger, T. S. Fisher, V. Mejia, J. J. Schuler, G. M. Scorziell, and Y. A. Katzenstein, Mutat. Res. 103:241-249 (1982).

56. J. Jäger, J. Chromatogr. 152:575-578 (1978).

57. J. N. Pitts, Jr., K. A. Van Cauwenberghe, D. Grosjean, J. P. Schmid, D. R. Fitz, W. L. Belser, Jr., G. B. Knudson, and P. M. Hynds, Science 202:515-519 (1978).

58. E. J. Rosenkranz, E. C. McCoy, R. Mermelstein, and H. S. Rosenkranz, Carcinogenesis 3:121-123 (1982).

59. D. Harvey and G. E. Penbeth, Analyst 82:498-503 (1957).

60. P. P. Fu, M. W. Chou, S. K. Yang, L. E. Unruh, F. A. Beland, F. F. Kadlubar, D. A. Casciano, R. H. Heflich, and F. E. Evans, In vitro metabolism of 6-nitrobenzo(a)pyrene: identification and mutagenicity of its metabolites, in Polynuclear Aromatic Hydrocarbons: Physical and Biological Chemistry, M. Cooke, A. J. Dennis, and G. L. Fisher (Eds.), Battelle Press, Columbus, Ohio, 1982, pp. 287-296.

61. M. C. MacLeod, G. M. Cohen, and J. K. Selkirk, Cancer Res. 39:3464-3470 (1979).

62. S. Tong and J. K. Selkirk, J. Environ. Health Toxicol, in press (1983).

63. C. N. Martin, C. A. Stanton, F. L. Chow, and R. C. Garner, Carcinogenesis 3:1423-1427 (1982).

64. K. El-Bayoumy, S. S. Hecht, and D. Hoffmann, Cancer Lett. 16:333-342 (1982).

65. J. K. Selkirk, R. G. Croy, F. J. Wiebel, and H. V. Gelboin, Cancer Res. *36*:4476-4479 (1976).

66. J. K. Selkirk, in *Carcinogenesis*, Vol. 5: *Modifiers of Chemical Carcinogenesis*, T. J. Slaga (Ed.), Raven Press, New York, 1980, pp. 1-30.

67. J. K. Selkirk, M. C. MacLeod, C. J. Moore, B. K. Mansfield, A. Nikbakht, and K. Dearstone, in *Mechanisms in Chemical Carcinogenesis*, C. C. Harris and P. A. Cerruti (Eds.), Alan R. Liss, New York, 1982, pp. 331-349.

9

Nitrogen-Containing Polycyclic Aromatic Compounds in Coal-Derived Materials

DOUGLAS W. LATER* / Battelle Northwest Laboratory, Richland, Washington

I.   Introduction   265

II. Nomenclature   266

    A. PAH nomenclature review   267
    B. N-PAC nomenclature   269
    C. N-PAC generic nomenclature   272

III. Physical, Chemical, and Spectroscopic Properties   272

IV. Biological Activity   283

V.   Isolation Methods   286

    A. Liquid-extraction chromatography   286
    B. Liquid-solid column chromatography   291
    C. Multiple-step chromatographic methods 301

VI. Analytical Methods   309

    A. Mass spectrometry   309
    B. Capillary column gas chromatography   311
    C. High-performance liquid chromatography   337

VII. Conclusions   340

    References   340

I. INTRODUCTION

Polycyclic aromatic compounds (PAC) are ubiquitous in the environment [1,2] and are prevalent in coal and products derived from coal [3]. Although the nonfunctional polycyclic aromatic hydrocarbons (PAH) are usually the most abundant class of PAC, nitrogen-containing polycyclic aromatic compounds (N-PAC) occur concomitantly with PAH in coal and coal-derived materials. Several classes of N-PAC, including pyridine- and pyrrole-type aza-PAH, amino-PAH, and cyano-PAH, have been extracted directly from coal [4,5]. N-PAC generally constitute a significant portion of materials derived from coal by combustion, coking, gasification, solvent refining, liquefaction, and other conversion and utilization processes [6,7].

There are many reasons for investigating the presence of N-PAC in coal products. Structural studies of these materials lead to a better understanding

---

*Current affiliation: The Scientific, Inc., Provo, Utah

of the chemistry of conversion processes and ultimately provide insights into more effective uses of the end products. N-PAC are known to have adverse effects on the activity and life span of catalyst used in conversion and refining processes [8,9]. The presence of N-PAC in stored fuels contributes to their instability [10]; Worstell et al. [11] reported significant increases in the deposition rate of liquid fuels that could be correlated with the levels of substituted quinolines, pyrroles, indoles, and pyridines. Carbazoles were identified by Wright et al. as the most concentrated species in the bottom sludge of stored solvent-refined coal liquids [12]. Materials rich in N-PAC produce high levels of noxious NO_x emission upon combustion [13]. As with the PAH, many N-PAC are mutagenic and/or carcinogenic and pose environmental as well as occupational hazards [1]. Conversely, coal-derived products could provide a source for N-PAC, such as quinolines, frequently used in the manufacture of pharmeceuticals, pesticides, dyes, and herbicides [7,14].

In this chapter, the analytical chemistry of N-PAC is reviewed. In addition, nomenclature and chemical, physical, and biological properties of the N-PAC are discussed briefly in an effort to provide background information in areas similar to those addressed for the PAH in both volumes of this handbook. Although a majority of the literature cited in this chapter is related to coal-derived materials, the principles discussed are applicable to the analysis of N-PAC in general. It is not within the scope of this chapter to provide an exhaustive or comparative listing of every aromatic nitrogen compound identified in coal or its conversion products. Rather, those aspects that will enable the analyst to understand the fundamental properties and analytical methods employed for the detailed characterization and identification of N-PAC in complex matrices such as coal products are emphasized.

II. NOMENCLATURE

The nomenclature used in the literature for the nitrogen-containing PAC is diversified and often confusing. Different preferences for naming generic classes of N-PAC, as well as individual nitrogen heterocycles, are common. The availability of prefixes, suffixes, and trivial names coupled with the bonding flexibility of the nitrogen atom offer several choices to the analyst for naming the nitrogen-functional PAC. Further confusion is added since the N-PAC are sister compounds to the nonfunctional PAH, and hence the nomenclature of this group of compounds necessarily influences the N-PAC nomenclature.

A synopsis of the current rules of nomenclature for the nitrogen-substituted PAC as published by the International Union of Pure and Applied Chemistry (IUPAC) [15] and the American Chemical Society [16] is presented here in an effort to clarify, validate, and unify the nomenclature used in this chapter. Furthermore, encouragement is offered to investigators and authors in this field of research to use these rules of nomenclature in their work and publications. The rules of nomenclature that are discussed in this section are taken mainly from Secs. A, B, and C of the IUPAC rules on the nomenclature of organic chemicals [15]. Other condensations of these rules, as they apply to PAH chemistry, have also been published and are recommended supplementary information [16-19].

Rules of nomenclature for the four major N-PAC classes are governed by either an extension of the Hantzsch-Widman system for monocyclic compounds (IUPAC, Sec. B) in conjunction with ring fusion nomenclature or replacement nomenclature (IUPAC, Sec. B4) for the endocyclic N-PAC, and substitutive nomenclature (IUPAC, Sec. C) for the exocyclic N-PAC. Fundamentals of nomenclature for the nonfunctional PAH that are applicable to naming the nitrogen-containing PAC are given in Secs. A21 and A22 of the IUPAC rules.

A. PAH Nomenclature Review

Rules of PAH nomenclature will not be reviewed in detail in this section. However, since the PAH nomenclature is fundamental to N-PAC nomenclature, a few key principles will be mentioned. A list of retained trivial names of selected PAH that are particularly applicable in N-PAC nomenclature is given in Table 1. Also shown are the accepted numbering schemes for each of the PAH. Fused-ring PAH that have no accepted trivial name are named by prefixing the names of the simplest component ring systems (e.g., benzo, dibenzo, naphtho, etc.) and joining them with the name of the base component, which contains as many rings as possible and occurs as high in Table 1 as possible (IUPAC Rules A21.3 and A21.4).

Spatial orientation of the polycyclic system requires (1) the greatest number of rings in a horizontal row, (2) a maximum number of rings above and to the right of the horizontal row (upper right-hand quadrant), and (3) as few rings as possible in the lower left-hand quadrant (IUPAC Rule A22.1). Numbering, in most cases, begins with the first carbon atom not engaged in ring fusion in the upper right-hand ring of the top row and proceeds clockwise around the molecule (IUPAC Rule A22.1). There are, however, notable exceptions, such as anthracene, phenanthrene, and others (Table 1). Isomers of substituted base compounds are denoted by lettering the peripheral sides of the base component a, b, c, and so on, beginning with a for the side 1,2, b for the side 2,3 or 2,2a, and continuing until every side is lettered around the periphery (IUPAC Rule A21.5). The numbering of the substituting component is used to identify points of attachment to the base component, if necessary, and should be done so as to provide the lowest possible numbering scheme. Naming of isomeric compounds is then achieved by using the prefix of the substituent component followed by the locants (numbers and letters) in square brackets and last the name of the base component. If both number and letter locants are required, numbers precede letters with a hyphen between them. The completed system is then renumbered as previously described (IUPAC A22). An example of the application of these rules is shown below for benz[a]anthracene, a benzolog of anthracene:

Anthracene   Benzo   Benz[a]anthracene

(1)

Another PAH nomenclature rule that is particularly important for the naming of amino-PAH and cyano-PAH compounds involves radical names obtained from the trivial names. For radicals derived from polycyclic hydrocarbons, the numbering of the hydrocarbon is retained, the point of attachment is given the lowest possible number, and the ending of the name is changed from "-ene" to "-enyl" (IUPAC Rules 24.1 and A24.2). Again, exceptions exist. The radical names of the trivial compounds are also listed in Table 1.

Table 1. IUPAC Structures, Numbering, and Trivial Names for Selected
PAH (IUPAC) A21.2

IUPAC No[a]	Structure[b]	Trivial name	Radical Name[b]
(2)		Indene	Indenyl-
(3)		Naphthalene	Naphthyl-
(9)		Acenaphthylene	Acenaphthyl-
		Acenaphthene	Acenaphthenyl-
		Biphenyl	Biphenyl-
(10)		Fluorene	Fluorenyl-
(12)		Phenanthrene	Phenanthryl-*
(13)		Anthracene	Anthryl-*
(14)		Fluoranthene	Fluoranthenyl-
(18)		Pyrene	Pyrenyl-
(19)		Chrysene	Chrysenyl-

[a] Listed in reverse order of preference.
[b] An asterisk indicates exceptions to numbering and radical naming rules.

B. N-PAC Nomenclature

1. Extension of Hantzsch-Widman System

Heterocyclic systems are fused-ring compounds containing atoms of at least two different elements as ring members. Nitrogen and carbon are the two elements for the N-PAC. An extension of the Hantzsch-Widman system is one of two methods applicable to the nomenclature of the two classes of endocyclic nitrogen heterocyclic PAC or aza-PAH. Table 2 provides a partial list of trivial names that are retained for the mononitrogen aza-PAC (IUPAC Rule B2.11). Unsaturated rings containing a nitrogen heteroatom have designated suffixes: compounds with a nitrogen heteroatom in a five-membered ring end in "-ole"; nitrogen in a six-membered ring has a compound name ending in "-ine" (IUPAC Rule B1.1, Table 2).

Fused-ring compounds containing a nitrogen heteroatom are named according to the fusion principles reviewed in Sec. II.A for the PAH (IUPAC Rule A21) except that the trivial names given in Table 2 are used; that is, the base component should be a nitrogen heterocyclic system with as many rings as possible and occurring as high as possible in Table 2 (IUPAC Rule B3). The numbering of fused-ring N-PAC systems should provide the nitrogen heteroatom with the lowest possible number while maintaining the orientation rules outlined for the PAH. Exceptions to this rule are carbazole and acridine. Lettering of peripheral sides for isomer differentiation is also done by the PAH method. When the name of a component in a fusion name contains locants that do not also apply to the numbering of the fused system, these locants are placed in brackets, as are the locants for fusion positions required by the PAH nomenclature (IUPAC Rule B3.1). For example, the structure obtained by fusing a benzene ring on the h side of isoquinoline shown in equation (2) has the systematic name benz[h]isoquinoline. Notice that the spatial orientation and numbering of this compound are in accordance with those rules reviewed for the PAH.

| Isoquinoline | Benzo | Benz[h]isoquinoline | (2) |

Other N-PAC with a single nitrogen heteroatom and three or more fused rings can be named by the application of these fundamental rules.

2. Replacement Nomenclature

Nitrogen-containing fused ring systems may also be named by prefixing the name of the corresponding nonfunctional PAH with the term "aza-" (IUPAC Rule B1.1, Table 1), preceded by the locant of the heteroatom (IUPAC Rule B4.2). The structures, numbering, and names of PAH in Table 1 can be used directly in naming endocyclic N-PAC. This method is known as replacement nomenclature, sometimes also referred to as "a" nomenclature. The numbering of the hydrocarbon, or PAH in the case of N-PAC, is retained irrespective of the heteroatom position. However, the lowest possible number consistent

Table 2. IUPAC Structures, Numbering, and Trivial Names for Selected N-PAC (IUPAC B2.11)

IUPAC No[a]	Structure	Trivial name[b]
(12)		Pyrrole
(15)		Pyridine
(22)		Indole
(26)		Isoquinoline
(27)		Quinoline
(35)		Carbazole*
(37)		Phenanthridine
(38)		Acridine*

[a]Listed in reverse order of preference.
[b]An asterisk indicates exceptions to IUPAC systematic numbering rules.

with the numbering of the PAH is assigned as the locant of the nitrogen hetero-atom. In these fusion names, the "aza" term always precedes the completed name of the parent hydrocarbon. Prefixes for ordinary substitution such as alkylation precede the "aza" term (IUPAC Rule B4.3).

A few examples of naming endocyclic N-PAC by the replacement and Hantzsch-Widman methods are given in Fig. 1, structures I-IV. Two additional points are noteworthy from the examples of Fig. 1. First, as shown for structure III, the location of a hydrogen atom in a polycyclic system not completely unsaturated is identified by using a capital "H" in conjunction with the numerical locant (IUPAC Rules A21.6 and B5.12). Further hydrogenation

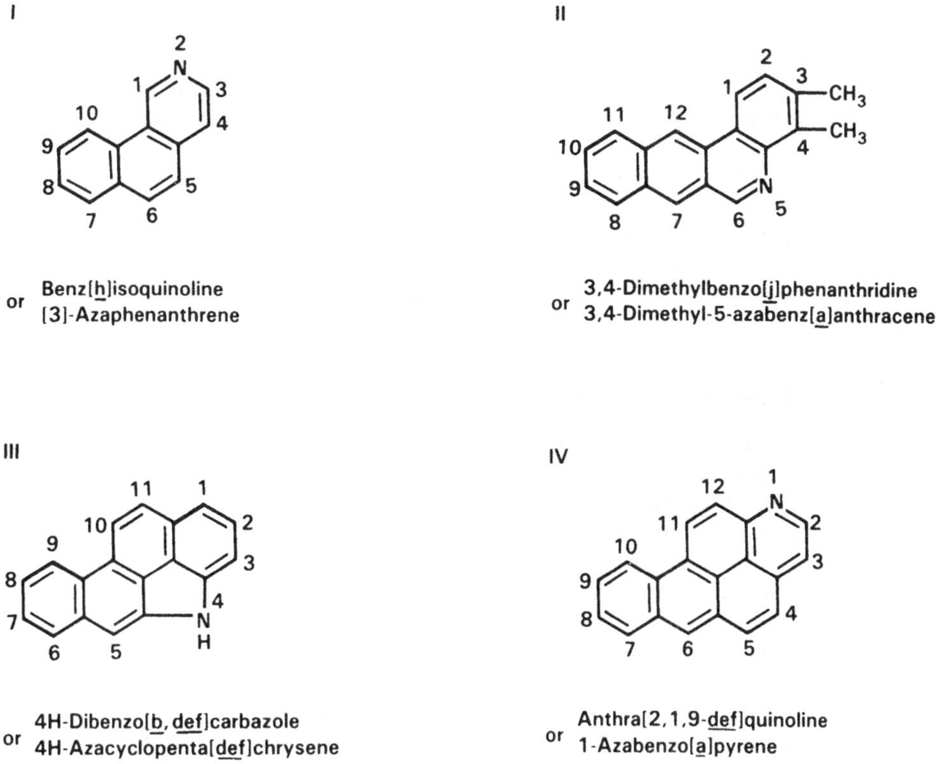

I

Benz[h̲]isoquinoline
or [3]-Azaphenanthrene

II

3,4-Dimethylbenzo[j̲]phenanthridine
or 3,4-Dimethyl-5-azabenz[a̲]anthracene

III

4H-Dibenzo[b̲, d̲e̲f̲]carbazole
or 4H-Azacyclopenta[d̲e̲f̲]chrysene

IV

Anthra[2,1,9-d̲e̲f̲]quinoline
or 1-Azabenzo[a̲]pyrene

Figure 1. Names of selected N-PAC.

is indicated by the prefixes "dihydro-," "tetrahydro-," and so on (IUPAC Rule B1.2). Second, from structure IV, the assignment of locants in naming complex fused-ring systems follows two additional rules: (1) locants are assigned the lowest possible numbering and lettering schemes; and (2) the attachment points numbering sequence for the substituting system(s) follows the direction of attachment as specified by the sequence of lettered locants on the base component system (IUPAC Rule A21.5).

3. Substitutive Nomenclature

The naming of exocyclic N-PAC or amino-PAH and cyano-PAH is achieved by use of substitutive nomenclature as contained in Section C of the IUPAC rules [15]. The cyano and amino functional groups may be cited as either suffixes or prefixes to the name of the parent compound (IUPAC Rule C10.2, Table III). The prefixes for the $-C \equiv N$ and $-NH_2$ functional groups of N-PAC are "cyano-" and "amino-," respectively, while the suffixes are "-carbonitrile" and "-amine," respectively.

Primary monoamines, RNH_2, where R is a PAC are named by adding (1) the suffix "-amine" to the name of the radical (Table 1), (2) the suffix "-amine" to the name of the parent compound with elision of the terminal "e," or (3) the prefix "amino-" to the name of the parent compound (IUPAC Rule C812). Use of locants of the lowest possible numbering are also required for isomer distinctions. Table 1 can be conveniently used for naming the amino-PAH (and the cyano-PAH). For example, phenanthrene with an amino functional group attached at the ninth position could be named 9-aminophenan-threne, 9-phenanthrylamine, or 9-phenanthrenamine.

Similar rules of nomenclature are used for naming the cyano functional PAH (IUPAC Rule C832). Monocyano PAC, R-C≡N, are named by adding (1) the suffix carbonitrile to the parent compound name (IUPAC Rule C832.2), (2) cyanide to the parent radical name (IUPAC Rule C832.4), or (3) the prefix cyano to the parent PAH compound (IUPAC Rule C832.5). "Carbonitrile" used as a suffix denotes the group -C≡N, including the carbon atom contained therein (IUPAC Rule C832.2). For the same example used for the amino functional phenanthrene, the corresponding cyano compound could be named 9-cyanophenanthrene, 9-phenanthrenecarbonitrile, or 9-phenanthryl cyanide.

C. N-PAC Generic Nomenclature

In this chapter the nitrogen-containing polycyclic aromatic compounds with a single nitrogen heteroatom will be referred to collectively as N-PAC, regardless of the chemical functionality of the nitrogen heteroatom. Other popular generic names used in the literature for the N-PAC include (but are not limited to) "amines," nitrogen bases, azaarenes, and aza-PAH. A plethora of corresponding acronyms have also been formulated and used in the literature for these compounds. The selection and use of generic names and acronyms is largely a matter of personal preference; arguments supporting the validity of their usages can be found in the different sections and rules of IUPAC, ACS, and *Chemical Abstracts* nomenclature guidelines.

In general, the N-PAC that are considered in this chapter occur in the complex matrices as subjugate species of the nonfunctional polycyclic aromatic hydrocarbons (PAH). Therefore, prefixing of the PAH term is preferred and will be used throughout this chapter. Aza-PAH will refer to N-PAC with an endocyclic nitrogen heteroatom. Both the benzologs of pyridine and pyrrole will be included in the aza-PAH class. The generic terms 3°-PANH (*tert*-polycyclic aromatic nitrogen heterocycles) for the pyridine benzologs and 2°-PANH (*sec*-polycyclic aromatic nitrogen heterocycles) or pyrrolic-PAH for the pyrrole benzologs may be used to distinguish between these two subclasses of aza-PAH when required. This terminology is based on the premise that the nitrogen heteroatom of the endocyclic N-PAC species can either be bound in a secondary configuration, -NH-, for the pyrrole-type aza-PAH or a tertiary configuration, -N=, for the pyridine-type aza-PAH [2,20]. Other generic terminology used in the literature for the aza-PAH compounds includes azaarenes, quinolines and benzoquinolines, indoles and carbazoles, and "amines."

The two possible exocyclic N-PAC classes will be referred to in this chapter as amino-PAH and cyano-PAH for the -NH₂ and -C≡N functional groups, respectively. The cyano-PAH are frequently given the generic name nitriles and/or carbonitriles, while primary aromatic amines (PAA) is another widespread generic term used for the amino-PAH class.

III. PHYSICAL, CHEMICAL, AND SPECTROSCOPIC PROPERTIES

Several classes of N-PAC, each with a different chemical functionality, are possible due to the versatile bonding properties of the nitrogen heteroatom in conjunction with cyclic aromatic ring systems. If the polycyclic aromatic compounds with a single nitrogen heteroatom are considered, there are four possible classes of N-PAC, as shown in Table 3. Those compounds with endocyclic nitrogen heteroatom substitution are generally the most abundant N-PAC in coal-derived materials: the pyridine type (e.g., quinoline and acridine) and the pyrrole type (e.g., indole and carbazole). There are two possible classes of N-PAC where substitution of the nitrogen heteroatom is exocyclic to the aromatic ring system: the amino-PAH (e.g., aminonaphthalenes and aminophenanthrenes) and the cyano-PAH (e.g., cyanonaphthalenes and cyanophenanthrenes). These classes of N-PAC are often much lower in concentration in coal products than either the pyrrole- or pyridine- type aza-PAH.

Overall, most coal-derived products have significant levels of N-PAC ranging from 5 to 40% of the total sample composition, depending on the material. Furthermore, the N-PAC content generally increases with boiling point. The relative abundance of these four classes of N-PAC in a coal product can vary widely and is generally a function of processing conditions, such as distillation, catalysis, conversion technology, and other process-related factors. For example, the cyano-PAH are more abundant in combustion process materials such as coal gasification condensates, while the amino-PAH are primarily found in noncatalytically processed direct coal liquefaction materials.

The implications with regard to the complexity and difficulty of chemical analysis of mixtures containing these four possible N-PAC classes are formidable compared to the nonfunctional PAH. As shown in Table 3, there are more than 30 possible isomers for the three-ring N-PAC as opposed to three possible PAH isomers: fluorene, anthracene, and phenanthrene. There are five possible cata-condensed four-ring PAH isomers (chrysene, benz[a]anthracene, naphthacene, triphenylene, benzo[c]phenanthrene), but there are 29 possible isomers of the pyridine-type aza-PAH alone for structurally equivalent N-PAC. If alkylation is considered, the possible number of isomers rapidly becomes quite staggering; for instance, there are 68 methylbenzoquinolines and 319 monomethylated four-ring pyridine-type aza-PAH. In general, N-PAC isolates of coal-derived materials can be composed of peri- and cata-condensed, multialkylated, partially hydrogenated ring structures of the four various classes of mono-nitrogen heterocycles and thus are extremely complex.

To complicate matters further, many N-PAC of different chemical functionality have the same elemental composition. Hence differentiation of molecular structure on the basis of molecular weight alone is not possible. Examples of isobaric N-PAC of different functionality include carbazole and the methylcyanonaphthalenes (m/z = 167), and the methylbenzoquinolines and aminophenanthrenes/anthracenes (m/z = 193). An N-PAC of molecular weight 217 could be a methylazapyrene, benzocarbazole, methylcyanophenanthrene, or aminopyrene. Unequivocable determination of individual N-PAC components in complex mixtures may require high-performance chromatographic separations as well as combinations of analytical techniques, such as high-resolution or tandem mass spectrometry, spectrophotometry, and capillary column gas chromatography.

Pyrrole, pyridine, aniline, and benzonitrile are the monocyclic aromatic compounds of the 2°-PANH, 3°-PANH, amino-PAH, and cyano-PAH N-PAC classes, respectively. The intrinsic physical properties and molecular structures of these four monomers are fundamentally different and have a strong influence on the nature of their respective benzologs. Chromatographic methods and analytical techniques take advantage of differences in their

Table 3. Possible Classes of Nitrogen-Containing Polycyclic Aromatic Compounds

Structure	Number of isomers	Name	Nitrogen functionality	Chemical class[a]
	8	Phenanthridine or 9-azaphenanthrene	Pyridine $-N=$	3°-PANH
	10	Carbazole or 9H-azafluorene	Pyrrole $-NH-$	2°-PANH
	8	9-Aminophenanthrene or 9-phenanthrylamine or 9-phenanthrenamine	Amino $-NH_2$	Amino-PAH
	8	9-Cyanophenanthrene or 9-phenanthrylcyanide or 9-phenanthrenecarbonitrile	Cyano $-C\equiv N$	Cyano-PAH

[a]PANH, polycyclic aromatic nitrogen heterocycle(s).

Table 4. Selected Physical Properties of Monocyclic Nitrogen Heterocyclic Compounds

Compound	Heat of combustion (kcal/mol)	Resonance stabilization energy (kcal/mol)	Dipole[a] moments (debye)
Aniline	811.7	~55[b]	-1.53
Benzonitrile	865.5	~54[b]	+4.18
Pyridine	567.7	~23	+2.19 to +2.26
Pyrrole	658.5	22-28	-1.81 to -1.84
Benzene	782	36-43	~0

[a]Positive indicates that the dipole moment direction (+ ⟶ -) is toward the nitrogen heteroatom; conversely, negative indicates that the direction of the dipole moment is away from the nitrogen heteroatom.
[b]Values calculated by Klages' method, as explained in Ref. 27.
Source: Values reported are from Ref. 21-23, 25, and 26.

acid-base qualities, dipole moments, molecular structure and bonding, chemical reactivities, and other physical/chemical properties to separate and distinguish among these four classes of N-PAC.

From thermodynamic information such as heats of combustion and heats of hydrogenation, the resonance stabilization energies (see Table 4) of the monocyclic, endocyclic nitrogen heterocycles are approximately 22-28 kcal/mol [22]. These values are somewhat less than 36-43 kcal/mol, the resonance energy of benzene, but much greater than the energy of conjugated dienes (3 kcal/mol). Hence the monocyclic, endocyclic, nitrogen heterocycles have a strong aromatic character, as do their benzologs. The exocyclic nitrogen heterocycles are also aromatic by virtue of substitution of the functional group onto an aromatic benzene ring. However, as shown in Table 4, their resonance energies are generally enhanced approximately 10-15 kcal/mol over benzene due to delocalization effects of the nitrogen heteroatom unshared electron pair and conjugation with the aromatic π orbitals of the ring system. In general, the fusion of additional benzene rings to the monocyclic nitrogen heterocycles increases the resonance stabilization energy, implying that the polycyclic nitrogen heterocycles are highly aromatic.

Pyrrole has a planar pentagonal structure in which the four carbon atoms and the nitrogen atoms are held by σ bonds with sp^2 orbital hybridization. A single electron from each carbon atom and two nitrogen heteroatom electrons reside in p orbitals and complete the electron sextet required for aromaticity. Since the nitrogen lone-pair electrons normally responsible for the basicity of amines are incorporated into the aromatic π orbitals and are not available for sharing, pyrrole, indole, carbazole, and other benzologs of these compounds are less basic than water [22-26]. The pK[a] values listed in Table 5 for the pyrrolic PAH as conjugate bases support this observation. Furthermore, it has been shown that protonation of pyrrole and its benzologs occurs on the pyrrole ring carbons rather than at the nitrogen heteroatom [24]. In fact, the pyrrolic PAH can be considered weak acids [23]; pK[a] =

Table 5. Literature pK_a Values for Selected N-PAC

Compound	$pK_a{}^a$	Compound	$pK_a{}^a$
2°-PANH[b]		3°-PANH	
Pyrrole	-0.3	Pyridine	5.23
Indole	-2.4	Quinoline	4.92
Carbazole	-1.9	Isoquinoline	5.40
Benzo[a]carbazole	-2.1	2-Azabiphenyl	4.48
Amino-PAH		4-Azabiphenyl	5.55
Aniline	4.63	Benzo[f]quinoline	5.11
5-Aminoindan	5.31	Benzo[h]quinoline	4.21
1-Aminonaphthalene	3.94	Acridine	5.58
2-Aminonaphthalene	4.15	Phenanthridine	4.61
2-Aminobiphenyl	3.82	Benz[a]acridine	3.95
4-Aminobiphenyl	4.35	Benz[c]acridine	4.70
4-Aminobiphenyl	4.35	1-Azapyrene	4.3
2-Aminofluorene	4.64	Cyano-PAH[b]	
1-Aminoanthracene	4.1	Benzonitrile	~ -10

[a] Sources of pK_a values include Ref. 21, 31-33.
[b] Values reported are for the compounds as bases.

13.5 for pyrrole as an acid (this compares with phenol, $pK_a = 10$, which is considered mildly acidic). Delocalization of the nitrogen electron pair into the pyrrole ring structure is responsible for these anomolous amine properties and is further substantiated by the direction of the dipole moment (Table 4), which places the positive portion of the charge distribution on the nitrogen heteroatom and the negative charge into the ring system (D $\cong$ -1.82) [24]. Hence those N-PAC with pyrrole as the basic structural component are fairly nonbasic, neutral compounds. These properties are used to separate the 2°-PANH from the other more polar, more basic N-PAC classes by liquid-liquid and liquid-solid chromatographic techniques, as discussed in Sec. V.

Pyridine is structurally analogous to benzene, except that one CH unit is replaced with a nitrogen heteroatom. The aromatic sextet of pyridine is composed of a single electron from each carbon and nitrogen ring atom. The nitrogen heteroatom has an unshared electron pair that resides in an sp^2 hydrid orbital in the same plane as the ring system. Hence this electron pair is available for sharing. The dipole moment of pyridine (see Table 4) is greater and in the reverse direction of pyrrole; the negative portion of the charge distribution is associated with the nitrogen heteroatom. Furthermore, the π and σ moments are in the same direction—toward the nitrogen heteroatom—and are thus additive, resulting in a surplus negative charge on the nitrogen. These properties are responsible for the more polar or basic character of the pyridine and its benzologs as compared to the pyrrolic-PAH. The pK_a values for the 3°-PANH are between 4 and 5 (see Table 5) and are indicative of mildly basic compounds. Quinoline and isoquinoline are benzopyridines and thus the orbital structures of both compounds are related to pyridine as well as naphthalene. The properties of the higher benzologs are also characteristic of the pyridine base component and the analogous nonfunctional PAH. All are weak bases (see Table 5).

Aniline can be considered as a substituted amine, $R-NH_2$, where R is benzene. Amines generally have an sp^3 pyramidal structure; the H-N-H bond angle in ammonia is 107.1°. The amino group in aniline is still pyramidal but has a larger H-N-H bond angle (113.9°) than ammonia. Furthermore, the H-N-H bonding plane intersects the plane of the benzene ring at a mean angle of 39.4° which places the nitrogen unshared electron pair in an orbital parallel (isovalent conjugation) to the ring p electron orbitals instead of in the preferred perpendicular orbital (heterovalent conjugation) like pyridine. This is due to the inductive effect of the phenyl group on the unshared electron pair of the nitrogen atom. These electrons are delocalized and isovalently conjugated into the aromatic π orbitals of the benzene ring and are therefore less accessible for bonding. The amino functional group is also resonance stabilized, as manifested by the direction and magnitude of the dipole moment and the high resonance energy of aniline compared to benzene, pyrrole, or pyridine (see Table 4). Isovalent conjugation of the unshared electron pair also decreases the basicity of the amino-substituted aromatic systems, such as aniline and its benzologs, compared to ammonia (pK_a = 9.5) or the alkylamines. The pK_a values shown for the amino-PAH in Table 5 range from 3 to 5 and are slightly less basic than pyridine and its benzologs.

Benzonitrile is considerably less basic than the other N-PAC classes (see Table 5). The protonated nitriles are some 15 powers of 10 more acidic than the amino-PAH or 3°-PANH. This is due to the powerful effect that the bonding hybridization has on the relative stability of the nitrogen lone-pair electrons. The orbital containing the unshared electrons in benzonitrile and its benzologs is approximately sp in character; electrons in aniline occupy sp^3 orbitals, while in pyridine and pyrrole electron pairs occupy sp^2 orbitals. The increased σ character of the orbital holding the unshared electron pair in nitriles increases its stability, decreases its energy, and in general, renders the unshared electron pair inaccessible. The net result of this effect is a far less basic functional group than would normally be anticipated for a nitrogen functional group. Interestingly, the dipole moment of benzonitrile (Table 4) is larger in magnitude than the other monocyclic N-PAC, with the negative charge directed toward the nitrile group. This seems to indicate that although the nitrilo group is comparatively nonbasic, it is polar. Much of the dipole is due to the nitrile group itself and is a consequence of the occupied orbitals. Finally, the p electrons of the triply bonded carbon and nitrogen atoms participate in resonance stabilization of the benzonitrile molecule, while the unshared nitrogen atom electron pair is heterovalently conjugated. The net result is a resonance energy for benzonitrile comparable to that of aniline (see Table 4). Like the pyrrolic-PAH, the cyano-PAH are a nonbasic, neutral chemical class. The distinguishing feature of these two classes is the apparent polar character of the cyano-PAH. This property is quite useful for separation of the cyano-PAH from other PAC and N-PAC by charge-transfer complexation chromatography, as discussed in Sec. V.

Variations in molecular structure, composition, and/or functionality are frequently reflected in the boiling points of organic compounds. The boiling points of the four monocyclic aromatic nitrogen heterocycles are: pyridine =

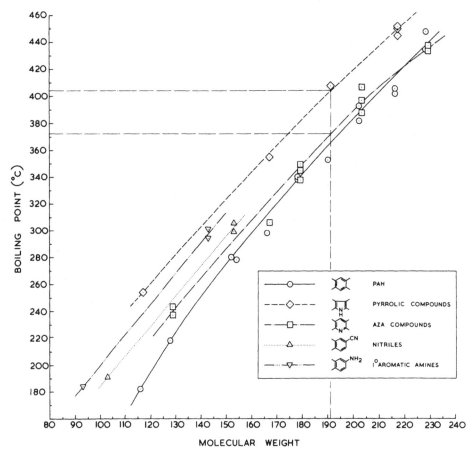

Figure 2. Graph of atmospheric boiling point versus molecular weight for N-PAC types and PAH. (From Ref. 34.)

116°C, pyrrole = 131°C, aniline = 185°C, and benzonitrile = 191°C. Burchill [34] plotted atmospheric boiling point against molecular weight for a range of N-PAC from each of the four classes and for the nonfunctional PAH as shown in Fig. 2. The particular classes of N-PAC lie on well-defined, approximately parallel lines. The pyridine-type aza-PAH have approximately the same boiling points as the PAH of similar molecular weight, but the pyrrolic-PAH have consistently higher boiling points for equivalent molecular weights. Burchill [34] utilized boiling-point characteristics to differentiate isomeric N-PAC of different chemical functionality based on their GC retention.

Molecular differences in the four functional N-PAC classes are manifested not only in the physical properties shown in Tables 4 and 5 and Fig. 2, but distinguishable differences also exist in their spectrophotometric behavior. For example, in the absence of N-substitution the pyrrolic-PAH have a sharp band at about 3430-3460 cm^{-1} in their infrared (IR) spectra, while the amino-PAH have strong IR absorption bands at 3220 and 3370 cm^{-1} [26,28,29]. A doublet IR absorption at 1598 and 1557 cm^{-1} due to C=N- stretch is characteristic of pyridine-type nitrogen [29]. Similarly, the cyano-PAH have a characteristic IR band at 2240-2260 cm^{-1} [22,30]. Specific details on the source and frequency of characteristic IR absorption bands for the various

functional N-PAC have been outlined by Conley [35]. The qualitative determination of N-PAC in coal-derived materials by IR was perhaps the earliest
used analytical approach. Karr and Chang [36] identified 51 aza-PAH and
amino-PAH in coal-derived tars in the late 1950s. Since then, other investigators have routinely used IR spectroscopy for the analytical determination
of nitrogen-functional PAC in coal products [37-40]. IR methods have also
been used to differentiate structural isomers. Caton et al. [37] distinguished
the eight benzoquinoline isomers by diffuse reflectance infrared (DRIFT)
spectroscopy. The most intense absorbance bands for each isomer are listed
from their work in Table 6.

In general, UV-visible absorption spectroscopy has not been as useful as
other spectroscopy techniques for differentiating and identifying N-PAC.
However, UV-visible spectroscopy does give a good indication of the size of
the aromatic conjugated system. The adsorption spectra of the polar N-PAC
typically exhibit less well resolved fine structure than their corresponding
nonfunctional PAH, with band intensities being very pH sensitive for the
more basic N-PAC. Wehry [41] illustrated these points with specific examples
for the aza- and amino-PAH in the first volume of this handbook.

Fluorescence spectroscopy is a well-established and sensitive method for
PAH analysis [41,42]. Kershaw [42,43] used this technique to identify N-PAC
in the basic fractions of coal-derived liquids. Tables 7 and 8 list the most intense fluorescence emission and excitation peaks of several aza-PAH standard
compounds in hexane at room temperature. Also shown are peak assignments
for components from coal-derived liquids identified by comparison with the
standard compound spectra. Comparison of the fluorescence data for structural isomers (see Tables 7 and 8) suggests that the position of the nitrogen
heteroatom does not significantly affect the emission and excitation bands
and hence this technique has not been useful for distinguishing structural
isomers [42]. McKay and coworkers [44,45] have also reported fluorescence
emission and excitation spectra for selected 2°-PANH and 3°-PANH standard
compounds and used this technique to identify the aza-PAH in crude oils.

The selectivity and resolution of fluorescence spectroscopy can be improved by low-temperature and matrix techniques. For example, Shpol'skii
spectroscopy has been used to identify isomeric aza-PAH in fossil fuel materials [43,46,47]. Fluorescence line narrowing spectroscopy in organic glasses
has been investigated for the determination of the amino-PAH [48]. Phosphorescence, another luminescence spectroscopy method, has similarly been
applied for the identification of N-PAC in the presence of special matrices
such as cyclodextrin [49] and silica gel [50]. Isomeric identifications not
achieved by normal room-temperature luminescence spectroscopy are possible
using these low-temperature and matrix techniques.

Nuclear magnetic resonance (NMR) spectroscopy has also been applied to
the determination of N-PAC in coal-derived products. Isomeric benzoquinolines have been distinguished by their ^{13}C chemical shifts by Buchanan and
Rubin [51]. ^{19}F NMR has been particularly useful for the analytical determination of derivatized amino-PAH [52].

In summary, although four classes can exist in a complex N-PAC isolate,
their differentiation is possible by analytical techniques which take advantage of fundamental differences in their physical properties. For example,
the amino-PAH and 3°-PANH are more basic than either the cyano-PAH or
2°-PANH, rendering them susceptible to separation according to basicity or
adsorptivity. The cyano-PAH are the most neutral class of N-PAC, although
the nitrile moiety is the most polar and/or polarizable of the N-PAC nitrogen

Table 6. Location of Most Intense Bands in DRIFT Spectra of Benzoquinolines[a]

Compound	>1500	Wavenumber region (cm^{-1}) 950 to 1500	850 to 950	800 to 850	750 to 800	680 to 750	Sets of hydrogen atoms
Benzo[b]quinoline (acridine)	1516[b]		860[c]				737	2
Benzo[c]quinoline (phenanthridine)		1458[b] 1246[b]			775[c]	748	2
Benzo[f]quinoline		1385 1493[b]		838 / 817		748 / 706[b]	3
Benzo[g]quinoline	1612[b]	1423[b]	887			736	3
Benzo[h]quinoline	1512	1443 1400		833 / 806		748 725	4
Benz[f]isoquinoline				833 814		748 718	4
Benz[g]isoquinoline			887 / 918[c]			641	4
Benz[h]isoquinoline	1608	1238		845 / 810[b]	741[c]	745	4

[a]All absorbance bands between 3800 and 600 cm^{-1} that have a maximum absorbance greater than 50% of the most intense band in the spectrum are listed. The most intense peak is underlined. The three most intense bands are always listed even if the maxima are less than 50% of the most intense band.
[b]Maximum absorbance of this band is greater than 25% but less than 50% of the most intense band.
[c]Maximum absorbance of this band is less than 25% of the most intense band.
Source: Ref. 37.

Table 7. Fluorescence Emission Spectra of Nitrogen Bases[a]

Compound	Fluorescence emission spectra wavelength (nm)			
	Model compound	Reference[b]	Coal-Derived Liquid	Sample[c]
Phenanthridine	353, 363, 382, 404	42	352, 368, 377(s), 384(s)	A
Benzo[f]quinoline	350, 362, 381, 402	42	351	B
Benzo[h]quinoline	347, 362, 381, 402	42		
Azapyrene	372, 380, 392, 402, 413	19	373, 382, 393, 401, 411	A
			375, 396, 403	B
Benz[a]acridine	382, 401, 425	26	381, 403, 422	A
Benz[c]acridine	384, 408, 430, 458	26	380, 406, 426, 464(s)	C
Phenaleno[1,9-gh]quinoline	410, 430, 462	26	410, 437, 461	C
(pyrenoline; azabenzo[a]pyrene)	410, 436, 463	26		
Dibenz[a,h]acridine	394, 405, 417	19	389, 418, 435	C
	395, 403, 418, 445, 470	18		
	390, 400, 410, 438, ~460(s)	26		
Dibenz[a,j]acridine	392, 403, 415, 426, 439	19		
	395, 400, 416, 440, 465	18		
	390, 400(s), 410, 434, 460(s)	26		

[a]The most intense peak in each spectrum is underlined. Shoulders are indicated by (s).
[b]See original paper for references.
[c]A, oil from hydrogenation of New Wakefield coal, South Africa; B, oil; C, asphaltene from a flash pyrolysis tar of Millmerran coal, Australia.
Source: Ref. 42.

Table 8. Fluorescence Excitation Spectra of Nitrogen Bases[a]

Compound	Fluorescence emission spectra wavelength (nm)			
	Model Compound	Reference[b]	Coal-Derived Liquid	Sample[c]
Phenanthridine	325, 340, 348	42	320(s), 335, 349	A
Benzo[f]quinoline	315, 322, 328, 336, 344	42		
Benzo[h]quinoline	315, 322, 329, 338, 346	42		
Azapyrene	236,[d] 260, 278, 300, 337, 352, 370	27	274, 337, 353, 372 338, 352, 373	A B
Benz[a]acridine	284, 330(s), 345, 360, 380	26	359, 378	A
Benz[c]acridine	273, 280, 300(s), 340, 356, 376	26	357, 378	C
Phenaleno[1,9-gh]quinoline	290, 297, 315(s), 340(s), 354, 368 384, 408	26	356, 367, 381, 408	C
Dibenz[a,h]acridine	290, 318, 330, 344, 350, 370, 380	26	348, 355, 388	C
Dibenz[a,j]acridine	264, 290, 320, 330, 352, 370, 388	26		
Benz[b]acridine	372,[d,e] 382, 391, 440, 465 543[d,f]	27 27	374(s),[e] 385, 397, 447(s), 468 543[f]	

[a]The most intense peak in each spectrum is underlined. Shoulders are indicated by (s).
[b]See original paper for references.
[c]A, oil from hydrogenation of New Wakefield coal, South Africa; B, oil; C, asphaltene from a flash pyrolysis tar of Millmerran coal, Australia.
[d]UV spectrum, not fluorescence excitation spectrum.
[e]In EtOH.
[f]In EtOH/HCl.
Source: Ref. 42.

functional groups. The amino-PAH are also susceptible to polarization and are probably more chemically reactive than the other N-PAC species. Differences in the physical properties of the N-PAC classes are manifest in their spectrophotometric behavior. Spectroscopy techniques have been used to distinguish the four N-PAC classes and identify structural isomers within a given N-PAC class. Other analytical techniques and methods that have been developed to isolate and identify the various components of the N-PAC classes based on the chemical and physical differences discussed in this section will be presented in Secs. V and VI.

IV. BIOLOGICAL ACTIVITY

Biodirected chemical analysis of coal-derived materials over the past several years has led to the identification of at least two general classes of compounds that are largely responsible for the biological response observed in laboratory systems upon exposure to these materials. The PAH are potent carcinogens in coal liquids, primarily those having from four to six aromatic rings. Kenneway, Cook, and others [53-56] in the early part of this century showed that certain coal-derived PAH could cause skin tumors in mice and rabbits. Later, Sexton [57] observed increased carcinogenic trends in human populations of workers in coal hydrogenation facilities. Mahlum and coworkers recently showed that the isolated PAH of coal liquefaction materials were responsible for a majority of the tumorigenicity in the initiation/promotion mouse skin assay, particularly isolates from high-boiling distillates [58-60].

The second class of compounds that have more recently been recognized as genotoxic constituents in coal products are the N-PAC. Table 9 lists selected N-PAC and PAH of representative one- to five-ring compounds along with their relative mutagenic and carcinogenic potencies. Several trends can be observed from the data presented in Table 9 [61-70]. First, the PAH with fewer than four aromatic fused rings are relatively inactive mutagens and/or carcinogens. Conversely, the nitrogen functional analogs display mutagenic and carcinogenic properties starting with two-ring systems. For example, indole, quinoline, and 2-aminonaphthalene are carcinogenic, but the PAH analog, naphthalene, is not carcinogenic. Similar trends are apparent when the mutagenicities of analogous two- and three-ring N-PAC and PAH are compared; the nitrogen functionality typically endows the inactive parent PAH structure with observable activity. When aromatic compound systems with four or more fused rings are considered, the carcinogenic potencies for the PAH are generally greater than for analogous N-PAC. Similar correlations for the mutagenicity of the PAH and N-PAC are not as straightforward, with the degree of mutagenic activity of the N-PAC being determined by the functionality of the nitrogen heteroatom. In general, the following mutagenic activity ranking for the N-PAC classes is accepted: amino-PAH >>> 3°-PANH > 2°-PANH > cyano-PAH. The PAH and 3°-PANH, in general, have comparable mutagenic activities [63,64].

Group-type comparisons like those shown in Table 9 are valid to a certain point in providing basic information about the relative biological activities of different classes of PAC. But variations in biological potency as a function of isomeric compound structure must also be considered. Structure-activity correlations are of increased importance for the N-PAC due to the increased number of possible isomers for a given structure as compared to their analogous nonfunctional PAH. Table 10 illustrates this point for the 10 isomers of aza- and aminophenanthrene. The degree of biological activity is a function

Table 9. Mutagenicity and Carcinogenicity of Selected N-PAC[a]

[a] o, Not active; +/o, weakly active; +, moderately active; ++, strongly active; ND, not determined.
Sources: Refs. 61-70.

Table 10. Structure-Activity Dependence of the Aminophenanthrenes and Azaphenanthrenes[a]

Compound	Mutagenic activity	Carcinogenic activity	References
Aza-PAH			
1-Azaphenanthrene/ benzo[f]quinoline	+/o	o	7,61,70
2-Azaphenanthrene/ benz[f]isoquinoline	o	ND	7,70
3-Azaphenanthrene/ benz[h]isoquinoline	o	ND	7,70
4-Azaphenanthrene/ benzo[h]quinoline	+/o	o	7,61,70
9-Azaphenanthrene/ phenanthridine	+/o	+	7,61,70
Amino-PAH			
1-Aminophenanthrene	+	+	62,65
2-Aminophenanthrene	++	++	62,65
3-Aminophenanthrene	++	++	62,65
4-Aminophenanthrene	o	ND	62,65
9-Aminophenanthrene	++	++	62,65

[a] o, Not active; +/o, weakly active; +, moderately active; ++, strongly active; ND, no data available.

of both chemical class and molecular structure. The aminophenanthrenes are more potent mutagens and carcinogens than the azaphenanthrenes. Furthermore, the activity of each isomer is not the same, but varies with structure. For example, 4-aminophenanthrene is not mutagenic, 1-aminophenanthrene is moderately mutagenic, and the other three isomers are potent mutagens. The fact that different N-PAC classes, as well as different isomers within the class, demonstrate varying biological potencies emphasizes the necessity for high-performance analytical methods and techniques that can be used not only to distinguish between the chemical classes, but also for the identification of structural isomers.

Chemical class separation of complex coal-derived products followed by microbial mutagenicity and mouse skin tumorgenicity assays, have revealed

that the N-PAC constituents are the primary microbial mutagens in these
materials [71-78]. The N-PAC also contribute to the overall tumorigenic
activity of the same materials, although the PAH class fractions are responsible for the majority of the tumorgenicity [60,72,79]. Furthermore, it has
been observed that the mutagenicity of the N-PAC isolates of coal conversion
products is determined primarily by the amino-PAH constituency [33,71,72,
78-82]. The 2°-PANH and 3°-PANH complex isolates are generally much less
mutagenic than the amino-PAH isolates. Methods for the class separation and
determination of specific compounds found in biologically active N-PAC isolates of coal-derived materials are reviewed in Secs. V and VI.

V. ISOLATION METHODS

Although the nitrogen-containing PAC chemical classes overall occur at significant levels in coal-derived materials, other classes of PAC, such as the
neutral PAH and oxygen-containing PAC, are in equal or greater concentrations. Hence isolation or fractionation methods are generally required for the
separation of the N-PAC from the other PAC components in coal-derived
complex mixtures prior to detailed, compound-specific chemical analysis.
Current methods for isolating the N-PAC employ liquid-extraction or liquid-solid
chromatography, or a combination of both separation techniques. In general,
many of the chromatographic methods currently used originated from the
analysis of complex petroleum and tobacco materials and have been adapted
for coal conversion products. In this section the application of liquid-liquid
and liquid-solid chromatographic techniques to the separation of homogeneous
N-PAC isolates will be discussed and examples of contemporary isolation procedures will be presented. Furthermore, chromatographic methods for isolating the various N-PAC classes, namely the aza-PAH (both 2°-PANH and
3°-PANH), amino-PAH, and cyano-PAH, will be reviewed.

A. Liquid Extraction Chromatography

Separations achieved by liquid extraction chromatography are based on the differential solubility of sample components in liquid solvents of different polari-
ties or chemical properties. There are two major divisions of liquid extraction
separations that have been widely used for either the isolation of chemical
classes or the gross characterization of coal-derived materials: solvent extraction and partition chromatography.

 Solvent extraction techniques have been widely used for the classification
of coal-derived products according to their solubility in organic solvents.
Solvents of increasing polarity or specific chemical property are used to sequentially extract solutes of similar polarity or chemical property from the
sample while leaving the insoluble or precipitated constituents. Several solvent extraction procedures for coal or coal oils have been published [83-90].
The most widely used liquid extraction method classifies constituents according to the following scheme: n-pentane or n-hexane soluble (oils); benzene
or toluene soluble/n-pentane or n-hexane insoluble (asphaltenes); pyridine
soluble/benzene or toluene insoluble (preasphaltenes); and pyridine insolu-
bles (residues) [88-90]. In one study [91], SRC process-derived distillable
liquids and nondistillable vacuum bottles were first separated into solvent
fractions by this method and then class fractionated by adsorption chromatography. Results indicated that the oils and asphaltenes of these materials
were composed of the same gross chemical classes: PAH, N-PAC, and hy-

droxyl-PAC. The main differences in the solvent fractions of these process materials were in the relative concentrations of these compound classes. It was found that the oils were rich in PAH while the asphaltenes had higher levels of N-PAC and hydroxyl-PAC than the oils. Solvent extraction techniques do not usually produce homogeneous N-PAC isolates and thus are not generally recommended for the isolation of an N-PAC class fraction.

Partition chromatography has been by far the most frequently used form of liquid extraction chromatography for isolating the N-PAC from coal products. This technique is based on the partitioning of sample solutes between two immiscible liquid phases. Usually, one phase is an organic solvent while the other liquid phase is aqueous. The pH of the aqueous phase can be sequentially adjusted so that compounds of different acidity, basicity, and/or chemical functionality partition into either the organic or aqueous phases, where they can be recovered in unique fractions. Hence partition chromatography is also frequently referred to as acid-base extraction chromatography and the extracted solutes are generally classified as acids, bases, or neutrals. The N-PAC are generally recovered in the basic and neutral isolates.

A generalized acid-base extraction scheme is depicted in Fig. 3. Organic solvents that are typically used to solvate the sample and in which the acid-base extractions are performed include methylene chloride, diethyl ether, isooctane, and benzene. These solvents are immiscible with the aqueous HCl and NaOH phases normally used to control the pH of the aqueous phase and extract the acidic and basic components from the crude sample material. These phase separations can be carried out in separatory funnels on the "macro scale" or on a "micro scale" in vials or reaction test tubes. After the basic components are extracted with aqueous HCl, the pH is readjusted to >10 with a NaOH solution, which deprotonates the nitrogen constituents, and the N-PAC are back-extracted into a fresh organic phase as uncharged neutral compounds with one of the previously mentioned organic solvents. Repetitive extractions are normally done to ensure efficient recoveries. Once the N-PAC are partitioned into an organic solvent, the sample volume can be reduced for secondary analytical steps.

Ho, Guerin, Rubin, and other coworkers [76,92-94] have described acid-base extraction procedures used as the initial isolation step in multiple-step chromatographic methods aimed at isolating chemically homogeneous N-PAC class fractions. Acid-base-neutral fractions of coal liquids were pro-

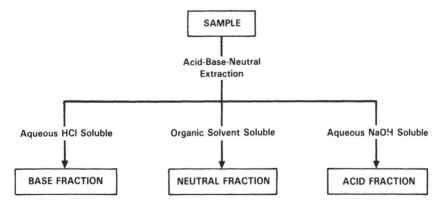

Figure 3. Generalized partition chromatography scheme for isolation of acid-base-neutral fractions of coal-derived materials.

duced by dissolving the crude materials in diethyl ether, extracting the
acids with 1 M NaOH, and then extracting the basic components with 1 M HCl
(see Fig. 3). This extraction procedure was adapted by Rubin [92] from a
method used to fractionate cigarette smoke condensate originally developed
by Swain et al. [95]. The more basic N-PAC, such as 3°-PANH and amino-
PAH, are extracted in the basic fraction, while the neutral aza-PAH, such
as the indoles and carbazoles, remain in the neutral fraction [94]. Additional
steps used to isolate the various N-PAC classes will be covered in other parts
of this section.

Another solvent fractionation method used for the separation of crude
coal liquefaction products has been described by Petersen et al. [96] and
applied by Wilson and coworkers [97,98] for the assessment of hydrotreated
SRC liquefaction materials. This version of liquid partition chromatography was carried out by dissolving the coal-derived liquid in isooctane
and then sequentially partitioning with 1 N HCl, 1 N NaOH, and dimethyl
sulfoxide (DMSO). The scheme is shown in Fig. 4 with the resulting fractions
designated as isooctane-insoluble tar, basic, acid-induced tar, acidic, base-

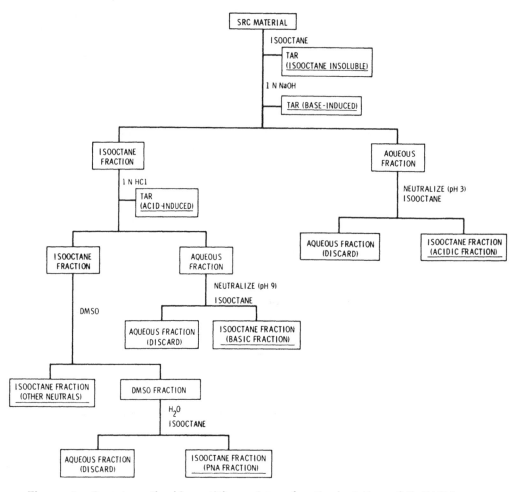

Figure 4. Isooctane liquid partition scheme for the isolation of N-PAC bases.

induced tar, neutral, and PAH. The N-PAC classes are concentrated primarily in the basic and basic-induced tar fraction.

Haugen [33,99] isolated the N-PAC constituency of gasification condensates by acid-base partition chromatography prior to subsequent class fractionation of amino-PAH and 3°-PANH by ion exchange HPLC. A solution of the nonvolatile condensate material in methylene chloride was extracted three times with 1 N HCl. The aqueous phases were combined and washed with diethyl ether. The pH was then adjusted to 11-12 by addition of NaOH and the N-PAC bases were extracted from the aqueous phase with three portions of methylene chloride. The back-extracted organic phase was dried over anhydrous Na_2SO_4 to remove residual water and the solvent was evaporated to obtain a purified basic N-PAC fraction.

Novotny and coworkers also used acid-base partition chromatography to isolate the N-PAC from both coal tar and SRC liquefaction materials [100-102]. Their scheme is shown in Fig. 5. The N-PAC or bases are isolated from the other PAC classes by extracting the methylene chloride-soluble portion of

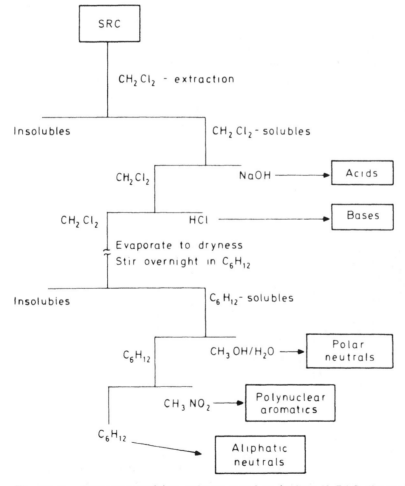

Figure 5. Solvent partition scheme used to isolate N-PAC (bases) from coal tar and SRC liquefaction materials. (From Ref. 101.)

290   Later

the crude material with 0.2 M HCl. The pH of the aqueous HCl extract is
then readjusted to pH 14 with base and the N-PAC are back-extracted with
methylene chloride.

Problems generally associated with acid-base extractions of complex coal-
derived materials are the formations of emulsions and tars at the liquid-liquid
interface and the low recoveries for the more hydrophobic three- and four-
ring N-PAC. Refinements in the acid-base extraction technique that enable
higher recoveries of the N-PAC in the basic fraction have been suggested by
Boparai [103]. The effect of HCl concentration on extraction efficiency was
investigated for representative N-PAC of the 3°-PANH and amino-PAH
classes. Figure 6 summarizes their results. Only 1-aminonaphthalene was extracted in high yield using 1 N HCl, with the recovery of the other N-PAC
bases shown being much less. The portion of the N-PAC standards not recovered in the base fraction were found in the neutral fraction. Increasing
the normality of the HCl did not enhance recoveries significantly. However,
when the N-PAC were extracted with a binary 3.4 N HCl/43% methanol phase,
the solubility of the hydrophobic bases in the aqueous phase was markedly
improved and higher recoveries for the three-, four-, and five-ring 3°-PANH
and amino-PAH in the basic fraction were obtained. Boparai [103] also pointed out that when methanol is used as a cosolvent, polar compounds, such as
the phenols, may be transferred to the aqueous phase simply by solvent partitioning. Therefore, it was recommended that samples be extracted first
with NaOH to remove the acidic compounds prior to isolation of the N-PAC
base fraction. Neutral PAH, such as methylnaphthalene, biphenyl, pyrene,
and diphenylanthracene, were not extracted by the methanol/HCl phase. The
results of this study strongly emphasize the necessity for verifying the recoveries of representative compounds in proposed extraction procedures and

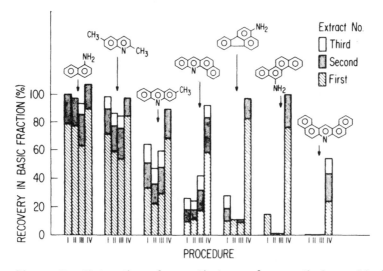

Figure 6. Extraction of aromatic bases from methylene chloride solutions
with aqueous phases containing different concentrations of HCl and methanol.
Mean values for duplicate experiments are shown. For procedures I, II, III,
and IV, the organic solutions were extracted with 1 N HCl, 3 N HCl, 6 N HCl,
or 3.4 N HCl in 43% methanol (v/v), respectively. Where results for the
second and third extracts do not appear, the values were too small to be included given the scale of the figure. (From Ref. 103.)

using cosolvents to enhance yields of N-PAC bases in liquid extraction chromatography.

A hybrid method of solvent extraction and acid-base partition chromatography, wherein the N-PAC components of coal are precipitated from coal conversion products with HCl, was developed and used by Sternberg [104] and White [14]. In this method, HCl gas is bubbled through a benzene solution of the crude material. The N-PAC bases form insoluble HCl salts that precipitate out of the benzene solvent. The precipitate is then centrifuged and separated from the mother liquor, slurred in an organic solvent, such as benzene or ether, and gaseous ammonia or aqueous NaOH is used to regenerate the original N-PAC bases, which are then soluble in the organic solvent. The resulting homogeneous N-PAC isolate can then be analyzed for compound-specific identifications. Finseth [105] reported that this HCl precipitation method favored the quantitative recovery of highly condensed aromatic bases but did not effectively concentrate the more highly hydrogenated species present in some coal-derived liquids. Furthermore, higher levels of residual acidic or phenolic compounds were detected in the N-PAC fractions isolated by the HCl precipitation method than those obtained using a cation exchange chromatographic method [105]. Nevertheless, the HCl precipitation method has been effectively used to isolate and characterize 2°-PANH, 3°-PANH, and amino-PAH in coal-derived liquids [14,106,107].

B. Liquid-Solid Chromatography

Separations based on the interaction of a solute between a liquid and solid phase have been widely used to separate N-PAC from other PAC classes in coal products. Examples of this technique include (but are not limited to) adsorption, gel permeation, sieving, and ion exchange chromatography. These separations are most commonly performed in open columns at ambient or low pressure; however, high-performance liquid chromatographic (HPLC) methods have also been used. A variety of materials are used as solid packings, including alumina, silicic acid, Sephadex LH-20, Bio-Beads, Florisil, XAD and Amberlyst resins, silica gel, cellulose acetate, and other gels and adsorbants. Solutes are separated based on (1) physiochemical interactions that occur between the compound and the solid surfaces of the packing, and (2) the solutes solubility in liquid solvents normally used as mobile phases. Inorganic metals and salts as well as chemically bonded functional groups such as amino and cyano can also be used as surface modifiers to enhance the selectivity of liquid-solid separations.

Compared to the neutral PAH, the increased electron density incorporated into N-PAC by the presence of the nitrogen heteroatom renders this class of compounds particularly susceptible to separations based on liquid-solid chromatography. This technique has been successfully used not only to isolate N-PAC from other PAC class, but also to separate the various classes of N-PAC.

The major advantages of liquid-solid chromatography over liquid-liquid chromatography include: (1) reproducible and precise class fractionations can be achieved; (2) formation of emulsions and tars encountered in liquid partition chromatography are not a problem in liquid-solid chromatography; (3) preparative-scale separations can be performed; and (4) overall, quantitative recoveries of PAC from most solid packings are usually obtained. In this section the major liquid-solid chromatographic methods used for the isolation of N-PAC from coal-derived products will be reviewed.

1. Gel Permeation Chromatography

A popular technique for separating N-PAC from other PAC classes has been liquid-solid chromatography on Sephadex LH-20 and Bio-Bead SX series. Sephadex LH-20, a lipophilic dextran-based material, and Bio-Beads, a styrene-divinylbenzene polymer, can be used in different chromatographic modes depending on the eluent solvent employed [108-110]. For example, if tetrahydrofuran is used as an eluent with LH-20, molecular sieving and hydrogen bonding retention mechanism prevail (sieving: nonpolar compounds are separated by size with the largest compounds eluting first; hydrogen bonding: polar compounds such as the N-PAC are eluted in reverse order of hydrogen bonding strengths). But if an alcohol like isopropanol is chosen as the eluent, the primary retention mechanism becomes π bonding or adsorption (the greater the aromatic π system, the greater the retention; compounds generally elute in order of number of condensed aromatic rings). While operating in the adsorption chromatographic mode, the presence of a nitrogen heteroatom in an aromatic compound increases its retention as compared to an analogous nonnitrogen compound. Various modes of gel permeation chromatography have been used for the enrichment of N-PAC; however, isolation of a homogeneous N-PAC fraction in a single step by this technique has not been demonstrated. The major advantages of this technique are the inertness of the gel materials, the quantitative elution of even polar functional PAC, and the relatively large loading capacities of gel columns [111].

Caution must be exercised when using gel permeation chromatography for the isolation of nitrogen functional PAC from complex mixtures. The polar moiety of the compound is generally subject to adsorption mechanisms regardless of the eluent solvent used. Thus the elution patterns of the N-PAC can be interdispersed with those compounds that elute strictly according to molecular size.

Snook studied the behavior of nitrogen heterocyclic PAC on Bio-Beads [112]. Dutcher, Royer, and coworkers [113-115] devised a separation scheme for coal gasification condensates using Sephadex LH-20 as the first step. Five fractions were collected with a tetrahydrofuran eluent. The retentions of selected standard compounds using this system are shown in Fig. 7. Although a complete separation of the N-PAC from the oxygen-containing PAC was not achieved, a significant separation of the N-PAC and neutral PAH was observed. Other investigators have also used Sephadex LH-20 column chromatography in multiple-step separation schemes that result in the isolation of the N-PAC constituents in coal-derived products [73,94,109,110,116].

2. Alumina/Silica Adsorption Chromatography

Open column, ambient pressure, liquid-solid chromatography on alumina and/or silica adsorbents has been demonstrated as an effective technique for the separation of N-PAC from other classes of PAC. The application of adsorption chromatography to fossil fuel separations experienced its genesis in the petroleum industry [117-119] and has been successfully adapted for the class separation of N-PAC in coal-derived materials. Alumina and silica are available in various grades, mesh sizes, pH levels, and activity (i.e., water content). Optimum class separation is achieved by sequential elution of the adsorbed sample material with organic solvents of increasing polarity.

A possible drawback to using alumina and/or silica adsorbents is that polar compounds such as N-PAC may be irreversibly adsorbed or undergo catalytic degradation on the adsorbent surface. Recovery results for several

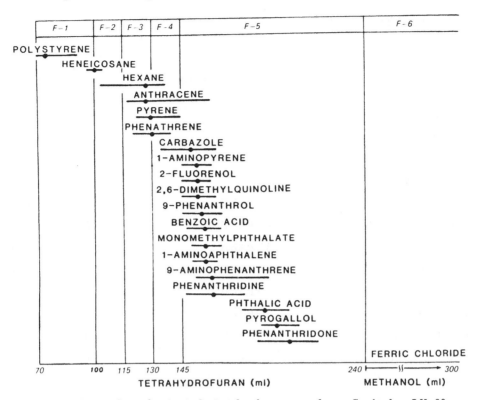

Figure 7. Retention of selected standard compounds on Sephadex LH-20. The bar indicates the width of the peak and the dot indicates the maximum. (From Ref. 113.)

two-, three-, and four-ring standard N-PAC on silicic acid and neutral alumina are shown in Table 11. Inspection of these data reveals that the aza-PAH recoveries are generally above 90% and are comparable to analogous nonfunctional PAH on both adsorbents. Overall yields for the amino-PAH are lower than the aza-PAH recoveries, with the best results obtained on alumina rather than silica. These factors must be carefully considered when quantitative assessments are required for N-PAC isolated by adsorption chromatography.

Farcasiu [120] reported a separation scheme developed for solvent-refined coal based on chromatographic fractionation by sequential elution with specific solvents on silica gel columns (SESC). The nonbasic and basic nitrogen heterocycles were concentrated in the third and fifth fractions, respectively, of a nine-fraction scheme. A similar gradient elution chromatographic method using alumina adsorbent was described by Callen and coworkers [121,122] for the separation of coal liquefaction materials into 12-14 fractions. These fractions were subsequently analyzed by spectrometric methods to determine the nature of functional groups present in each fraction. The nitrogen functional PAC were separated from the PAH and generally eluted in fractions beyond the fourth fraction.

An alumina adsorption chromatographic method initially developed by Schiller and Mathiason [123,124] and modified versions described by others [20,38,125,126] results in the rapid isolation of homogeneous N-PAC frac-

Table 11. Recoveries of Selected PAC from Neutral Alumina and Silicic Acid

Compound	Chemical class	Recoveries (%) Al$_2$O$_3$	SiO$_2$
n-C$_{18}$	Aliphatic hydrocarbon	~100	—
Phenanthrene	PAH	97.6[a]	—
Triphenylene	PAH	94.0[b]	—
Benzo[a]pyrene	PAH	96.8	—
Carbazole	2°-PANH	87.0[a]	87.8[b]
Benzo[a]carbazole	2°-PANH	84.2[b]	62.5
Quinoline	3°-PANH	99.8	81.1[a]
Acridine	3°-PANH	99.2[a]	90.4[a]
Phenanthridine	3°-PANH	94.4[b]	97.7[a]
1-Azapyrene	3°-PANH	92.8[b]	94.6[a]
Benz[c]acridine	3°-PANH	97.8[b]	93.5[a]
1-Aminonaphthalene	Amino-PAH	83.7[a]	73.8[a]
9-Aminophenanthrene	Amino-PAH	56.5[b]	38.9[a]
3-Aminofluoranthene	Amino-PAH	64.1[a]	64.0[a]
6-Aminochrysene	Amino-PAH	73.7[a]	57.5[b]
1-Acenaphthenol	Hydroxy-PAH	85.0	—
8-Hydroxybenzo[a]pyrene	HPAH	93.6	—

[a]Average of three or more determinations.
[b]Average of two determinations.
Source: Ref. 125.

tions from a wide range of coal conversion materials in a single chromatographic step. Figure 8 is a schematic diagram of one version of this method [20, 125]. Either basic or neutral alumina can be used for this separation with the N-PAC eluted following the neutral PAH. Popl et al. [32] discussed the mechanisms and advantages of using pH-controlled alumina for separating aromatic hydrocarbons and nitrogen heterocycles. Table 12 and Figs. 9 and 10 demonstrate the selectivity of this alumina adsorption chromatographic system for separating the N-PAC components of complex mixtures into discrete, class-specific fraction. Klemm [127] described adsorption mechanisms and proposed a general set of rules for predicting the adsorbabilities of N-PAC on alumina. The reproducibility of this separation technique was addressed by Later and Lee [125], who found that excellent reproducibility could be achieved if the activity (i.e., water content) of the alumina adsorbent was optimized for the separation and closely monitored to ensure that the adsor-activity remained within optimal parameters. The detailed chemical analyses

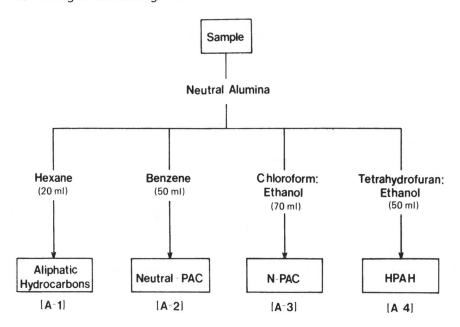

Figure 8. Schematic diagram for the chemical class separation of PAC on neutral alumina. (From Ref. 125.)

Table 12. Fractionation of Selected Standard PAC on Neutral Alumina

	Recovery (%)		
Compound	A-2	A-3	A-4
Naphthalene	44.8	0	0
Phenanthrene	103.8	0	0
Triphenylene	91.9	0	0
Dibenzofuran	93.9	0	0
Indole	0	81.4	0
Dimethylquinoline	0	109.1	0
Carbazole	0	83.6	0
Acridine	0	95.6	0
1-Azapyrene	0	85.6	0
Benz[c]acridine	0	95.6	0
Benzo[a]carbazole	0	68.4	0
Dibenz[a,h]acridine	0	40.7	0
Phenol	0	0	21.9
Naphthol	0	0	42.7

Source: Ref. 131.

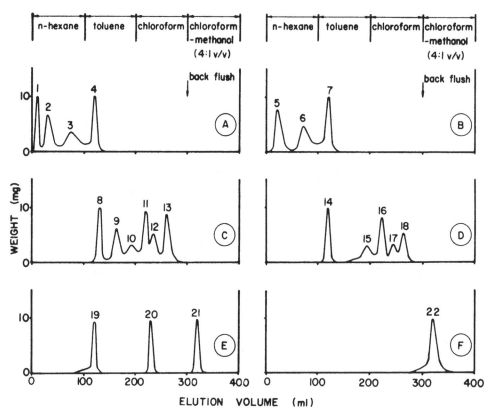

Figure 9. Composite elution curves of model compounds on basic alumina: (A) saturated and hydroaromatic hydrocarbons; (B) polycyclic aromatic hydrocarbons; (C) basic nitrogen heterocycles; (D) nonbasic nitrogen heterocycles; (E) oxygen-containing compounds; (F) hydroxyl aromatics. *Note*: Peak numbers and corresponding model compound structures are given in Fig. 10. (From Ref. 38.)

of alumina N-PAC isolates of coal-derived products by analytical instrumentation methods are discussed in Sec. VI.

Alumina and silica adsorption chromatography have also been used to separate classes of N-PAC. Ho [128] found that the unsubstituted amino-PAH were more strongly retained on basic alumina with benzene as an eluent than the N-substituted amino-PAH and aza-PAH (3°-PANH). The separation of these N-PAC classes by their method is shown in Fig. 11. After the first two classes of N-PAC were eluted with benzene, the solvent system was changed to acetone, which immediately eluted the adsorbed amino-PAH. Ho and coworkers have also described the separation of pyrrolic-PAH and neutral PAH by silicic acid adsorption chromatography [94,129].

Figure 12 demonstrates the feasibility of separating the 2°-PANH, 3°-PANH, and amino-PAH classes by gradient elution chromatography on silicic acid [20,71,125,130,131]. Except for minimal overlap of the 3°-PANH with the amino-PAH, a class-specific separation of the various N-PAC is achieved. The aza-PAH that elute prematurely are compounds such as benzo[h]quinoline and benz[c]acridine in which the nitrogen heteroatom electron lone pair

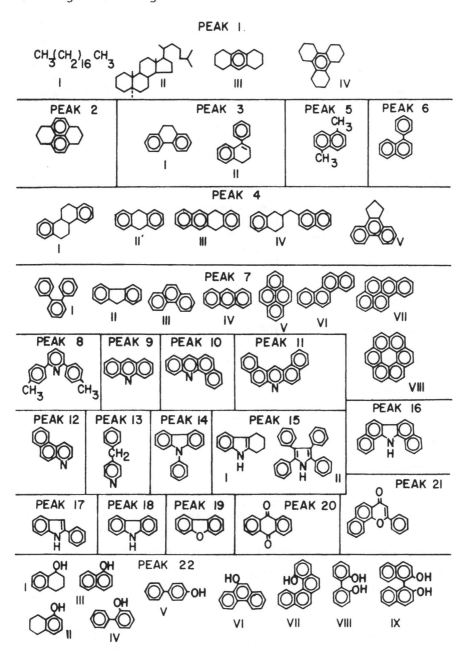

Figure 10. Model compounds eluted from basic alumina column. Key: see Fig. 9. (From Ref. 38.)

is sterically shielded. This structural effect results in a lowered adsorptivity for these compounds and partial elution with the amino-PAH [71,127,130, 131].

The cyano-PAH have also been isolated from the other N-PAC classes by alumina adsorption chromatography. Dubay and Hites used neutral alumina

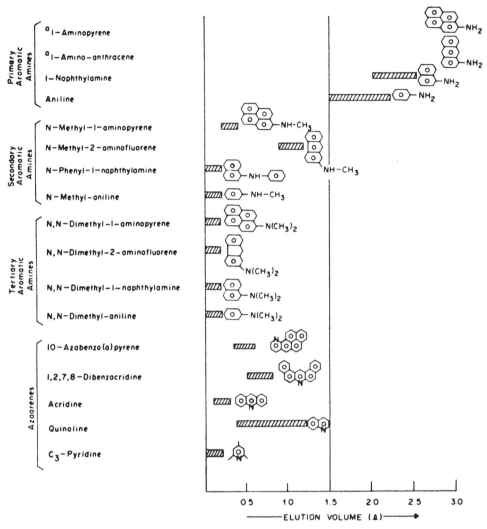

ᵃTHESE COMPOUNDS WERE NOT ELUTED WITH 3 LITERS OF BENZENE

Figure 11. Preparative-scale elution of selected N-PAC from basic alumina using benzene. (From Ref. 128.)

(activity I) column chromatography to partially separate the cyano-PAH from the neutral PAH and aza-PAH [132]. An isolate enriched in cyano-PAH was eluted from the alumina with benzene-hexane (7:3) prior to elution of the aza-PAH with benzene. In another method, the cyano-PAH were isolated from coal gasification condensates by a two-step alumina method [67,133]. In the first step, the activity of the alumina adsorbent was decreased by increasing the water content from about 1 to 2%. This allowed the cyano-PAH class to elute with the neutral PAH in the benzene eluate; the cyano-PAH are least basic of the N-PAC classes and hence have the lowest adsorptivities. The amino-PAH, 2°-PANH, and 3°-PANH remained adsorbed on the column. Next the PAH/cyano-PAH fraction was separated on a second neutral alumina column where the front segment of the column had been doped with

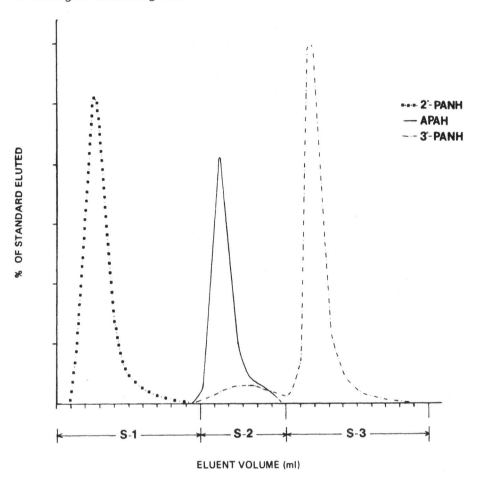

ELUENT VOLUME (ml)

Figure 12. Elution profile of several N-PAC standard compounds on silicic acid. Fractions: S-1, hexane-benzene, 1:1, 50 ml; S-2, benzene 30 ml; S-3, benzene-ether, 1:1, 50 ml. The following compounds were used for each class: (1) 2°-PANH: indole, 3-methylindole, 2,3-dimethylindole, carbazole, benzo[def]carbazole, benzo[a]carbazole, and dibenzo[a,i]carbazole; (2) APAH:1-aminonaphthalene, 2-aminofluorene, 1-aminoanthracene, 2-aminoanthracene, 9-aminophenanthrene, 3-aminofluoranthene, and 1-aminopyrene; (3) 3°-PANH:quinoline, isoquinoline, 2,6-dimethylquinoline, 4-azafluorene, benzo[h]quinoline, aciridine, phenthridine, benzo[f]quinoline, 3-methyl-benzo[f]quinoline, 2-methylacridine, 2-azafluoranthene, 1-azapyrene, benz[c]acridine, benz[a]acridine, and dibenz[a,h]acridine. (From Ref. 131.)

picric acid, a charge-transfer medium. The PAH, which do not form strong picrates, were eluted with benzene. The polarizable nitrile functional group of the cyano-PAH readily forms picrates by complexing with this charge transfer agent, and thus the cyano-PAH remained adsorbed on the column. Finally, the cyano-PAH picrate complexes where disrupted by the use of a more polar chloroform solvent system and the cyano-PAH were eluted in a class-specific fraction. Detailed compound-specific analyses of the cyano-PAH was achieved by capillary column gas chromatography and an example is shown later in this chapter (see Fig. 30).

3. Ion Exchange Chromatography

The application of ion exchange chromatography for the removal of neutral and basic N-PAC from complex fossil fuel materials was also initiated in the petroleum industry [134,135] and has been adapted for use with coal products [106,107]. Coordination or complexation by use of cation exchange resins or other agents such as metal salts have been used for the separation of the N-PAC from other acidic and neutral PAC. Formation of complexes between the solutes and functional sites of the solid phase by ionic and/or coordination mechanisms results in the selective retention of the nitrogen-functional compounds on the solid phase. After eluting the nonnitrogen functional compounds with an appropriate nonpolar solvent, a polar mobile phase such as dichloroethane, dichloromethane, methanol, or chloroform is used to disrupt the complex and elute the N-PAC. Amberlyst and XAD resins are the predominant ion exchange solids used for the basic nitrogen compounds, while ferric chloride adsorbed on a kaolin or clay substrate complexes the neutral or nonbasic nitrogen compounds [106,107,134,135].

Schweighardt and coworkers [106,107] separated several pentane-soluble solvent-refined coal liquids on a combination column of cation/anion exchange resins, ferric chloride-impregnated clay, and silica gel adsorbent. This scheme was used to obtain basic and neutral nitrogen isolates and is known as the SARA method (saturates-aromatics-resins-asphaltenes). Monocyclic and two- to six-ring aza-PAH were identified in these isolates as well as mono- and dicyclic amino-PAH. Klimisch and Beiss described a method for the separation of the 2°-PANH (carbazole type) and 3°-PANH (acridine type) by chromatography on strong acidic and strong basic ion exchangers of the Sephadex gels [136,137]. Aza-PAH compounds sorbed on the ion exchangers were readily eluted by addition of acids and bases to the solvents. Although their method was demonstrated with a basic isolate of cigarette smoke condensate, this method could probably be adapted to coal products.

Both the usual XAD resins and a novel sorbent were used by Conditt et al. [138] to selectively complex the N-PAC from shale and coal conversion effluents. Selective adsorption of nitrogen compounds was achieved using a lanthanide metal chelating agent (i.e., a europium complex). Compounds like the N-PAC which have electron-donating functional groups form complexes with the europium-based chelating agent. Both solvent and thermal desorption of the retained compounds was feasible.

4. High-Performance Liquid Chromatography

The liquid-solid chromatographic techniques discussed in the three preceding sections dealt mainly with separations at ambient or low pressure. Gross separations of crude materials according to chemical class using the same techniques but at higher pressures have also been reported using HPLC. Examples of these methods will be given in this subsection.

Matsunaga [139] studied the retention of standard PAC on several commercially available packings, including chemically bonded phases (NO_2, NH_2, CN, and sulfonic acid), silica, alumina, and porous polystyrene gels. He reported capacity factors and compared the selectivity of these phases for the class-type separation of standard PAC and coal tars. The chromatographic separation of compound classes on a plain silica adsorption column by HPLC with aprotic, dipolar solvents was investigated by Chmielowiec [140]. Aprotic dipolar solvents such as DMSO strongly moderate the adsorbent solid surface and selectively complex with various PAC functional groups to determine their retention. Evaluation of moderators was based on their ability to discriminate among chemically distinct compound classes with specific functional groups. Capacity factors, ln k', for several PAC classes are presented in Fig. 13. This chart illustrates the class grouping tendencies, system-resolving ability, and possible overlaps between compound classes. In general, resolution of PAH, 3°-PANH, and 2°-PAH was achieved. The amino-PAH elute partially with the pyrrolic-PAH. This HPLC method was also used to separate coal-derived naphtha into chemically distinct classes [140].

An HPLC method that has been widely used for the class-specific separation of PAH and N-PAC in alternative fuels employs a normal-phase NH_2 column with multiple-solvent mobile-phase systems [73,116,141-143]. Wise and coworkers [141,142] used a preparative-scale aminosilane column and modified the mobile-phase composition from 100% hexane to 100% methylene chloride to effect a separation between nonpolar PAH and N-PAC. Similarly, Toste and coworkers [73,116] used an aminosilane column HPLC separation scheme for the isolation of chemically distinct N-PAC fractions from coal liquefaction materials. They employed a ternary solvent system of hexane/methylene chloride/isopropanol to elute the samples. Three elution regions were defined; the N-PAC eluted efficiently in the moderately polar region. The retentions of several standard PAC on this NH_2-HPLC system are shown in Fig. 14. This HPLC method has also been recommended as a standard analysis method to be performed along with CHNOS elemental analysis, neutron activation analysis, and simulated distillation analysis for the comparative characterization of coal-derived process materials in terms of mass balance and compound class distribution [144].

As with the low-pressure techniques, HPLC has also been used to separate N-PAC classes. Haugen and coworkers [33,145] described a cation exchange HPLC method for the separation of two classes of N-PAC: 3°-PANH and amino-PAH. These N-PAC classes were readily resolved except for unusual cases in which structural properties were responsible for similar basicities. The HPLC system consisted of a silica-based cation exchange column with a 40% acetonitrile/10 mM aqueous sodium citrate mobile phase. The elution order of those N-PAC investigated appeared to be primarily a function of their pK_a values. Figure 15 shows an HPLC separation of several standard amino-PAH and 3°-PANH by this method. This HPLC scheme was used to separate the basic fractions from coal gasification and liquefaction materials. Several subfractions were collected during the HPLC separation of the basic fraction and high-resolution gas chromatographic techniques were applied for the detailed characterization of each subfraction.

C. Multiple-Step Chromatographic Methods

The complete isolation of N-PAC classes from complex matrices, such as coal-derived materials, in a single chromatographic step is not generally feasilble, especially if 3°-PANH, 2°-PANH, amino-PAH, and cyano-PAH class-specific fractions are to be obtained. Hence multiple-step procedures have been designed that use a combination of the various techniques discussed in Secs. V.A and V.B. The primary aim of these multiple-step procedures is to isolate class-specific fractions that are as homogeneous as possible and which can be subsequently characterized in detail by higher-performance analytical techniques.

Separation of the N-PAC according to the chemical functionality of the nitrogen heteroatom is advantageous for several reasons. First, different

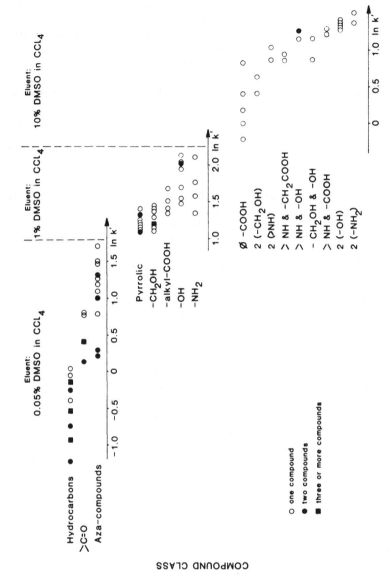

Figure 13. Chart of retentions of model compounds on a silica HPLC column. (From Ref. 140.)

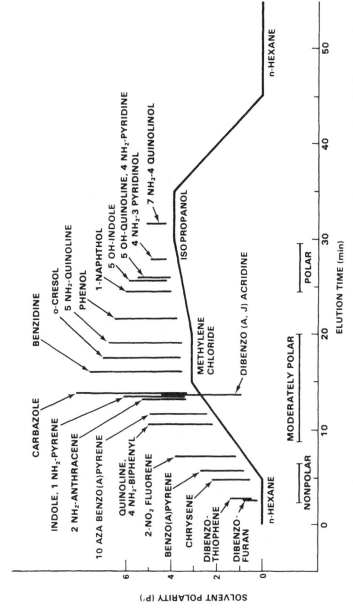

Figure 14. HPLC of standards. Each standard (1 mg/ml in methanol or chloroform) was chromatographed on a μBondapak NH₂ column using the three-solvent mobile-phase system depicted above at a flow rate of 5 ml/min. The organic species in the synfuel samples studied typically eluted in three distinct regions of polarity: nonpolar, moderately polar, and polar. (From Ref. 73.)

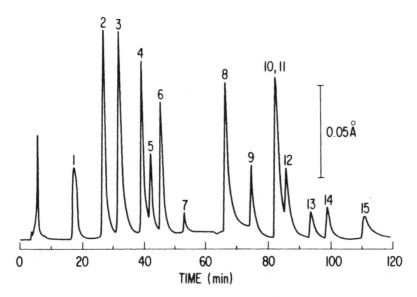

Figure 15. Separation of a standard mixture of amino-PAH and aza-PAH by HPLC. The standards were as follows: 1, 1-aminopyrene; 2, 1-aminoanthracene; 3, 2-aminoanthracene; 4, 4-aminobiphenyl; 5, 2-aminonaphthalene; 6, 2-aminofluorene; 7, aniline; 8, benzo[f]quinoline; 9, quinoline; 10, acridine; 11, isoquinoline; 12, pyridine; 13, 3-methylpyridine; 14, 2-methylpyridine; 15, 2,4-dimethylpyridine. (From Ref. 33.)

classes of N-PAC in coal products are typically present at different concentration levels. Thus detection interferences and obscurring of certain classes of N-PAC can be minimized. Next, as discussed in Sec. III, many compounds from the four N-PAC classes have overlapping molecular weights, making differentiation of molecular structure and functionality by conventional analytical methods, such as mass spectrometry, very difficult. Furthermore, as discussed in Sec. IV, the biological activities of the various N-PAC classes are different depending on the functionality of the nitrogen heteroatom. Finally, other reasons for class separation of the N-PAC include understanding the role and effect of these compounds on coal product stability, processing technologies, end-product usefulness, coal conversion chemistry, coal genesis and origin, and so on. For these reasons, multiple-step chromatographic methods have been developed for the isolation of the different N-PAC classes.

In the first example shown in Figs. 16 and 17, a combination of acid-base extraction, Sephadex LH-20 gel chromatography, and alumina and silica adsorption chromatography is used to isolate the N-PAC classes. Initial separation by acid-base extraction yields two fractions that contain N-PAC; the basic N-PAC (3°-PANH and amino-PAH) are isolated in the ether-soluble base fraction, while the neutral or nonbasic N-PAC (pyrrolic-PAH) are obtained in the neutral fraction [76,92-94]. Subsequent separation of the neutral fraction, first on Sephadex LH-20 and then by silicic acid adsorption chromatography, results in the isolation of a neutral N-PAH fraction (Fig. 16; fraction II) and a polar fraction [94]. Chemical characterization of fraction II revealed that the primary components isolated from synthetic crude oils were the pyrrolic-PAH: indoles, phenylpyrroles, carbazoles, phenylin-

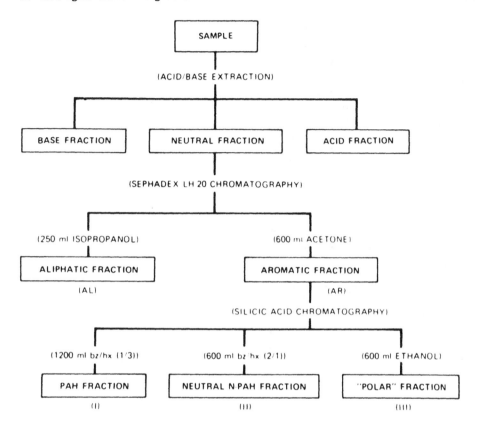

Figure 16. Isolation scheme for neutral N-PAH (2°-PANH) from coal-derived crude oils. (From Ref. 94.)

doles, benzocarbazoles and naphthylindoles, and so on [146,147]. Fraction III contained mostly hydroxy-PAH; however, trace levels of 3°-PANH and amino-PAH (aminophenylnaphthalenes, etc.) were also detected in the polar fraction of the neutral extract [129,146,147]. The presence of the N-PAC in the polar fraction is presumably due to incomplete partitioning of these compounds in the initial acid-base extraction step. Figure 17 outlines the secondary separation of the ether-soluble base (ESB) fraction, which contained mostly the 3°-PANH and amino-PAH. Purification of these N-PAC was achieved in two steps by chromatography on basic alumina followed by Sephadex LH-20 [94,128,129]. The elution parameters of the basic alumina were determined as discussed and shown in Fig. 11. The benzene fraction was composed mainly of N-substituted amino-PAH and strongly basic 3°-PANH, such as pyridines, quinolines, acridines, benzacridines, and alkylated homologs of these compounds [40,128,129]. The ethanol fraction was further purified on Sephadex LH-20 (see Fig. 17) to give an isopropanol subfraction, which contained partially hydrogenated one- and two-ring aza-PAH, and an acetone subfraction that was composed primarily of multiring 3°-PANH and amino-PAH ranging from one to four aromatic rings [40,129,147]. Ho et al. [129] added a third step to this scheme and separated the 3°-PANH and amino-PAH by basic alumina adsorption chromatography. Detailed compound-specific analyses were performed on each of these subfractions by

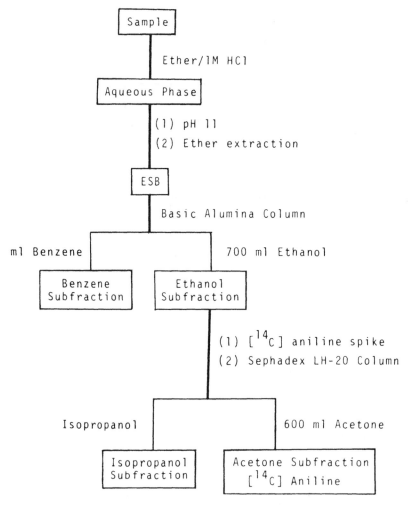

Figure 17. Isolation of alkaline N-PAC. (From Ref. 129.)

gas chromatographic, mass spectrometric, and spectrophotometric (UV and IR) techniques.

Haugen and coworkers [33,99] also used an initial acid-base extraction step in conjunction with ion exchange HPLC of the basic fraction to separate the 3°-PANH and amino-PAH from coal gasification and liquefaction process materials. They, however, incorporated the modifications in the acid-base extraction step suggested by Boparai [103] to ensure quantitative recoveries of the N-PAC in the basic fraction. (See Figs. 6 and 15 and related discussions.)

Multiple-step separation methods that employ Sephadex LH-20 gel permeation chromatography as an initial step have also been reported [73,113-116]. Toste and coworkers [73,116] performed initial fractionations of solvent-refined coal liquids by LH-20 chromatography, then subfractionated each of three eluates from the gel column using an aminosilane column and a ternary solvent mobile phase HPLC system. The N-PAC were eluted in the moderately polar region (see Fig. 14). Another combination scheme in which Sephadex

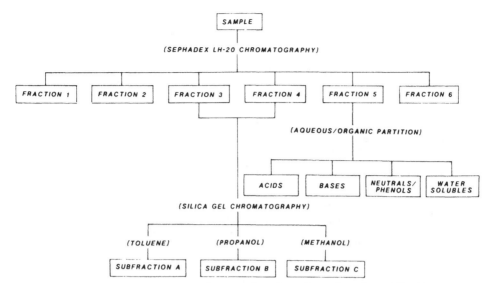

Figure 18. Separation scheme for the fractionation of coal gasification condensate tars. (From Ref. 113.)

LH-20 chromatography was used as the initial step is shown in Fig. 18 and was reported by Dutcher and coworkers [113-115]. The quinolines/benzoquinolines and other 3°-PANH were isolated in the basic isolate of the fifth LH-20 gel fraction while the pyrrolic-PAH were fractionated into the neutral/phenol isolate by acid-base extraction chromatography.

Multiple-step column chromatographic schemes have been developed and reported for the chemical class separation of not only the N-PAC as a unique isolate, but also individual, homogeneous isolates of amino-PAH, 2°-PANH, and 3°-PANH. An example of such a separation scheme is shown in Fig. 19 [20,71,125,130,148]. The N-PAC were initially isolated from the other PAC by adsorption chromatography on neutral alumina. Next, the pyrrolic-PAH were isolated from the other N-PAC and a partial separation of the amino-PAH and 3°-PANH was achieved by silicic acid adsorption chromatography. In the third step, the amino-PAH-rich isolate was derivatized with pentafluoropropionic anhydride (this procedure results in the quantitative formation of the fluoroalkylamide derivatives of only the amino-PAH) and separated by gel permeation chromatography. The derivatized amino-PAH and remaining 3°-PANH were completely separated in this step [71,125,148]. The derivatized amino-PAH isolate was then hydrolyzed to regenerate the free amino compounds [148]. The main advantage of this separation procedure is that each of the three N-PAC classes shown are isolated in three homogeneous fractions that can be chemically characterized in detail with minimal isomeric interferences from other closely related N-PAC classes. Furthermore, the relative chemical class distribution of the N-PAC classes in coal products can easily be obtained by gravimetric measurements. The detailed analysis of the amino-PAH and aza-PAH isolates was performed by capillary column gas chromatography [71,125,148]; examples are shown in Sec. VI.

In summary, the isolation of nitrogen-functional PAC has received considerable attention. Methods are currently available that allow class-specific separations of N-PAC classes from complex matrices such as coal liquefac-

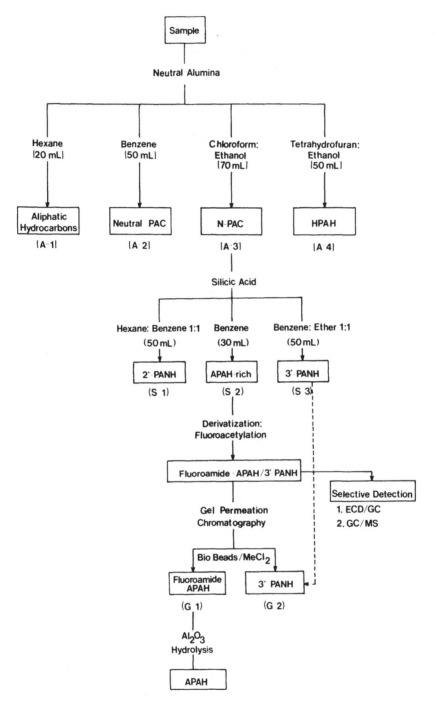

Figure 19. Schematic diagram of the procedure used for the isolation of the N-PAC classes from coal-derived materials. (From Ref. 148.)

tion, gasification, and carbonization materials. An overview of the chromatographic techniques and procedures required for such separations has been presented in this section. In the next section, analytical techniques that have been used to provide detailed, compound-specific characterizations of the isolates obtained by these separation methods are discussed.

VI. ANALYTICAL METHODS

The isolation methods discussed in Sec. V provide N-PAC isolates from coal-derived materials that are very complex and that may be composed of hundreds of compounds. Hence compound-specific identifications require further chemical analysis. In this section analytical methods used for the identification of N-PAC in complex mixture will be discussed. Particular emphasis will be given to the analysis of N-PAC by mass spectrometry (MS) and capillary column gas chromatography (GC) since these methods have been widely used and generally provide the analyst with the most useful and detailed qualitative and quantitative information.

A. Mass Spectrometry

Mass spectrometry (MS) has been widely used for the analysis of PAC in coal and coal-derived products for many years. Josefsson [149] reviewed mass spectrometry methods for PAC and White [150] covered both early and contemporary MS techniques for the analysis of PAC in coal-derived materials in the first volume of this handbook. Although many of the techniques discussed in these two chapters are directly applicable to N-PAC analysis, both authors dealt primarily with the nonfunctional PAH and only briefly mentioned MS analysis of N-PAC. In this part of the analytical methods section, MS methods that have been used for the analysis of coal-derived N-PAC will be discussed briefly with emphasis on methods developed for the separation of the N-PAC classes.

Direct inlet probe MS of N-PAC isolates can be routinely used for qualitative and quantitative mass-specific analysis. Although isomeric N-PAC cannot be differentiated by this method, it is useful for obtaining more general information such as mass distribution, total amounts of structural isomers, or relative chemical composition. This MS technique is particularly useful for the analysis of high molecular weight N-PAC that might not be amenable to gas chromatographic analysis. Since nitrogen-containing compounds adhere to the nitrogen rule [151], most aromatic compounds with a single or odd number of nitrogen heteroatom(s) have odd molecular masses. Hence probe inlet mass spectra of coal-derived N-PAC isolates are composed predominantly of peaks at odd masses.

Both high- and low-resolution probe inlet MS have been used for the analysis of N-PAC in coal and its conversion products [149,150]. The main advantage of high-resolution MS is, of course, that accurate mass data are acquired from which molecular formulas and Z numbers can be assigned. Table 13 shows excerpts from two published tables [152,153] and demonstrates the utility of high-resolution MS for the generic determination of mono- and dinitrogen PAC in coal products. The Z number (C_nH_{2n-z}) is a measure of hydrogen deficiency and/or aromaticity and is often used for the gross characterization of the principal functional components of complex mixtures. Z numbers for aromatics are sometimes listed as negative values to emphasis their hydrogen deficiency. Furthermore, N-PAC Z numbers are frequently

Table 13. (a) N-PAC Identified in High-Volatile Bituminous Coal by High-Resolution MS

Z	First member of homologous series	Carbon numbers experimentally determined	Possible structures
1	2	3	4

Nitrogen compounds: CHN

Z	First member of homologous series	Carbon numbers experimentally determined	Possible structures
-3	C_4H_5N	7-8, 11	Pyrrole
-5	C_5H_5N	7-15	Pyridine, aniline
-7	C_7H_7N	7-9, 14	Azaindane, tetrahydroquinoline
-9	C_8H_7N	8-15	Indole, dihydroquinoline, methylbenzonitrile
-11	C_9H_7N	9-10,12-15	Quinoline, naphthylamine
-13	$C_{11}H_9N$	11-16	Phenylpyridine, phenylaniline, tetrahydroacridine
-15	$C_{11}H_7N$	11-15	Carbazole, dihydroacridine
-17	$C_{13}H_9N$	13-17, 19	Acridine, phenylindole, benzoquinoline
-19	$C_{14}H_9N$	13-19	Benzo[def]carbazole, methylenephenanthridine, fluorenenitrile
-21	$C_{15}H_9N$	15-21	Azapyrene, benzocarbazole, azafluoranthene
-23	$C_{17}H_{11}N$	17-22	Benzacridine
-25	$C_{17}H_9N$	18-23	Azabenzofluoranthene, benzofluorenenitrile
-27	$C_{19}H_{11}N$	19-22	Azabenzopyrene
-29	$C_{21}H_{13}N$	22-23	Azapicene
-31	$C_{21}H_{11}N$	21-24	Azabenzoperylene
-33	$C_{23}H_{13}N$	24	Azadibenzopyrene

Source: Ref. 152.

Z	First member of homologous series	Carbon numbers experimentally determined	Possible structures
1	2	3	4

Nitrogen compounds: CHN_2

Z	First member of homologous series	Carbon numbers experimentally determined	Possible structures
-6	$C_8H_{10}N$	9-10, 13	Diazatetrahydronaphthalene, azatetrahydroquinoline
-8	$C_7H_6N_2$	9-10	Diazaindene
-10	$C_8H_6N_2$	10	Diazanaphthalene
-12	$C_{10}H_8N_2$	10, 13	Bipyridine
-14	$C_{11}H_8N_2$	11-14	Diazafluorene
-16	$C_{12}H_8N_2$	11-14	Diazanthracene/phenanthrene
-18	$C_{13}H_8N_2$	19	Diazabenzofluorene, diazaphenylnaphthalene
-20	$C_{14}H_8N_2$	15-18	Diazapyrene, diazaaceanthrylene
-22	$C_{16}H_{10}N_2$	19-20	Diazachrysene, diazanaphthacene
-24	$C_{16}H_8N_2$	20	Diazabenzofluoranthene
-26	$C_{18}H_{10}N_2$	20, 22	Diazabenzopyrene

Table 13. (b) N-PAC Detected by High-Resolution MS in a Coal Tar Pitch

Hydrocarbon class	Example of structural type	Z	Nominal mass of species				
CHN	Methylquinoline	11	143				
	Dimethylphenylpyridine	13	183	197			
	Acridine	17	179	193	207	221	235
	Benzo[def]carbazole	19	191	205	219	233	247
	Azapyrene	21	203	217	231	245	259
	Benzacridine	23	229	243	257	271	285
	Azabenzofluoranthene	25	227	241	255	269	283
	Azabenzopyrene	27	253	267	281	295	309
	Dibenzacridine	29	279	293	307		
	Azabenzoperylene	31	—	291	305	319	
CHN_2	Dimethylazaindole	8	146				
	Azacarbazole	14	168	182	196		
	Azabenzo[def]carbazole	18	192				
		24	270				

Source: Ref. 153.

reported with a capital N to denote the nitrogen functionality. For example, an accepted designation for azapyrene would be Z = -21N. N-PAC with a single or odd number of nitrogen heteroatoms have odd Z numbers and can be differentiated on this basis from the other PAH and PAC that have even Z numbers. Furthermore, generic structural assignments can be made from the molecular formula and Z number. For example, if the four-ring aza-PAH are considered, the benzocarbazoles, azapyrenes, and azafluoranthenes have a Z number of 21, but Z = 23 for the benzacridines and naphthoquinolines; alkylated N-PAC analogs have the same Z numbers as their parent compounds.

A specific high-resolution mass spectrometry (HRMS) application where molecular formula data were used to distinguish the amino-PAH from other N-PAC was reported by Later et al. [130]. Selective conversion of the amino-PAH in coal liquid isolates to their corresponding pentafluoropropyl amides by derivatization with pentafluoropropionic anhydride followed by probe inlet, high-resolution MS enabled the determination of two- to six-ring amino-PAH. Considering nitrogen and fluorine as possible heteroatoms, the ions of the high-resolution mass spectrum were assigned molecular formulas by computerized data processing techniques. Those compounds with a 5:1 ratio of fluorine and nitrogen atoms and deviations between the measured and calculated masses of less than 5 millimass units were identified as derivatized amino-PAH.

Fluoroalkylamide derivatization of amino-PAH also facilitates their identification by gas chromatography mass spectrometry at low resolution [130,

154]. Fragmentation of the amino-PAH derivatives under normal 70 eV electron-impact ionization conditions are characterized by loss of the derivatization group from the parent compound; m/z 146 for pentafluoropropyl amide and m/z 96 for trifluoroacetyl amide. Furthermore, the increased molecular weight of the parent ions compared to that of their analogous underivatized aza-PAH is also useful for identification of the derivatized amino-PAH.

Different ionization techniques have been useful for the MS analysis of N-PAC in coal products. Field ionization (FI) and field desorption (FD) MS are particularly effective for analysis of high molecular weight N-PAC. Many N-PAC, as well as multiple-heteroatomic nitrogen, sulfur, and oxygen PAC, have been identified in heavy-end coal products by high-resolution FI- and FD-MS [91,152,155]. Fast atom bombardment (FAB) MS has also been used for the analysis of fossil fuel materials. Grigsby et al. [156] evaluated a number of N-PAC standard compounds and a base fraction from an anthracene oil by FAB-MS. Comparison by these workers of the FAB and FI spectrum of the same N-PAC fraction revealed that many molecular ions appearing in the FI spectrum were shifted to $(M + H)^+$ ions in the FAB spectrum. The M^+ and $(M + H)^+$ ions in the FAB mass spectrum were also used to classify sample constituents by nominal mass Z numbers and were found to be in good agreement with high-resolution MS data [156].

Low-voltage ionization techniques have been widely used at both low and high resolution for group-type MS identification of N-PAC in coal products [123,124,153,157,158]. The main advantage of using a low ionization voltage is that fragmentization is minimized, with most ions detected representing molecular masses, particularly in the case of the PAH and N-PAC. Figure 20 is an example of a probe inlet, low-voltage, low-resolution mass spectrum of the N-PAC isolate from a high-boiling SRC distillate. The fraction was obtained by alumina adsorption chromatography as shown in Fig. 8. As expected, the majority of ions in this spectrum are molecular ions and have odd masses. Representative structures for the components of this particular sample include isomeric parent and alkylated aza- and amino-PAH with four to six aromatic rings (see Fig. 20).

Electron-impact (EI) ionization MS at 70 eV is frequently used for the identification of N-PAC, particularly in conjunction with gas chromatography. Comparison of sample mass spectra with published standard compound spectra can be found in the literature or in mass spectral libraries such as the *Eight Peak Index of Mass Spectra* or the *EPA/NIH Mass Spectral Data Base* [159,160]. Olerich and Buchanan have also compiled a library of mass spectra for 161 N-PAC commonly found in coal-derived materials [161].

Typically, EI mass spectra of isomeric aza-PAH, amino-PAH, and cyano-PAH are very similar; the molecular ion is usually the base peak and normally very few fragmentation ions are observed. However, special MS techniques have been developed for differentiation of isomeric N-PAC of different chemical functionality. Buchanan [162] developed a chemical ionization MS technique using ammonia-d$_3$ as a reagent gas for the differentiation of N-PAC with primary, secondary, and tertiary nitrogen heteroatom functionalities. This technique has been used to characterize N-PAC fractions of many coal-derived liquids by gas chromatographic mass spectrometry [147, 162-165]. As reviewed by Bartle in Chap. 6 [Eqs. (7) and (8) and Table 6], the exchange of the ammonia deuterium atoms with the nitrogen-bound hydrogen of the target N-PAC results in a characteristic mass increase that can be used to distinguish the different isomeric nitrogen compounds.

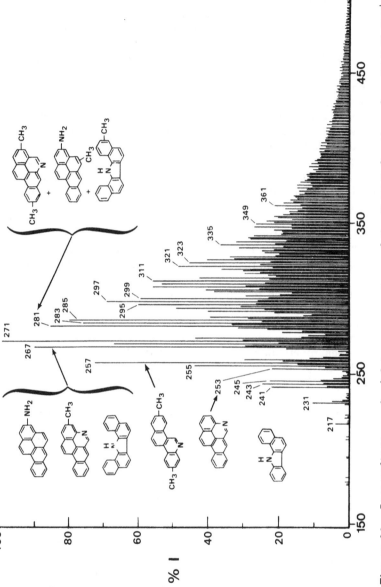

Figure 20. Low-voltage, probe inlet mass spectrum of the N-PAC fraction of an SRC II 850+°F distillate cut. (From Ref. 79.)

Amino-PAH, 2°-PANH, and 3°-PANH in complex mixtures have been successfully differentiated by this chemical ionization MS technique.

More sophisticated MS techniques have also been used to differentiate isomeric N-PAC of different chemical functionality. Both mass-analyzed ion kinetic energy spectrometry (MIKES) and collision-induced dissociation (CID) mass spectrometry techniques have been used to identify different N-PAC classes in coal liquids [166-168]. Figure 21 compares the high-energy CID/ MIKES and low-energy CID/tandem quadrapole chemical ionization mass spectra of isomeric 2-methylbenzo[f]quinoline and 1-aminoanthracene (193 amu). At high energy (7 keV) under chemical ionization conditions, protonated methyl aza-PAH fragment to give a cluster of ions with prominent losses of 15 and 16 amu, CH_3 and CH_4, (Fig. 21B), while the corresponding protonated amino-PAH shows prominent losses of 17 and 18 amu, NH_3 and NH_4 (Fig. 21D). Low-energy (20 eV) CID MS/MS spectra show major losses of 15 and 16 amu for methylaza-PAH and 16 and 17 amu, NH_2 and NH_3, for amino-PAH. The spectra shown in Fig. 21 for the aza- and amino-functional 193 amu isomers are adequately different and provide a basis for class differentiation of these N-PAC. Zakett et al. [166] further suggested that nitrogen-containing ions could be recognized and speciated by their enhanced tendency for charge stripping as determined by the relative intensities of the sharp doubly charged ion peaks in the MIKES spectra of the molecular ions. Structural isomers such as acridine and benzo[f]quinoline have been differentiated by this technique.

Wilson and coworkers [168] used EI metastable MS techniques to distinguish the aza and amino N-PAC. They reported that differences in the relative intensities of the M-15 (CH_3), M-17 (NH_3), and M-27 (HCN) molecular metastable ions could be used to speciate the amino-PAH, 3°-PANH, and 2°-PANH. The M-15 ion was generally the most intense of the three metastable ions for the 3°-PANH, with relatively less intense M-17 and M-27 ions. The M-17 and M-27 ions were the most intense for the amino-PAH. The 2°-PANH typically lacked the M-17 ion, with the M-27 ion being more intense than the M-15 metastable ion. This technique has also been used to distinguish the cyano-PAH from the other N-PAC classes [169]. Figure 22 shows an example where the intensity of the M-27 ion (176 amu) with respect to the other metastable ions can be used to differentiate isomeric cyano- and aza-PAH compounds.

One of the most powerful analytical tools used for the identification of N-PAC in coal-derived products couples the mass spectrometer with a high-resolution, capillary column gas chromatographic system. Most of the MS techniques discussed in this section are applicable to gas chromatographic/ mass spectrometry analysis techniques. The separation of N-PAC by capillary column gas chromatography with on-line MS identification of the separated components has been a widely used analytical technique.

B. Capillary Column Gas Chromatography

The demonstrated ability of capillary column gas chromatography (GC) for resolving components in complex mixtures of PAC makes this analytical technique vital for the detailed characterization of complex N-PAC isolates of coal-derived materials. The theory and mechanics of capillary column GC as they apply to PAC analysis, including the N-PAC, have been discussed in detail by Lee et al. [170], reviewed by Bartle in Chap. 6 and by Olufsen and Bjørseth in Chap. 6 of the first volume [172]. GC techniques and applica-

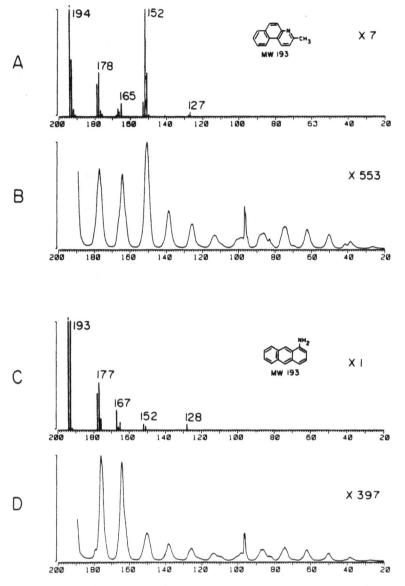

Figure 21. Comparison of low-energy (triple quadrupple) and high-energy
(MIKE) MS/MS spectra: (A) low-energy MS/MS spectrum of 2-methylbenzo-
[f]quinoline; (B) high-energy MS/MS spectrum of 2-methylbenzo[f]quino-
line; (C) low-energy MS/MS spectrum of 1-aminoanthacene; (D) high-energy
MS/MS spectrum of 1-aminoanthacene. These spectra have been normalized
by the scaling factor shown such that the second most intense ion equals the
molecular ion 194^+. (From Ref. 167.)

tions that relate specifically to the compositional elucidation of complex N-
PAC isolates of coal-derived products will be reviewed in this subsection.

Applications in the literature for the analysis of N-PAC by capillary
column GC have relied on a variety of stationary phases of a wide polarity

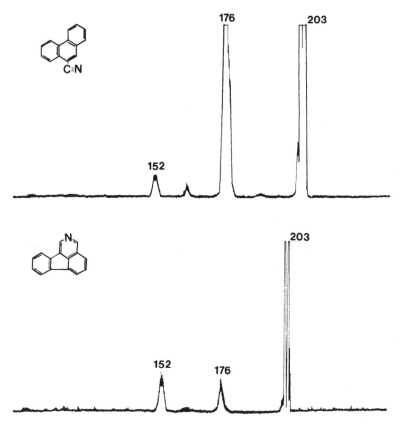

Figure 22. MIKES spectra of isomeric 2-azafluoranthene and 9-cyanophenan-threne (203 amu).

range. Table 14 lists stationary phases that have been used for the GC analyses of N-PAC standard compounds and complex isolates of coal-derived materials. Ignatidis [179] has reviewed methods for the preparation of non-polar, moderately polar, and polar capillary columns for the analysis of N-PAC. Parameters that affect deactivation, efficiency, and thermal stability of both glass and fused silica capillary columns have also been considered and reviewed [170,171,179,183].

In general, the more polar or basic stationary phases, such as Carbowax, Superox, and Amine 220, have been traditionally used to separate the one- and two-ring N-PAC, mainly because enhanced selectivity and resolution are obtained with these phases due to the relatively strong influence of the nitrogen heteroatom in the "smaller" aromatic systems. These phases have also been used because of the higher solubility of the nitrogen compounds in basic or polar phases and their sensitivity to subtle differences in mole-cular structure and/or polarity that are a result of the enhanced selectivity of the polar stationary phases and which ultimately lead to the separation of parent and alkylated one- and two-ring N-PAC. A concomitant deactivation of the capillary wall also occurs and is another advantage of using polar stationary phases [170,188]. These factors result in more symmetrical peak shapes and, in some cases, better resolution of the nitrogen-containing components. However, in earlier work lower efficiencies were normally ob-

Table 14. Selected Gas Chromatagraphic Stationary Phases Used for the Separation of N-PAC

Stationary phase[a]	Description	Selected references	Matrix[b]	Polarity[c] Σ ΔI
Apiezon L&M	Long-chain hydrocarbon	181, 186	STC, CTP	~140
OV-1, OV-101, SE-30, CPSil5, SP-2100	100% Methylpolysiloxane	33, 99, 102, 164, 165, 177, 178, 179, 194, 195, 201	PC, CTP, STC, CLM, SRC, CGM	~215-230
OV-73	5.5% Phenyl methylpolysiloxane	177, 179	PC, CTP, STC	~335
SE-52	5% Phenyl methylpolysiloxane	20, 71, 102, 125, 130, 131, 147, 148, 170-175, 177, 202, 207	CTP, STC, SRC, CGM, CLM, PC	334
SE-54, DB-5	5% Phenyl 1% Vinyl methylpolysiloxane	179, 183, 184, 210	CTP, SRC, STC, CGM, CLM, PC	337
OV-7	20% Phenyl methylpolysiloxane	183	CTP	592
OV-61	33% Phenyl methylpolysiloxane	177, 179, 183	PC, CTP, STC	773
SP-2250	50% Phenyl methylpolysiloxane	34, 177, 179, 200	PC, CTP, STC, CGM, CLM	~880
OV-17	50% Phenyl 1% Vinyl methylpolysiloxane	185	SRC, STC	884
OV-225	25% Cyano 25% Phenyl methylpolysiloxane	177, 179, 185	PC, CTP, STC	1813
SP-2340	75% Cyano 25%Phenyl methylpolysiloxane	177, 179	PC, CTP, STC	3642
N.P.	76% Cyanopropyl 20% tolyl 4% Vinyl polysiloxane	185, 192, 193	STC, SRC	—
N.P.	25% Biphenyl 1% Vinyl methylpolysiloxane	185, 191	STC, SRC	—
Carbowax 20M	Polyethylene glycol	14, 186, 189, 190, 197	CLM, SRC, STC	2308

Superox	Polyethylene glycol	184, 185	CTP, SRC	2309
Pluronic L64	Polypropylene-polyethylene Glycol block copolymer	177, 179	PC, CTP, STC	—
Pluronic L68	Polypropylene-polyethylene Glycol block copolymer	177, 179	PC, CTP, STC	—
PEG 400	Polyethylene glycol	181	CTP	—
PEG 1500	Polyethylene glycol	187	STC	2587
Reoplex 400	Polypropylene glycol/adipate	181	CTP	2750
UCON 50-HB-200	Polypropylene-polyethylene Glycol block copolymer	102, 176	SRC, STC, CTP	1582
Triton X-305	Octylphenoxypolyethoxyethanol	186	STC	1961
Amine 220	1-Ethanol-2-heptadecenyl-2-isoimidazole	180, 181	CTP, STC	—
Solvamin 20	Octadecylamine/ethylene oxide	181	CTP, STC	—
Versamid 900	Polyamide resin	188	STC	—
FFAP	Free fatty acid phase	196	STC	2546
BMBT liquid crystal	N,N'-bis(p-methoxybenzylidene)-α,α'-bis-p-toluidine	182	STC	—
Dexsil 400	Carborane methylphenylsilicone	80, 94, 179, 198	STC, CGM, SRC, CLM	673
Poly S 179	Polyphenyl/sulfonyl ether	183	CTP, STC	—

[a] PC, petroleum crudes and products; SRC, solvent-refined coal; CTP, coal tar products; CLM, coal liquefaction material; CGM, coal gasification materials; STC, standard compounds.

[b] The McReynolds values, $\Sigma \Delta I$, are a measure of polarity and are obtained by summing the retention indices differences of a phase versus squalane for the following compounds: benzene, 1-butanol, 2-pentanone; nitropropane, pyridine, 2-methyl-2-pentanol, and 2-octyne (i.e., $\Sigma \Delta I$ for squalane = 0).

[c] N.P., new phase with no name.

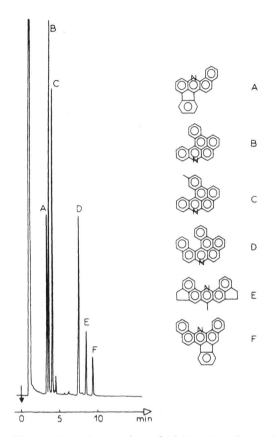

Figure 23. Separation of high molecular weight standard aza-PAH by capillary column GC. Stationary phase, SE-52; column length, 30 m x 0.3 mm i.d.; film thickness, 0.15 μm; isothermal at 280°C. (From Ref. 179.)

served for polar and/or polarizable phase capillary columns than for analogous apolar phase columns. Furthermore, nonbonded polar phases could not generally be operated at the high temperatures required to elute higher molecular weight N-PAC frequently present in coal products.

Capillary column GC analysis of N-PAC with more than three rings have been performed on apolar or slightly polar gum phases, such as the phenyl-methylpolysiloxanes. This is probably due in part to the increased physical similarities of the higher molecular weight N-PAC and the PAH which have been successfully analyzed on these apolar or slightly polar phases; the nitrogen heteroatom contributes relatively less to the overall molecular properties of the "larger" N-PAC systems than do one- and two-ring nitrogen compounds. Figure 23 demonstrates that efficient separations can be obtained for high molecular weight aza-PAH on SE-52.

To date, a majority of the applications in the literature have employed the methylpolysiloxane and phenylmethylpolysiloxane phases (SE-30 and SE-52 or comparable phases) for the analysis of complex N-PAC isolates from coal liquefaction, coking, and gasification materials. However, with the advent of immobilization and cross-linking techniques, the more polar and/or polarizable phases with high phenyl, biphenyl, and cyano content have been advantageously used for highly efficient and selective separations of isomeric

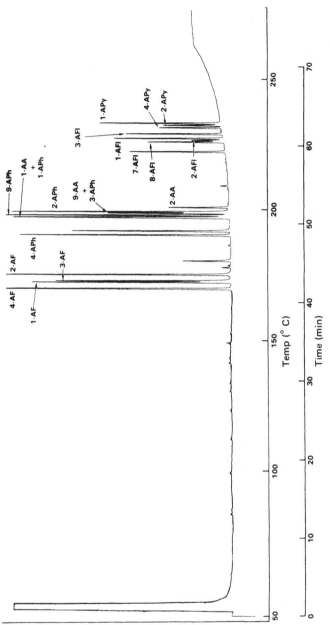

Figure 24. Capillary gas chromatogram of a temperature-programmed separation of amino-PAH isomer groups on a 25% biphenyl methylpolysiloxane stationary phase. Conditions: capillary column 16 m × 0.20 mm i.d.; temperature programmed from 50 to 275°C at 4°C/min; H₂ carrier gas at 100 cm/sec linear velocity. AF, aminofluorene; APh, aminophenanthrene; AA, aminoanthracene; AFl, aminofluoranthene; APy, aminopyrene. (From Ref. 185.)

Table 15. Values of Base Indices of Reference Aza-PAH, Measured on SE-52 Stationary Phase

No.	Compound	BI	Standard deviation	Number of determinations	Boiling point (°C)
1	Quinoline	200.00	—	—	237.1
2	Isoquinoline	203.47	0.37	5	243.2
3	2-Methylquinoline	212.56	0.15	6	246.5
4	2,6-Dimethylquinoline	233.00	0.13	3	266
5	2-Benzylpyridine	243.13	0.19	4	276 (742 mmHg)
6	5H-Indeno[1,2,-b]pyridine	272.81	0.39	3	314
7	Benzo[h]quinoline	298.50	0.11	4	338 (719 mmHg)
8	Acridine	300.00	—	—	345
9	Phenanthridine	303.55	0.30	4	360
10	Benzo[f]quinoline	304.40	0.17	4	350 (721 mmHg)
11	2-Methylbenzo[h]quinoline	312.08	0.27	5	324
12	9-Methylacridine	332.33	0.05	2	
13	2,4-Dimethylbenzo[h]quinoline	334.30	0.14	6	
14	2,3-Dimethylbenzo[h]quinoline	335.48	0.17	8	
15	1,3-Dimethylbenzo[f]quinoline	341.00	0.35	5	
16	Indeno[1,2,3-ij]isoquinoline	345.82	0.15	4	394

17	Acenaphtho[1,2-b]pyridine	348.70	0.03	2	396
18	2,4,6-Trimethylbenzo[h]quinoline	356.10	0.11	5	
19	7,6-Ethylenebenzo[h]quinoline	356.94	0.28	7	
20	Benzo[1,m,n]phenanthridine	357.40	0.44	3	407
21	2-Tolyl-3-methylquinoline	362.10	0.21	3	
22	2,3,4-Trimethylbenzo[h]quinoline	363.34	0.32	5	
23	11H-Indeno[1,2-b]quinoline	372.19	0.11	4	410
24	Benz[c]acridine	393.75	0.09	4	434
25	Benz[a]acridine	400.00	—	—	438
26	Naphtho[2,1-f]quinoline	409.50	0.20	5	
27	8-Methylbenz[a]acridine	411.54	0.08	9	
28	10-Methylbenz[a]acridine	420.94	0.04	2	
29	8-Ethylbenz[a]acridine	423.09	0.11	3	
30	12-Methylbenz[a]acridine	427.39	0.36	2	
31	8,9,10-Trimethylbenz[a]acridine	431.14	0.08	4	
32	Indeno[1,2-de]benzo[h]quinoline	448.19	0.16	3	
33	7-Methylindeno[7,1-bc]acridine	481.81	0.37	7	
34	Dibenz[a,j]acridine	500.00	—	—	

Source: Ref. 177.

N-PAC [185,192,193]. Furthermore, bonded or cross-linked polar phases are thermally stable and can be operated at high temperatures (>275°C) and thus can be used for the analysis of high molecular weight N-PAC. Polarizable phases, such as biphenyl and cyanopropyl methylpolysiloxane, have been used for the selective and efficient separation of isomeric three- and four-ring N-PAC [185,191]. An example of the separation achieved on a biphenyl phase capillary column for the three- and four-ring amino-PAH is shown in Fig. 24; nearly all of these structurally similar isomers were separated. Such separations are important in light of the different biological activities exhibited by the different amino-PAH isomers (see Tables 9 and 10) and the need for quantitative determinations of these compounds in coal products.

Retention data from GC analyses performed on different GC phases can be used as supportive information for the positive identification of compounds separated by capillary column gas chromatography. Retention measurements and indexing techniques have been developed and used to facilitate the unambiguous identification of N-PAC in complex mixtures [173-177]. Although absolute retention times of standard N-PAC can be matched with sample constituents for their identification, utilization of a retention index systems is preferred.

Kováts-type retention index values have been reported for selected N-PAC, mostly the 3°-PANH, using an improved linear interpolation method [173]. A more reliable retention index system specifically used for indexing PAC was proposed by Lee et al. [174]. Retention indices of compounds are calculated from the equation

$$I = 100Z + 100\left(\frac{t_{R_x} - t_{R_Z}}{t_{R_{Z+1}} - t_{R_Z}}\right) \tag{3}$$

based on their absolute retention time with respect to those of naphthalene (I = 200.00), phenanthrene (I = 300.00), chrysene (I = 400.00), and picene (I = 500.00). Although only a few index values for the N-PAC were reported in the initial publication on this retention system, a subsequent paper reported retention index values for more than 100 additional aza-PAH, amino-PAH, and cyano-PAH [175]. This retention index system has assisted in the identification of N-PAC isolated from coal-derived products; retention indices for many N-PAC are given in Tables 18 to 20. It is important to note that these retention index values are usually obtained on phenylmethylpolysiloxane gum phases, such as SE-52, and that values can vary if other capillary column stationary phases are used.

Retention index systems similar to Lee's method have been developed specifically for the identification of N-PAC. Novotny et al. [176] used pyridene (I = 100.00), quinoline (I = 200.00), and acridine (I = 300.00) as the designated retention standards for an index system and reported values for approximately 60 one- and two-ring pyridines and quinolines on a UCON stationary phase. This retention system was used to support the identification of several 3°-PANH in the basic fraction of a solvent-refined coal recycle oil [176]. Schmitter et al. [177] added benz[a]acridine (I = 400.00) and dibenz[a,j]acridine (I = 500.00) to Novotny's system and reported the retention indices shown in Table 15. They also showed that a nearly linear correlation existed between Kováts retention indices and the indices obtained by this method designed specifically for the N-PAC. These workers applied this system for the identification of N-PAC in crude petroleums and coking gas oils

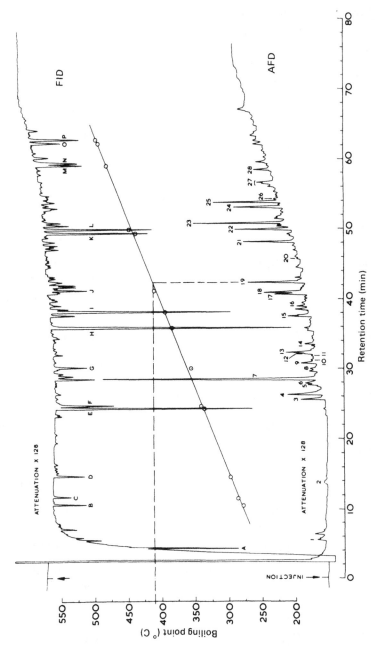

Figure 25. FID-GC and NPD-GC chromatograms of anthracene oil. Peak numbers refer to compounds listed in Table 16. Conditions: 40 m SP-2250 glass SCOT capillary column temperature programmed from 135 to 285°C at 3°C/min. (From Ref. 34.)

Table 16. Nitrogen compounds Identified in Anthracene Oil

Peak no.[a]	m/z	Composition	Z number	Possible type
1	129	C_9H_7N	-11.N	Quinoline
2	129	C_9H_7N	-11.N	Isoquinoline
3	143	$C_{10}H_9N$	-11.N	Methylquinoline
4	143	$C_{10}H_9N$	-11.N	Methylquinoline
5	117	C_8H_7N	-9.N	Indole
6	153	$C_{11}H_7N$	-15.N	Cyanonaphthalene or Aza-acenaphthylene
6	155	$C_{11}H_9N$	-13.N	Aza-acenaphthene or phenylpyridine
7	153	$C_{11}H_7N$	-15.N	Cyanonaphthalene or Aza-acenaphthylene
8	167	$C_{12}H_9N$	-15.N	Azafluorene
9	181	$C_{12}H_7NO$	-17.NO	Azafluorenone
10	181	$C_{12}H_7NO$	-17.NO	Azafluorenone
11	179	$C_{13}H_9N$	-17.N	Benzo[h]quinoline
12	179	$C_{13}H_9N$	-17.N	Acridine
13	193	$C_{14}H_{11}N$	-17.N	Methylbenzoquinoline
14	179	$C_{13}H_9N$	-17.N	Phenanthridine
15	179	$C_{13}H_9N$	-17.N	Benzo[f]quinoline
16	179	$C_{13}H_9N$	-17.N	Benzoquinoline or benzoisoquinoline
17	167	$C_{12}H_9N$	-15.N	Carbazole
18	179	$C_{13}H_9N$	-17.N	Benzoquinoline or benzoisoquinoline
19	193	$C_{14}H_{11}N$	-17.N	Methylbenzoquinoline
20	193	$C_{14}H_{11}N$	-17.N	Methylbenzoquinoline
21	181	$C_{13}H_{11}N$	-15.N	Methylcarbazole
22	193	$C_{14}H_{11}N$	-17.N	Methylbenzoquinoline

Table 16. Continued

Peak no.[a]	m/z	Composition	Z number	Possible type
23	181	$C_{13}H_{11}N$	-15.N	Methylcarbazole
23	193	$C_{14}H_{11}N$	-17.N	Methylbenzoquinoline
24	193	$C_{14}H_{11}N$	-17.N	Methylbenzoquinoline
25	207	$C_{15}H_{13}N$	-17.N	C_2-Alkylbenzoquino-line
26	181	$C_{13}H_{11}N$	-15.N	Methylcarbazole
27	203	$C_{15}H_9N$	-21.N	Azafluoranthene
28	203	$C_{15}H_9N$	-21.N	Azafluoranthene
29	203	$C_{15}H_9N$	-21.N	Azapyrene
30	203	$C_{15}H_9N$	-21.N	Azapyrene
31	191	$C_{14}H_9N$	-19.N	Phenanthro(bcd)-pyrrole
32	229	$C_{17}H_{11}N$	-23.N	Benz[c]acridine
33	229	$C_{17}H_{11}N$	-23.N	Dibenzoquinoline
34	217	$C_{16}H_{11}N$	-21.N	Benzocarbazole
35	217	$C_{16}H_{11}N$	-21.N	Benzo[b]carbazole
36	217	$C_{16}H_{11}N$	-21.N	Benzocarbazole

[a]Peak numbers refer to NPD-GC chromatogram in Fig. 25.
Source: Ref. 34.

using several different stationary phases [179]. One advantage of using the N-PAC reference standards for GC retention index analyses, as opposed to the PAH, is that these compounds are typically present in complex N-PAC isolates of coal-derived materials and hence their identification by retention index methods can be achieved more readily.

Several gas chromatographic detectors other than the universal flame ionization detector (FID) have been particularly useful for the qualitative and quantitative analysis of N-PAC in coal-derived products. Selective and/ or nitrogen-specific GC detectors that have been used for the determination of N-PAC in complex mixtures include the nitrogen-phosphorous detector, NPD (also referred to as the alkali flame ionization detector or thermionic emission detector); the Hall electrolytic conductivity detector, HECD (responds to both nitrogen- and sulfur-containing compounds); the electron capture detector, ECD; and the glow discharge detector, GDD. The principals of operation of these detectors were recently reviewed by Drushel [199].

The NPD has been applied to the analysis of N-PAC in both unfractionated and class-fractionated coal products [34,198-202]. The design, theory of operation, and optimization of this detector have been described by Aue [203], Hartigan [204], Rubin and Bayne [205], Kolb and Bischoff [206], and Patterson and Howe [207]. Albert [208] compared the NPD response for different types of nitrogen compounds; those N-PAC having molecular structures most favorable for the production of cyano radicals when combusted had larger response factors. The response rank order observed was carbazoles > indoles > quinolines > pyridines. A selective response ratio of 1:1000 is generally obtainable with the NPD for N-PAC versus nonnitrogen compounds. Splitting of a capillary column GC effluent between the universal FID and selective NPD can provide information relative to the overall composition of the sample and the presence and comparative concentrations of nitrogen-containing compounds in a single analysis step [34,198-202]. Figure 25 shows a dual trace NPD/FID capillary column GC analysis of an unfractionated anthracene oil [34]; the identifications reported for the N-PAC constituents are listed in Table 16. A plethora of quantitative and qualitative data are available from this single analysis. In a companion paper, Burchill et al. [200] isolated the basic fraction of the same anthracene oil by partition chromatography and, without the interfering hydrocarbons, identified over 100 N-PAC using this NPD/FID dual-detector analysis technique. Most of the compounds identified were aza-PAH, with only a few possible cyano- and amino-PAH suggested. Other investigators have used NPD/FID dual detection for the analysis of N-PAC in coal tar [198], solvent-refined coal liquids [202], coal gasification tars [201], and for engine oil fingerprinting [209].

The HECD can also be used as a nitrogen-specific detector. This detector responds to the presence of NH_3 in the chromatographic effluent that is derived from the pyrolytic reduction of nitrogen-containing compounds. The HECD is highly sensitive, with detection limits around 5-10 ng under optimal conditions and a selectivity factor of approximately 10^6:1 for nitrogen compared to hydrocarbon compounds [210]. Westerman et al. [210] analyzed solvent-refined coal liquids by HECD/FID dual-detection capillary GC and studied the conversion kinetics of representative aza-PAH in catalytically hydroprocessed coal liquids.

Although the ECD and GDD have not been used for the direct determination of N-PAC, these GC detectors have been applied to the determination of the amino-PAH after chemical modification. As discussed in Sec. VI.A, derivatization of the amino-PAH with fluorinated acid anhydrides selectively converts the amino group to a fluoroalkylamide functionality; the fluorine atoms can then be selectively detected by either the ECD [71,130,154] or GDD [211]. Figure 26 and Table 17 demonstrate the qualitative and quantitative data obtained using this selective method, with compounds identified by ECD-GC retention time and GC/MS. Campbell and Grimsrud [212,213] extended this GC method to selectively determine isomeric amino-PAH by doping the ECD with trace amounts of oxygen. They reported that derivatized isomeric amino-PAH of different molecular structure produced adequately different O_2-doped ECD responses and could be differentiated on this basis. For example, all eight aminophenanthrene/aminoanthracene isomers could be distinguished by this technique.

Interfacing the gas chromatograph to the mass spectrometer (GC/MS) provides perhaps the most versatile and useful detection system for the qualitative analysis of N-PAC in complex mixtures. The high-resolution sepa-

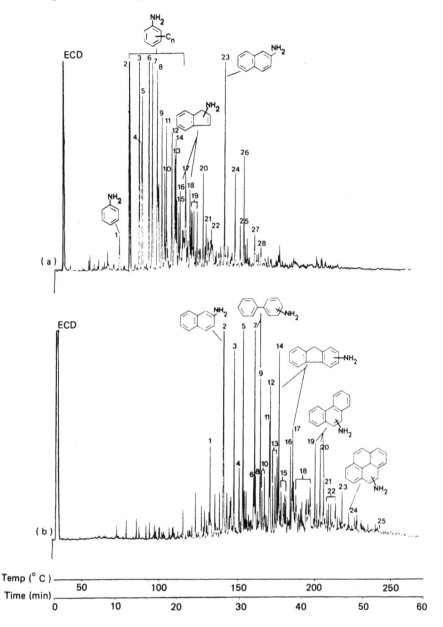

Figure 26. ECD capillary column ECD gas chromatograms of the pentafluoro-propyl derivatized amino-PAH isolates from (A) an SRC I process solvent and (B) an SRC II heavy distillate. Peak numbers refer to the compounds listed in Table 17. (From Ref. 71.)

rations achieved by capillary column GC coupled with the mass-specific detection provided by the MS make this a very powerful analytical tool. However, as discussed in Sec. VI.A, only isomer group-type identifications are possible using MS. Hence specialized capillary column GC techniques, such as retention index systems, selective stationary phases, and selective detec-

Table 17. Identification of and Semiquantitation of Amino-PAH in an SRC I Process Solvent and an SRC II Heavy Distillate

Compound[a]	Peak no.	Approximate concentration[b] ($\mu g/g$)
A. SRC I PS		
Aniline	1	4.3
C_1-Anilines	2, 3, 4	115.3
C_2-Anilines	5-10	119.2
C_3-Anilines	11-15	59.9
C_4-Anilines	17	18.0
Aminoindans	16, 18	16.0
C_1-Aminoindan	19, 20	32.8
C_2-Aminoindans	21	5.6
1-Aminonaphthalene	22	2.8
2-Aminonaphthalene	23	29.9
C_1-Aminonaphthalenes	24, 25, 26	39.1
C_2-Aminonaphthalenes	27, 28	17.3
B. SRC II HD		
1-Aminonaphthalene	1	42.9
2-Aminonaphthalene	2	161.4
C_1-Aminonaphthalenes	3, 4, 5	226.9
C_2-Aminonaphthalenes	6, 8	86.3
3-Aminobiphenyl	7	141.5
4-Aminobiphenyl	9	95.8
C_1-Aminobiphenyls	10-13	327.1
C_2-Aminobiphenyls	15	96.5
Aminofluorenes	14, 16, 17	121.2
C_1-Aminofluorenes	18	125.3
Aminoanthracenes/ phenanthrenes	19, 20, 21	153.2
C_1-Aminoanthracenes/ phenanthrenes	22	104.1

Table 17. Continued

Compound[a]	Peak no.	Approximate concentration[b] (μg/g)
Aminofluoranthene/ aminophenylnaphthalene	23	31.0
Aminopyrene	24	18.0
Aminochrysene	25	18.7

[a]Compounds identifed by GC/MS and ECD-GC retention time.
[b]Semiquantitation was accomplished using 1-aminonaphthalene as an internal standard. The results reported are the average of two determinations.
Source: Ref. 71.

tors, aid in providing detailed characterizations of N-PAC in coal-derived materials. Furthermore, quantitative analysis can be achieved by standard addition, internal standard, or external standard methods by capillary column GC [170].

Although direct analysis of unfractionated coal products by these techniques is feasible, a greater amount of useful information is obtained if analyses are performed with class-specific isolates of the various N-PAC classes. Figures 27 to 30 and Tables 18 to 20 show examples of the detailed analyses that can be achieved by correlated use of GC, GC/MS, and class-specific isolation techniques. From this information it is obvious that a wide variety of N-PAC at varying concentrations can be found in coal-derived materials. Of particular interest in these class-specific fractions is the occurrence of several series of nitrogen heterocyclic compounds that are structurally analogous to the nonfunctional PAH present in coal-derived products. For example, if the three-ring compounds are considered, there are N-PAC from each class that correspond to the PAH structures of fluorene, phenanthrene, and anthracene: carbazoles for the pyrrole-type aza-PAH class; aza-fluorenes, acridine, and benzoquinolines for the pyridine-type aza-PAH class; aminofluorenes, aminophenanthrenes, and aminoanthracenes for the amino-PAH class; and for the cyano-PAH class there are the cyanofluorenes, cyanoanthracenes, and cyanophenanthrenes. In fact, the nature of N-PAC constituents found in coal products correlates closely to the types and quantities of PAH in the same materials.

In summary, detailed qualitative and quantitative information is obtained when capillary column gas chromatographic techniques are applied to the analysis of the N-PAC constituency of coal-derived materials. N-PAC of different chemical functionality, as well as isomeric N-PAC, can be analyzed and differentiated with the aid of capillary column GC. The data obtained from GC analyses can be used for the compositional elucidation of N-PAC in coal products, which in turn can be related to process conditions, coal type and source, formation mechanisms of N-PAC during conversion or combustion of coal, environmental assessments of utilizing coal-derived materials containing N-PAC, and many other factors.

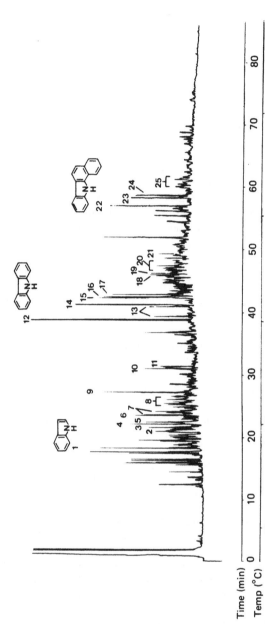

Figure 27. FID-GC capillary column chromatogram of the 2°-PANH fraction (Fig. 19) of a coal gasification tar. Peak numbers refer to compounds listed in Table 18. Chromatographic conditions: 20 m x 0.20 mm i.d. fused silica capillary coated with SE-52; temperature programmed from 40 to 265°C at 3°C/min; H_2 carrier gas at 100 cm/sec linear velocity; splitless injections. (From Ref. 125.)

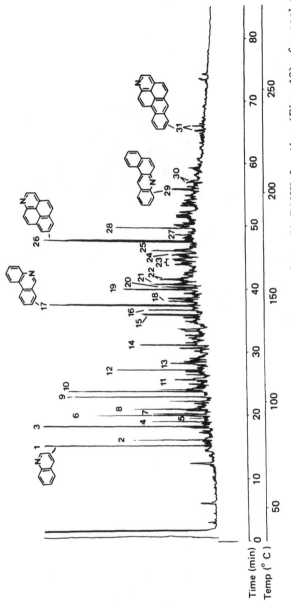

Figure 28. FID-GC capillary column chromatogram of the 3°-PANH fraction (Fig. 19) of a coal gasification tar. Peak numbers correspond to compounds listed in Table 19. Conditions as described in Fig. 27. (From Ref. 125.)

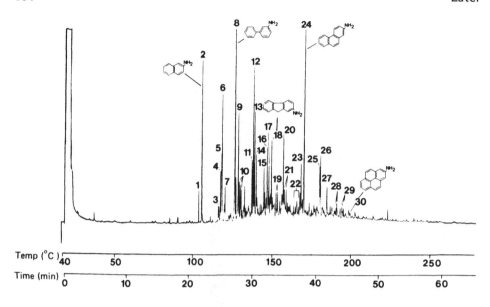

Figure 29. FID-GC capillary column chromatogram of hydrolyzed amino-PAH isolate (Fig. 19) of a solvent-refined coal heavy distillate. Peak identification are listed in Table 20. Conditions: 25 m x 0.20 mm i.d. fused silica capillary column coated with SE-52; temperature programmed from 40 to 265°C at 4°C/ min. (From Ref. 148.)

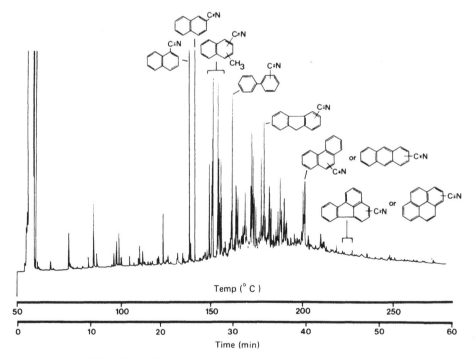

Figure 30. FID-GC capillary column chromatogram of the cyano-PAH fraction of a gasifier medium oil. Compound structures shown were assigned based on GC retention and GC/MS data. Chromatographic conditions: 25 m x 0.25 mm i.d. fused silica DB-5 capillary column; temperature programmed from 50 to 265°C at 4°C/min. (From Ref. 67.)

Table 18. 2°-PANH Identified by GC and GC/MS in a Coal Gasification Tar (Fig. 27)

Peak no.	Compound	RI[a]	Concentration[b] ($\mu g/g$)
1	Indole	222.15	104
2	7-Methylindole	235.42	15
3	C_1-Indole	238.62	74
4	3-Methylindole	239.18	104
5	2-Methylindole	240.01	57
6	1,2-Dimethylindole	244.53	73
7	C_2-Indoles	—	—
8	C_3-Indoles	—	—
	1-Cyanonaphthalene	256.96	13
9	2-Cyanonaphthalene	260.52	123
10	C_1-Cyanonaphthalene	276.83	53
11	C_1-Cyanonaphthalene	278.48	61
12	Carbazole	311.94	516
13	Carbazole isomer	314.27	54
	Carbazole isomer	322.17	50
14	1-Methylcarbazole	323.98	132
15	3-Methylcarbazole	329.12	142
16	2-Methylcarbazole	329.95	107
17	4-Methylcarbazole	331.76	94
18	1,2-Dimethylcarbazole	347.30	52
19	1,3-Dimethylcarbazole	348.44	63
20	C_2-Carbazoles	—	—
	9-Cyanoanthracene	350.42	39
	9-Cyanophenanthrene	351.52	44
21	Benzo[def]carbazole	363.89	40
22	Benzo[a]carbazole	401.94	91
23	Benzo[b]carbazole	409.41	62
24	Benzo[c]carbazole	411.61	56
25	C_1-Benzocarbazoles	—	—

[a]RI, retention index by method of Lee et al. [174].
[b]Approximate concentration of solute in the crude product from semiquantitative capillary gas chromatographic analysis using naphthalene, phenanthrene, chrysene, and picene as internal standards.
Source: Ref. 125.

Table 19. 3°-PANH Identified by GC and GC/MS in a Coal Gasification Tar (Fig. 28)

Peak no.	Compound	RI[a]	Concentration[b] (μg/g)
1	Quinoline	210.02	147
2	Isoquinoline	214.02	55
3	C_1-Quinoline/isoquinoline	223.45	171
4	C_1-Quinoline/isoquinoline	226.82	46
5	1-Methylisoquinoline	229.02	13
6	7-Methylquinoline	231.01	94
7	3-Methylquinoline	232.09	35
8	4-Methylquinoline	235.35	63
9	2,7-Dimethylquinoline	243.77	99
10	2,4-Dimethylquinoline	247.48	122
11	4-Azabiphenyl	252.56	38
12	C_3-Quinoline/isoquinoline	262.31	79
13	C_3-Quinoline/isoquinoline	266.64	25
14	4-Azafluorene	279.59	40
15	Benzo[h]quinoline	301.72	36
16	Acridine	303.87	35
17	Benzo[f]quinoline and/or phenanthridine	307.65	134
18	Benzoquinoline	310.70	22
19	3-Methylbenzo[f]quinoline	320.07	58
20	C_1-Methylbenzoquinoline	321.76	25
21	2-Methylacridine	324.07	42
22	C_1-Benzoquinolines	—	~58
23	C_2-Benzoquinolines	—	—
24	2-Azafluoranthene	347.09	17
25	Azafluoranthene	350.01	39
26	1-Azapyrene and/or 4-azapyrene	358.01	198
27	2-Azapyrene	362.71	4

Table 19. Continued

Peak no.	Compound	RI[a]	Concentration[b] (μg/g)
28	C_1-Azapyrene/fluoranthene and/or azabenzofluorenes	367.90	81
29	Benz[a]acridine	398.32	29
30	Naphthoquinolines	—	~25
31	Azabenzofluoranthenes/ azabenzopyrenes	452.79 456.87	19 18

[a]RI, retention index by method of Lee et al. [174].
[b]Approximate concentration of solute in the crude product from semiquantitative capillary gas chromatographic analysis using naphthalene, phenanthrene, chrysene, and picene as internal standards.
Source: Ref. 125.

C. High-Performance Liquid Chromatography

High-performance liquid chromatographic (HPLC) techniques have been used less frequently than either GC or MS techniques for the detailed analytical determination of N-PAC in coal-derived materials. However, when HPLC separations are used in conjunction with on-line UV, IR, or fluorescence spectrophotometric detection systems, HPLC is particularly valuable for functional and/or isomeric identifications. For example, Tomkins et al. [214] used a sequential two-step HPLC procedure for the rapid isolation and determination of 2-aminonaphthalene in coal oil crudes. They used fluorescence detection, which permitted a high level of discrimination (at least 100-fold) between the 1- and 2-aminonaphthalene isomers.

Wise has reviewed theory, operational considerations, and applications of HPLC for PAC analyses in both volumes of this handbook. Furthermore, he also reviewed HPLC techniques for the separation and determination of N-PAC [2]. Many HPLC methods have been investigated and proposed for the separation of N-PAC standard compounds, mostly the aza-PAH. Both normal and reversed phase as well as adsorption HPLC techniques have been employed. Reversed-phase HPLC on C_8 or C_{18} phases separates N-PAC according to the number of carbon atoms in the molecule, with little selectivity toward positional isomers [215-217]. Conversely, separation of N-PAC is governed by the basicity and accessibility of the nitrogen heteroatom in the normal-phase mode with NH_2- or CN-bonded HPLC phases [139,215-217].

Sonnefeld reported capacity factors for 19 amino-PAH and 36 aza-PAH on five different stationary phases [Partisil-PAC, Nucleosil-NO_2, Nucleosil-NH_2, Chromegabond-Diamine, and μBondapak-NH_2) with various mobile phases [218]. Ray and Frei [219] used a p-nitrophenylisocyanate-bonded HPLC phase to selectively separate the benzoquinoline isomers. The effects of organic mobile phase modifiers, such as acetic acid [220], methanol [215], and dimethyl sulfoxide [139], on the retention properties of N-PAC on various HPLC phases have also been investigated. Vivilecchia et al. [221] sepa-

Table 20. Peak Assignments for Amino-PAH Fraction (Fig. 29) of a Solvent-Refined Coal Heavy Distillate

Peak no.	Compound	Retention index[a]	Mol. wt.[b]
1	1-Aminonaphthalene	262.61	143
2	2-Aminonaphthalene	265.32	143
3	Methylaminonaphthalene		157
4	Methylaminonaphthalene		157
5	Methylaminonaphthalene		157
6	3-Methyl-2-aminonaphthalene	283.58	157
7	Methylaminonaphthalenes		157
8	3-Aminobiphenyl		169
9	4-Aminobiphenyl	297.80	169
10	C_2-Aminonaphthalenes		171
11	Methylaminobiphenyl		183
12	Methylaminobiphenyl		183
13	Methylaminobiphenyl		183
14	Methylaminobiphenyl		183
15	Methylaminobiphenyl		183
16	1-Aminofluorene	327.11	181
17	3-Aminofluorene	329.00	181
18	2-Aminofluorene	331.63	181
19	Methylaminofluorenes		195
20	Methylaminofluorenes		195
21	Methylaminofluorenes		195
22	C_2-Aminofluorenes		209
23	1-Aminoanthracene and/or 1-aminophenanthrene and/or 9-aminophenanthrene	362.12	193
24	2-Aminophenanthrene and/or 3-aminophenanthrene	365.04	193
25	Methylaminophenanthrene/anthracene		207
26	Methylaminophenanthrene/anthracene		207

Table 20. Continued

Peak no.	Compound	Retention index[a]	Mol. wt.[b]
27	Methylaminophenanthrene/ anthracene		207
28	C_2-Aminophenanthrenes/ anthracenes		221
29	Aminofluoranthenes		217
30	2-Aminopyrene	413.45	217

[a]Peaks identified by comparing retention indices of sample solutes and authentic standards [174].
[b]Molecular weight as determined by capillary GC/MS.
Source: Ref. 148.

rated aza-PAH with a silver ion-impregnated HPLC packing. The compounds were separated by donor-acceptor complexing between the adsorbed silver ion and the nitrogen heteroatom, with the elution order being determined by basicity and steric accessibility of the nitrogen unshared electron pair. Haugen [33,145] used ion exchange HPLC to separate the 3°-PANH and amino-PAH in coal gasification condensate oils (see Fig. 15).

Other than spectrophotometric HPLC detectors, electrochemical detectors, such as the HECD, have been modified for use as HPLC detectors for the selective detection of N-PAC [222,223]. Combined HPLC-MS has also been applied to the analysis of coal-derived products [224]. However, as mentioned earlier, the spectrophotometric HPLC detection systems have proved to be the most useful in differentiating structural N-PAC isomers and providing functional group information.

Leary et al. [126] distinguished the methylcarbazoles from the isomeric aminofluorenes, both of molecular weight 181, by HPLC separation coupled with on-line, scanning UV detection. Scanning UV-HPLC has also been employed to provide positive identifications of parent and alkylated benzoquinolines in petroleum crudes [225] and cyano-PAH in coal gasification condensates [131]. Brown and Taylor [226] analyzed coal-derived materials using an HPLC system coupled to a Fourier transform IR spectrophotometer and showed that the N-PAC in the intermediate polar fraction were predominantly 2°-PANH. Garrigues and coworkers [46,227] separated the nitrogen base fraction of crude oils by reversed-phase HPLC and analyzed the subfractions by off-line high-resolution spectrofluorimetry in Shpol'skii matrices. Structural isomers of the dimethyl- and trimethylbenzoquinolines were identified by emission and excitation spectra in both the fluorescence and phosphorescence modes.

In summary, HPLC techniques are not as widely used for the detailed chemical analysis of coal products and do not provide the same level of resolution as GC techniques. However, HPLC does provide definite advantages for differentiating N-PAC according to structural and functional features when coupled to spectrophotometric detection systems or when special mobile and stationary phases are employed.

VII. CONCLUSIONS

The N-PAC are known to occur concomitantly with the nonfunctional PAH in coal-derived materials. In this chapter the analytical chemistry of this chemical class of polycyclic aromatic compounds was reviewed. The nomenclature as well as the chemical, physical, and biological properties of the four N-PAC subclasses have also been reviewed. Pyrrole- and pyridine-type aza-PAH, amino-PAH, and cyano-PAH have all been identified as either major or minor constituents of coal products from various conversion technologies, such as gasification, liquefaction, coking, or other processing methods. As disucssed in the introduction, the N-PAC are of concern for many reasons, including their biological activity, adverse effects on refining, upgrading and storage of coal-derived products, and their potential usefulness as chemical feedstocks.

Methods for the isolation and analytical determination of specific compounds from the different N-PAC subclasses were particularly emphasized. Chromatographic methods that have been employed for the class-specific isolation of N-PAC, as well as techniques aimed at separating the 2°-PANH, 3°-PANH, amino-PAH, and cyano-PAH subclasses, were reviewed in detail. Instrumental analytical methods for compound-specific identifications and differentiation of the four subclasses of N-PAC were also discussed. Although the primary purposes of this chapter did not include providing an unabridged list of N-PAC identified in coal liquids, the information presented in the tables in this chapter provided an accurate assessment of the types of nitrogen-containing compounds, as well as the specific N-PAC that have been found and reported in coal-derived materials.

Future work in this area will certainly be focused in one or more areas. Developing better chromatographic methods for the separation and resolution of N-PAC in complex matrices will be important. Gaining a better understanding of the chemical, physical, and spectroscopic properties of N-PAC, particularly those compounds with three or more aromatic rings and alkylated N-PAC, will be of interest. Developing new and improved capabilities in the instrumental analytical areas of spectrophotometry, mass spectrometry, and liquid and gas chromatography for isomeric determinations and class differentiation of N-PAC will receive attention. Finally, further investigations into the environmental and biological effects of this chemical class of polycyclic aromatic compounds are required.

REFERENCES

1. M. L. Lee, M. V. Novotny, and K. D. Bartle, in *Analytical Chemistry of Polycyclic Aromatic Compounds*, Academic Press, New York, 1981, p. 17.
2. K. D. Bartle, M. L. Lee, and S. A. Wise, Chem. Soc. Rev. (London) *10*:113 (1981).
3. C. M. White, in *Handbook of Polycyclic Aromatic Hydrocarbons*, A. Bjorseth (Ed.), Marcel Dekker, New York, 1983, p. 525.
4. K. F. Lang and I. Eigen, Fortschr. Chem. Forsch. *8*:91 (1967).
5. R. Hayatsu, R. E. Winans, R. G. Scott, L. P. Moore, and M. H. Studier, Fuel, *57*:541 (1978).
6. D. W. Koppenaal and S. E. Manahan, Environ. Sci. Technol. *10*:1104 (1976).
7. M. Dong, I. Schmeltz, E. LaVoie, and D. Hoffman, in *Carcinogenesis*

Vol. 3: *Polynuclear Aromatic Hydrocarbons*, P. W. Jones and R. I. Freudenthal (Eds.), Raven Press, New York, 1978, p. 97.

8. G. K. Hartung and D. M. Jewell, Anal. Chim. Acta *26*:514 (1962).

9. S. E. Moschopedis, R. W. Hawkins, and J. G. Speight, Fuel *60*:397 (1981).

10. H. V. Drushel and A. L. Sommers, Anal. Chem. *38*:19 (1966).

11. J. H. Worstell, S. R. Daniel, and G. Frauenhoff, Fuel *60*:485 (1981).

12. C. W. Wright, E. K. Chess, D. L. Stewart, and W. C. Weimer, Fuel, *64*:443 (1985).

13. A. E. Axworthy, G. R. Schnelder, M. D. Shuman, and V. H. Dahan, PB 250373, U.S. National Technical Information Service Rep., 1976, p. 35.

14. C. M. White, F. K. Schweighardt, and J. L. Schultz, Fuel Process. Technol. *1*:209 (1978).

15. International Union of Pure and Applied Chemistry, in *Nomenclature of Organic Chemistry*, Pergamon Press, Oxford, 1979, Secs. A, B, C, D, E, F, and H.

16. J. H. Fletcher, O. C. Dermer, and R. B. Fox, in *Nomenclature of Organic Compounds*, Advances in Chemistry Series 126, American Chemical Society, Washington, D.C., 1974.

17. J. Jacob and G. Grimmer, Int. Agency Res. Cancer *3*(29):15 (1977).

18. M. L. Lee, M. Novotny, and K. D. Bartle, in *Analytical Chemistry of Polycyclic Aromatic Hydrocarbons*, Academic Press, New York, 1981, Chap. 1.

19. K. L. Loening and J. E. Merritt, in *Polynuclear Aromatic Hydrocarbons: Formation, Metabolism, and Measurement*, M. Cooke and A. J. Dennis (Eds.), Battelle Press, Columbus, Ohio, 1983, p. 819.

20. D. W. Later, M. L. Lee, K. D. Bartle, R. C. Kong, and D. L. Vassilaros, Anal. Chem. *53*:1612 (1981).

21. R. C. Weast (Ed.), *Handbook of Chemistry and Physics*, Chemical Rubber Co., Cleveland, Ohio, 1971, p. D-119.

22. R. T. Morrison and R. N. Boyd, in *Organic Chemistry*, 3rd ed., Allyn and Bacon, Boston, 1973, p. 1004.

23. A. R. Gennaro and J. G. Nikelly, in *The Analytical Chemistry of Nitrogen and Its Compounds*, Vol. 28, Pt. 2, C. A. Streuli and P. R. Averell (Eds.), Wiley-Interscience, New York, 1969, p. 557.

24. A. C. Capomacchia and S. G. Schulman, Anal. Chim. Acta *59*:471 (1972).

25. A. Streitwieser and C. H. Heathcock, *Introduction to Organic Chemistry*, Macmillan, New York, 1976, pp. 467 and 1065.

26. J. B. Hendrickson, D. J. Cram, and G. S. Hammond, *Organic Chemistry*, McGraw-Hill, New York, 1970, p. 143.

27. R. D. Gilliom, *Introduction to Physical Organic Chemistry*, Addison-Wesley, Reading, Mass., 1970, p. 68.

28. C.-H. Ho, B. R. Clark, M. R. Guerin, C. Y. Ma, and T. K. Rao, ACS, Fuel Chem. Div., Prepr. *24*:281 (1979).

29. C. D. Ford, S. A. Holmes, L. F. Thompson, and D. R. Latham, Anal. Chem. *53*:831 (1981).

30. C. Izard, P. Morée-Testa, and R. Gautheret, C.R. Acad. Sci., Paris *272*:2581 (1971).

31. L. R. Snyder and B. E. Buel, J. Chem. Eng. Data *11*:545 (1966).

32. M. Popl, V. Dolansky, and J. Mostecky, J. Chromatogr. *74*:51 (1972).

33. D. A. Haugen, M. J. Peak, K. M. Suhrbier, and V. C. Stamoudis, Anal. Chem. 54:32 (1982).
34. P. Burchill, A. A. Herod, and E. Pritchard, Fuel 62:11 (1983).
35. R. T. Conley, in Infrared Spectroscopy, 2nd ed., Allyn and Bacon, Boston, 1972, p. 139.
36. C. Karr and T.-C. L. Chang, J. Inst. Fuel 31:522 (1958).
37. J. E. Caton, C. Y. Ma, and C.-H. Ho, Oak Ridge National Laboratory, personal communication, 1984.
38. M. M. Boduszynski, R. J. Hurtubise, and H. F. Silver, Anal. Chem. 54:375 (1982).
39. I. Schwager and T. F. Yen, Anal. Chem. 51:569 (1979).
40. C.-H. Ho, B. R. Clark, M. R. Guerin, C. Y. Ma, and T. K. Rao, ACS, Fuel Chem. Div., Prepr. 24:281 (1979).
41. E. L. Wehry, in Handbook of Polycyclic Aromatic Hydrocarbons, A. Bjorseth (Ed.), Marcel Dekker, New York, 1983, p. 323.
42. J. R. Kershaw, Fuel 62:1430 (1983).
43. J. R. Kershaw, Anal. Proc. 18:155 (1981).
44. J. F. McKay, J. H. Weber, and D. R. Latham, Anal. Chem. 48:891 (1976).
45. J. F. McKay, T. E. Cogswell, J. H. Weber, and D. R. Latham, Fuel 54:50 (1975).
46. P. Garrigues, R. DeVazelhes, M. Ewald, J. Joussot-Dubien, J.-M. Schmitter, and G. Guiochon, Anal. Chem. 55:138 (1983).
47. C. S. Woo, A. P. D'Silva, and V. A. Fassel, Anal. Chem. 52:159 (1980).
48. I. Chiang, J. M. Hayes, and G. J. Small, Anal. Chem. 54:318 (1982).
49. S. Scypinski and L. J. C. Love, Anal. Chem. 56:331 (1984).
50. C. D. Ford and R. J. Hurtubise, Anal. Lett. 13:485 (1980).
51. I. Rubin and M. V. Buchanan, Org. Mass Resolut., in press (1985).
52. K. D. Bartle, R. S. Matthews, and J. W. Stadelhofer, Fuel 60:1172 (1981).
53. J. W. Cook, Proc. R. Soc. B. 11:455 (1932).
54. N. M. Kennaway and E. L. Kennaway, J. Hyg. 36:236 (1936).
55. J. W. Cook, C. L. Hewett, and I. Hieger, J. Chem. Soc. 395 (1933).
56. K. Yamagiwa and K. Ichikawa, Mitt. Med. Fak Tokio 15:295 (1916).
57. R. J. Sexton, Arch. Environ. Health 1:208 (1960).
58. D. D. Mahlum, J. Appl. Toxicol. 3:31 (1983).
59. D. D. Mahlum, J. Appl. Toxicol. 3:254 (1983).
60. D. D. Mahlum, C. W. Wright, E. K. Chess, and B. W. Wilson, Cancer Res. 44:5176 (1984).
61. M. L. Lee, M. V. Novotny, and K. D. Bartle, in Analytical Chemistry of Polycyclic Aromatic Compounds, Academic Press, New York, 1981, p. 441.
62. J. C. Arcos and M. F. Argus, in Chemical Induction of Cancer, Vol. 2B, Academic Press, New York, 1974.
63. J. McCann, E. Choi, E. Yamasaki, and B. N. Ames, Proc. Natl. Acad. Sci. USA 72:5135 (1975).
64. C.-H. Ho, B. R. Clark, M. R. Guerin, B. D. Barkenbus, T. K. Rao, and J. L. Epler, Mutat. Res. 85:335 (1981).
65. D. W. Later, R. A. Pelroy, D. L. Stewart, T. McFall, G. M. Booth, M. L. Lee, M. Tedjamulia, and R. N. Castle, Environ. Mutat., in 6:497 (1984).
66. W. Karcher, J. Dubois, R. Fordham, Ph. Glaude, R. Barale, and D.

Zucconi, in *Polynuclear Aromatic Hydrocarbons: Mechanisms, Methods and Metabolism*, Battelle Press, Columbus, Ohio, 1985, p. 685.

67. D. W. Later, S. A. Barraclough, R. B. Lucke, R. A. Pelroy, P. Stoker, and G. M. Booth, in *Polynuclear Aromatic Hydrocarbons: Mechanisms, Methods and Metabolism*, Battelle Press, Columbus, Ohio, 1985, p. 827.

68. T. Ramdahl and G. Becher, Anal. Chim. Acta *144*:83 (1982).

69. S. Krishnan, D. A. Kaden, W. G. Thilly, and R. A. Hites, Environ. Sci. Technol. *13*:1532 (1979).

70. E. A. Adams, E. J. LaVoie, and D. Hoffman, in *Polynuclear Aromatic Hydrocarbons: Formation, Metabolism, and Measurement*, M. Cooke and A. J. Dennis (Eds.), Battelle Press, Columbus, Ohio, 1983, p. 73.

71. D. W. Later, M. L. Lee, R. A. Pelroy, and B. W. Wilson, in *Polynuclear Aromatic Hydrocarbons: Physical and Biological Chemistry*, M. Cooke, A. J. Dennis, and G. L. Fisher (Eds.), Battelle Press, Columbus, Ohio, 1982, p. 427.

72. D. W. Later, R. A. Pelroy, D. D. Mahlum, C. W. Wright, M. L. Lee, W. C. Weimer, and B. W. Wilson, in *Polynuclear Aromatic Hydrocarbons: Formation, Metabolism, and Measurement*, M. W. Cooke and A. J. Dennis (Eds.), Battelle Press, Columbus, Ohio, 1983, p. 771.

73. A. P. Toste, D. S. Sklarew, and R. A. Pelroy, J. Chromatogr. *249*: 267 (1982).

74. T. Kosuge, H. Zenda, H. Nukaya, A. Terada, T. Okamoto, K. Shudo, K. Yamaguchi, Y. Iitaka, T. Sugimura, M. Nagao, K. Wakabayashi, A. Kosugi, and H. Saito, Chem. Pharm. Bull. *30*:1535 (1982).

75. W. B. Greenley, W. D. Barta, and R. T. Eddinger, Fuel Process. Technol. *6*:9 (1982).

76. M. R. Guerin, I. B. Rubin, T. K. Rao, B. R. Clark, and J. L. Epler, Fuel *60*:282 (1981).

77. B. W. Wilson, M. R. Petersen, R. A. Pelroy, and J. T. Cresto, Fuel *60*:289 (1981).

78. D. W. Later, R. A. Pelroy, M. L. Lee, and B. W. Wilson, Mutat. Res., in press (1985).

79. R. A. Pelroy, D. L. Stewart, D. D. Mahlum, E. K. Chess, and D. W. Later, Mutat. Res., in press (1985).

80. M. R. Guerin, C.-H. Ho, T. K. Rao, B. R. Clark, and J. L. Epler, Environ. Res. *23*:42 (1980).

81. R. A. Pelroy and D. L. Stewart, Mutat. Res. *92*:297 (1981).

82. B. W. Wilson, R. A. Pelroy, and J. T. Cresto, Mutat. Res. *79*:193 (1980).

83. M. J. Mima, H. Schultz, and W. E. McKinstry, in *Analytical Methods for Coal and Coal Products*, Vol. 1, C. Karr (Ed.), Chap. 19, Academic Press, New York, 1978, p. 557.

84. F. P. Burke, R. A. Winschel, and D. L. Wooton, Fuel *58*:539 (1979).

85. I. Schwager and T. F. Yen, Fuel *57*:100 (1976).

86. F. W. Steffgen, K. T. Schroeder, and B. C. Bockrath, Anal. Chem. *51*:1164 (1979).

87. H. Schultz and M. J. Mima, ACS, Div. Fuel Chem. *25*:18 (1980).

88. M. M. Boduszynski, R. J. Hurtubise, and H. F. Silver, Anal. Chem. *54*:372 (1982).

89. F. K. Schweighardt and B. M. Thames, Anal. Chem. *50*:1381 (1978).

90. R. G. Ruberto and D. C. Cronaur, Fuel Process. Technol. 3:215 (1979).

91. M. M. Boduszynski, R. J. Hurtubise, and H. F. Silver, Fuel 63:93 (1984).

92. I. B. Rubin, M. R. Guerin, A. A. Hardigree, and J. L. Epler, Environ. Res. 12:358 (1976).

93. M. R. Guerin, C.-H. Ho, T. K. Rao, B. R. Clark, and J. L. Epler, Int. J. Environ. Anal. Chem. 8:217 (1980).

94. C.-H. Ho, C. Y. Ma, B. R. Clark, M. R. Guerin, T. K. Rao, and J. L. Epler, Environ. Res. 22:412 (1980).

95. A. P. Swain, J. E. Cooper, and R. L. Stedman, Cancer Res. 29:579 (1969).

96. M. R. Petersen, J. Fruchter, and J. C. Laul, in *Characterization of Substances in Products Effluents, and Wastes from Synthetic Fuel Production Tests*, PNL-2131, Pacific Northwest Laboratory, Richland, Wash., 1976.

97. B. W. Wilson, M. R. Petersen, R. A. Pelroy, and J. T. Cresto, Fuel 60:289 (1981).

98. B. W. Wilson, R. A. Pelroy, and J. T. Cresto, Mutat. Res. 79:193 (1980).

99. D. A. Haugen, V. C. Stamoudis, M. J. Peak, and A. S. Boparai, in *Polynuclear Aromatic Hydrocarbons: Physical and Biological Chemistry*, M. Cooke, A. J. Dennis, and G. L. Fisher (Eds.), Battelle Press, Columbus, Ohio, 1982, p. 347.

100. M. Novotny, M. L. Lee, and K. D. Bartle, J. Chromatogr. Sci. 12:606 (1974).

101. R. V. Schultz, J. W. Jorgenson, M. P. Maskarinec, M. Novotny, and L. J. Todd, Fuel 58:783 (1979).

102. M. Novotny, J. W. Strand, S. L. Smith, D. Wiesler, and F. J. Schwende, Fuel 60:213 (1981).

103. A. S. Boparai, D. A. Haugen, K. M. Suhrbier, and J. F. Schneider, in *Advanced Techniques in Synthetic Fuels Analysis*, C. W. Wright, W. C. Weimer, and W. D. Felix (Eds.), National Technical Information Service, U.S. Department of Commerce, Springfield, Va., 1983, p. 3.

104. H. W. Sternberg, R. Raymond, and F. K. Schweighardt, Science 188:49 (1975).

105. D. H. Finseth, Z. T. Przybylski, and C. E. Schmidt, Fuel 61:1175 (1982).

106. F. K. Schweighardt, C. M. White, S. Friedman, and J. L. Schultz, in *Organic Chemistry of Coals*, J. W. Larsen (Ed.), ACS Symposium Series 71, American Chemical Society, Washington, D.C., 1978, p. 240.

107. J. L. Schultz, C. M. White, F. K. Schweighardt, and A. G. Sharkey, in *Characterization of the Heterocyclic Compounds in Coal Liquefaction Products, Part I: Nitrogen Compounds*, PERC/RI-77/7, National Technical Information Service, U.S. Department of Commerce, Springfield, Va., 1977.

108. C. A. Streuli, J. Chromatogr. 56:225 (1971).

109. R. A. Jones, M. R. Gurin, and B. R. Clark, Anal. Chem. 49:1766 (1977).

110. T. K. Rao, B. E. Allen, D. W. Ramey, J. L. Epler, I. B. Rubin, M. R. Guerin, and B. R. Clark, Mutat. Res. 85:29 (1981).

111. H.-J. Klimisch and L. Stader, J. Chromatogr. 67:291 (1972).

112. M. E. Snook, Anal. Chim. Acta 99:299 (1978).

113. R. E. Royer, C. E. Mitchell, R. L. Hanson, J. S. Dutcher, and W. E. Bechtold, Environ. Res. 31:460 (1983).

114. J. S. Dutcher, R. E. Royer, C. E. Mitchell, and A. R. Dahl, in Advanced Techniques in Synthetic Fuels Analysis, C. W. Wright, W. C. Weimer, and W. D. Felix (Eds.), National Technical Information Service, U.S. Department of Commerce, Springfield, Va., 1983, p. 12.

115. J. S. Dutcher, R. E. Royer, J. D. Hill, C. E. Mitchell, and R. L. Hanson, in Advanced Techniques in Synthetic Fuels Analysis, C. W. Wright, W. C. Weimer, and W. D. Felix (Eds.), National Technical Information Service, U.S. Department of Commerce, Springfield, Va., 1983, p. 133.

116. A. P. Toste, D. S. Sklarew, and R. A. Pelroy, in Advanced Techniques in Synthetic Fuels Analysis, C. W. Wright, W. C. Weimer, and W. D. Felix (Eds.), National Technical Information Service, U.S. Department of Commerce, Springfield, Va., 1983, p. 74.

117. A. A. Rosen and F. M. Middleton, Anal. Chem. 27:790 (1955).

118. L. R. Snyder and B. E. Buell, Anal. Chem. 40:1295 (1968).

119. L. R. Snyder, B. E. Buell, and A. E. Howard, Anal. Chem. 40:1303 (1968).

120. M. Farcasiu, Fuel 56:9 (1977).

121. R. B. Callen, C. A. Simpson, J. G. Bendoraitis, and S. E. Votz, Ind. Eng. Chem. Prod. Res. Dev. 15:222 (1976).

122. R. B. Callen, C. A. Simpson, and J. G. Bendoraitis, in Analytical Chemistry of Liquid Fuel Sources, P. C. Uden, S. Siggia, and H. B. Jensen (Eds.), Advances in Chemistry Series 170, American Chemical Society, Washington, D.C., 1978, p. 307.

123. J. E. Schiller and D. R. Mathiason, Anal. Chem. 49:1225 (1977).

124. J. E. Schiller, Anal. Chem. 49:2292 (1977).

125. D. W. Later and M. L. Lee, in Advanced Techniques in Synthetic Fuels Analysis, C. W. Wright, W. C. Weimer, and W. D. Felix (Eds.), National Technical Information Service, U.S. Department of Commerce, Springfield, Va., 1983, p. 44.

126. J. A. Leary, A. L. Lafluer, J. P. Longwell, W. A. Peters, E. L. Kruzel, and K. Biemann, in Polynuclear Aromatic Hydrocarbons: Formation, Metabolism, and Measurement, M. Cooke and A. J. Dennis (Eds.), Battelle Press, Columbus, Ohio, 1983, p. 799.

127. L. H. Klemm, C. E. Klopfenstein, and H. P. Kelly, J. Chromatogr. 23:428 (1966).

128. C.-H. Ho, M. R. Guerin, B. R. Clark, T. K. Rao, and J. L. Epler, J. Anal. Toxicol. 5:143 (1981).

129. C.-H. Ho, M. R. Guerin, and M. V. Buchanan, in Advanced Techniques in Synthetic Fuels Analysis, C. W. Wright, W. C. Weimer, and W. D. Felix (Eds.), National Technical Information Service, U.S. Department of Commerce, Springfield, Va., 1983, p. 24.

130. D. W. Later, M. L. Lee, and B. W. Wilson, Anal. Chem. 54:117 (1982).

131. D. W. Later, The determination of nitrogen-containing polycyclic aromatic compounds in coal-derived materials. Dissertation, Brigham Young University, Provo, Utah, 1982.

132. G. R. Dubay and R. A. Hites, Environ. Sci. Technol. 12:965 (1978).

133. D. W. Later, submitted to Anal. Chem. (1985).

134. D. M. Jewell and R. E. Snyder, J. Chromatogr. *38*:351 (1968).

135. D. M. Jewell, J. H. Weber, J. W. Bunger, H. Plancher, and D. R. Latham, Anal. Chem. *44*:1391 (1972).

136. H.-J. Klimisch and A. Beiss, J. Chromatogr. *128*:117 (1976).

137. H.-J. Klimisch and K. Fox, J. Chromatogr. *120*:482 (1976).

138. M. K. Conditt, S. B. Hawthorne, E. J. Williams, and R. E. Seivers, in *Advanced Techniques in Synthetic Fuels Analysis*, C. W. Wright, W. C. Weimer, and W. D. Felix (Eds.), National Technical Information Service, U.S. Department of Commerce, Springfield, Va., 1983, p. 299.

139. A. Matsunaga, Anal Chem. *55*:1375 (1983).

140. J. Chmielowiec, Anal. Chem. *55*:2367 (1983).

141. S. A. Wise, S. N. Chesler, H. S. Hertz, L. R. Hilpert, and W. E. May, Anal Chem. *49*:2306 (1977).

142. H. S. Hertz, J. M. Brown, S. N. Chesler, F. R. Guenther, L. R. Hilpert, W. E. May, R. M. Parris, and S. A. Wise, Anal. Chem. *52*:1650 (1980).

143. B. A. Tomkins, W. H. Griest, J. E. Caton, R. R. Reagan, in *Polynuclear Aromatic Hydrocarbons: Physical and Biological Chemistry*, M. Cooke, A. J. Dennis, and G. L. Fisher (Eds.), Battelle Press, Columbus, Ohio, 1982, p. 813.

144. M. L. Lee, B. Bechtold, A. S. Boparai, D. W. Later, C.-H. Ho, D. Spall, A. P. Toste, and T. J. Wozniak, in *Advanced Techniques in Synthetic Fuels Analysis*, C. W. Wright, W. C. Weimer, and W. D. Felix (Eds.), National Technical Information Service, U.S. Department of Commerce, Springfield, Va., 1983, p. 95.

145. D. A. Haugen, V. C. Stamoudis, M. J. Peak, and A. S. Boparai, in *Polynuclear Aromatic Hydrocarbons: Physical and Biological Chemistry*, M. Cooke, A. J. Dennis, and G. L. Fisher (Eds.), Battelle Press, Columbus, Ohio, 1982, p. 347.

146. C.-H. Ho, M. V. Buchanan, B. R. Clark, and M. R. Guerin, in *Coal Conversion and the Environment*, D. D. Mahlum, R. H. Gray, and W. D. Felix (Eds.), National Technical Information Service, U.S. Department of Commerce, DOE Symposium Series 5, Springfield, Va., 1981, p. 34.

147. M. V. Buchanan, C.-H. Ho, M. R. Guerin, and B. R. Clark, in *Polynuclear Aromatic Hydrocarbons: Chemical Analysis and Biological Fate*, M. Cooke and A. J. Dennis (Eds.), Battelle Press, Columbus, Ohio, 1981, p. 133.

148. D. W. Later, T. G. Andros, and M. L. Lee, Anal. Chem. *55*:2126 (1983).

149. B. Josefsson, in *Handbook of Polycyclic Aromatic Hydrocarbons*, A. Bjorseth (Ed.), Marcel Dekker, New York, 1983, p. 301.

150. C. M. White, in *Handbook of Polycyclic Aromatic Hydrocarbons*, A. Bjorseth (Ed.), Marcel Dekker, New York, 1983, pp. 543, 587.

151. F. W. McLafferty, in *Interpretation of Mass Spectra*, 2nd ed., W. A. Benjamin, Menlo Park, Calif., 1973, p. 33.

152. D. Bodzek and A. Marzec, Fuel *60*:47 (1981).

153. J. L. Schultz, T. Kessler, R. A. Friedel, and A. G. Sharkey, Fuel *51*:242 (1972).

154. B. A. Tomkins and C.-H. Ho, Anal. Chem. *54*:91 (1982).

155. D. Bodzek, T. Krzyzanowska, and A. Marzec, Fuel *58*:196 (1979).

156. R. D. Grigsby, S. E. Scheppele, Q. G. Grindstaff, G. P. Strum, L. C. E. Taylor, H. Tudge, C. Wakefield, and S. Evans, Anal. Chem. *54*:1108 (1982).

157. J. L. Schultz, R. A. Friedel, and A. G. Sharkey, Fuel *44*:55 (1965).

158. V. H. Pichler and A. Herlan, Erdoel Kohle Erdgas *26*:401 (1973).

159. *Eight Peak Index of Mass Spectra,* Mass Spectrometry Data Centre, Awre, U.K., 1974.

160. S. R. Heller and G. W. A. Milne, *EPA/NIH Mass Spectral Data Base,* U.S. Department of Commerce, NSRDS-NBS/63, U.S. Government Printing Office, Washington, D.C., 1978.

161. G. Olerich and M. V. Buchanan, *Compilation of Mass Spectra of Nitrogen-Containing Aromatics,* ORNL/TM-8855, Oak Ridge National Laboratory, Oak Ridge, Tenn., Sept. 1983.

162. M. V. Buchanan, Anal. Chem. *54*:571 (1982).

163. M. V. Buchanan, G. L. Kao, B. D. Barkenbus, C.-H. Ho, and M. R. Guerin, Fuel *62*:1177 (1983).

164. M. V. Buchanan, J. Flanagan, I. B. Rubin, and M. R. Guerin, in *Polynuclear Aromatic Hydrocarbons: Formation, Metabolism, and Measurement,* M. Cooke and A. J. Dennis (Eds.), Battelle Press, Columbus, Ohio, 1983, p. 211.

165. M. V. Buchanan, M. R. Guerin, G. L. Kao, I. B. Rubin, and J. E. Caton, in *Advanced Techniques in Synthetic Fuels Analysis,* C. W. Wright, W. C. Weimer, and W. D. Felix (Eds.), National Technical Information Service, U.S. Department of Commerce, Springfield, Va., 1983, p. 286.

166. D. Zakett, V. M. Shaddock, and R. G. Cooks, Anal. Chem. *51*:1849 (1979).

167. J. D. Clupek, D. Zakett, R. G. Cooks, and K. V. Wood, Anal. Chem. *54*:2215 (1982).

168. B. W. Wilson, A. P. Toste, R. A. Pelroy, B. Vieux, and D. Wood, in *Coal Conversion and the Environment,* D. D. Mahlum, R. H. Gray, and W. D. Felix (Eds.), National Technical Information Service, U.S. Department of Commerce, Springfield, Va., 1981, DOE Symposium Series 5, p. 148.

169. E. K. Chess and D. W. Later, unpublished data.

170. M. L. Lee, F. J. Yang, and K. D. Bartle, in *Open Tubular Column Gas Chromatography: Theory and Practice,* Wiley-Interscience, New York, 1984.

171. K. D. Bartle, Chapter 6 of this volume.

172. B. S. Olufsen and A. Bjorseth, in *Handbook of Polycyclic Aromatic Hydrocarbons,* A. Bjorseth (Ed.), Marcel Dekker, New York, 1983, p. 257.

173. H. Beernaert, J. Chromatogr. *173*:109 (1979).

174. M. L. Lee, D. L. Vassilaros, C. M. White, and M. Novotny, Anal. Chem. *51*:768 (1979).

175. D. L. Vassilaros, R. C. Kong, D. W. Later, and M. L. Lee, J. Chromatogr. *252*:1 (1982).

176. M. Novotny, R. Kump, F. Merli, and L. J. Todd, Anal. Chem. *52*:401 (1980).

177. J. M. Schmitter, I. Ignatiadis, and G. Guiochon, J. Chromatogr. *248*:203 (1982).

178. D. H. Stuermer, D. J. Ng, and C. J. Morris, Environ. Sci. Technol. *16*:582 (1982).

179. I. Ignatiadis, J. M. Schmitter, and G. Guiochon, J. Chromatogr. *246*:23 (1982).

180. J. Macak, V. M. Nabivach, P. Buryan, and J. S. Berlizou, J. Chromatogr. *209*:472 (1981).

181. K. Tesarik and S. Ghyczy, J. Chromatogr. *91*:723 (1974).

182. M. Pailer and V. Hlozek, J. Chromatogr. *128*:163 (1976).

183. H. Borwitzky and G. Schomburg, J. Chromatogr. *170*:99 (1979).

184. M. Novotny, D. Wiesler, and F. Merli, Chromatographia *15*:374 (1982).

185. D. W. Later and B. W. Wright, J. Chromatogr. *289*:183 (1984).

186. R. E. Poulson, J. Chromatogr. Sci. *7*:152 (1969).

187. G. Goretti, M. Clardl, and C. DiPalo, J. High Resolut. Chromatogr. Chromatogr. Commun. *3*:523 (1980).

188. D. Brocco, A. Cimmino, and M. Possanzini, J. Chromatogr. *84*:371 (1973).

189. G. Becher, J. Chromatogr. *211*:103 (1981).

190. B. Olufsen, J. Chromatogr. *179*:97 (1979).

191. J. C. Kuei, J. I. Shelton, L. W. Castle, R. C. Kong, B. E. Richter, J. S. Bradshaw, and M. L. Lee, J. High Resolut. Chromatogr. Chromatogr. Commun. *7*:13 (1984).

192. B. E. Richter, J. C. Kuei, J. I. Shelton, L. W. Castle, J. S. Bradshaw, and M. L. Lee, J. Chromatogr. *229*:21 (1983).

193. B. E. Richter, J. C. Kuei, L. W. Castle, B. A. Jones, J. S. Bradshaw, and M. L. Lee, Chromatographia 17:570 (1983).

194. T. Nakazawa, M. Kuroki, and Y. Tsunashima, J. Chromatogr. *211*:388 (1981).

195. G. Grimmer, J. Jacob, and K.-W. Naujack, Fresenius' Z. Anal. Chem. *314*:29 (1983).

196. P. C. Uden, A. P. Carpenter, H. M. Hackett, D. E. Henderson, and S. Siggia, Anal. Chem. *51*:38 (1979).

197. W. W. Paudler and M. Cheplen, Fuel *58*:775 (1979).

198. G. Schomburg, H. Husmann, and H. Borwitzky, Chromatographia *12*:651 (1979).

199. H. V. Drushel, J. Chromatogr. Sci. *21*:375 (1983).

200. P. Burchill, A. A. Herod, and E. Pritchard, Fuel *62*:20 (1983).

201. S. K. Gangwal, J. Chromatogr. *204*:439 (1981).

202. D. W. Later, B. W. Wright, and M. L. Lee, J. High Resolut. Chromatogr. Chromatogr. Commun. *4*:406 (1981).

203. W. A. Aue, C. W. Gehrke, R. C. Tindle, D. L. Stalling, and C. D. Ruyle, J. Gas Chromatogr. *5*:381 (1967).

204. M. J. Hartigan, J. E. Purcell, M. Novotny, M. L. McConnell, and M. L. Lee, J. Chromatogr. *99*:339 (1974).

205. I. B. Rubin and C. K. Bayne, Anal. Chem. *51*:541 (1979).

206. B. Kolb and J. Bischoff, J. Chromatogr. Sci. *12*:625 (1974).

207. P. L. Patterson and R. L. Howe, J. Chromatogr. Sci. *16*:275 (1978).

208. D. K. Albert, Anal. Chem. *50*:1822 (1978).

209. M. L. Lee, K. D. Bartle, and M. V. Novotny, Anal. Chem. 47:540 (1975).

210. D. W. B. Westerman, S. S. Katti, M. W. Vogelzang, C.-L. Li, B. C. Gates, and L. Petrakis, Fuel *62*:1376 (1983).

211. B. A. Tomkins and C. Feldman, Anal. Chim. Acta *119*:283 (1980).

212. J. A. Campbell and E. P. Grimsrud, J. Chromatogr. *284*:27 (1984).

213. J. A. Campbell, E. P. Grimsrud, and L. R. Hageman, Anal. Chem. 55:1335 (1983).

214. B. A. Tomkins, V. H. Ostrum, and J. E. Caton, Anal. Chim. Acta *134*:301 (1982).

215. H. Colin, J.-M. Schmitter, and G. Guiochon, Anal. Chem. *53*:625 (1981).

216. L.-A. Truedsson and B. E. F. Smith, J. Chromatogr. *214*:291 (1981).

217. M. Dong, D. C. Locke, and D. Hoffman, J. Chromatogr. Sci. *15*:32 (1977).

218. B. Sonnefeld, High performance liquid chromatogrpahy: vapor density measurements in multi-dimensional analysis of polynuclear aromatic hydrocarbons, Dissertation, University of Maryland, College Park, Md., 1982.

219. S. Ray and R. W. Frei, J. Chromatogr. *71*:451 (1972).

220. E. Soczewinski and M. Waksmundzka-Hajnos, J. Liq. Chromatogr. *3*:1625 (1980).

221. R. Vivilecchia, M. Thiebaud, and R. W. Frie, J. Chromatogr. Sci. *10*:411 (1972).

222. R. J. Lloyd, J. Chromatogr. *216*:127 (1981).

223. I. Mefford, R. W. Keller, R. N. Adams, L. A. Sternson, and M. S. Yllo, Anal. Chem. *49*:683 (1977).

224. R. D. Smith, in *Advanced Techniques in Synthetic Fuels Analysis*, C. W. Wright, W. C. Weimer, and W. D. Felix (Eds.), National Technical Information Service, U.S. Department of Commerce, Springfield, Va., 1983, p. 332.

225. J.-M. Schmitter, H. Colin, J.-L. Excoffier, P. Arpino, and G. Guiochon, Anal. Chem. *54*:769 (1982).

226. R. S. Brown and L. T. Taylor, Anal. Chem. *55*:723 (1983).

227. P. Garrigues, R. DeVazelhes, J. -M. Schmitter, and M. Ewald, in *Polynuclear Aromatic Hydrocarbons: Formation, Metabolism, and Measurement*, M. Cooke and A. J. Dennis (Eds.), Battelle Press, Columbus, Ohio, 1983.

10
Atmospheric Reactions of PAH

KAREL A. VAN CAUWENBERGHE / Chemistry Department, University of
Antwerp, Antwerp, Belgium

I.   Introduction   351

II.   Reactivity of PAH: Theoretical Aspects   353

III. The Reactive Gaseous Species in Polluted Air   354

IV.   Reactions of PAH with Molecular Oxygen   357

V.   Reactions of PAH with Ozone   362

VI.   Reactions of PAH with Free Radicals   367

VII. Reactions of PAH with Sulfur Oxides   369

VIII. Reactions of PAH with Nitrogen Oxides   370

IX.   Sampling Artifacts in High-Volume Filtration   374

X.   Parameters Involved in Reactivity of PAH   375

XI.   Conclusions   378

      References   380

I. INTRODUCTION

Polycyclic aromatic hydrocarbons (PAH) constitute an important class of organic aerosol components, which are generated in a variety of combustion
processes and are emitted into the atmosphere preferentially associated with
the submicron size particles [1,2]. It has been known for more than three
decades that organic extracts of the fine particulate matter (POM) collected
in ambient urban air is carcinogenic when administered subcutaneously to
mice [3,4]. Subsequently, this effect was also observed in experimental animals administered extracts from ambient particulates collected during a photochemical smog episode in Los Angeles [5] and in seven other U.S. cities [6].
Similar results have been found with ambient samples collected in major urban
centers throughout the world.

   The carcinogenicity of the organic fraction of particulates has been
mainly attributed to the presence of several PAH and aza-heterocyclic analogs, which are known carcinogens in animals: for example, benzo[a]pyrene
(BaP), benz[a]anthracene, benz[c]acridine, and several dibenzopyrene,
dibenzanthracene, and dibenzacridine isomers. Using fluorescence techniques
[7] and, more recently, combined gas chromatography/mass spectrometry,
an impressive list of PAH has been identified, both in ambient particulates
[8] and in spark ignition and diesel engine exhaust [9].

However, it is noteworthy that the observed carcinogenicity in animals or transformation in cell cultures is significantly greater than can be accounted for by the concentrations of carcinogenic PAH determined in those samples [5,6,10-13], both for ambient particulates and for automobile exhaust. Thus Gordon et al. [14] reported that the benzene extract of airborne particles collected in the Los Angeles area exhibited a cell transformation activity 10^2-10^3 times stronger than that attributable to its BaP content. Furthermore, the methanol extract, containing the more polar material, had activity comparable to the benzene extract, while containing only 1/30 of the BaP present in the total sample. This problem is often referred to as the "excess" carcinogenicity of particulate organic matter extracts.

Recently, a rapid and relatively inexpensive microbiological assay for mutagenic activity has been developed by Ames and coworkers [15,16]. It is a reverse mutation system employing histidine-requiring mutants of the bacteria *Salmonella typhimurium*. The general application of the Ames microbiological assay to environmental samples has demonstrated the mutagenicity of the organic extracts of ambient particulate matter on several occasions [17-21]. The mutagenicity of the frame shift type detected in those samples (TA 98 is usually the most sensitive strain) was often not enhanced by the addition of liver microsomes, thus indicating the presence of direct-acting mutagens, which do not require metabolic activation and therefore could not be simple PAH or aza-heterocyclic analogs. Direct-acting mutagens were also shown to be present in the exhaust of spark ignition and diesel engines [22, 23]. Fly-ash and soot extracts also exhibit mutagenic activity [24,25]. Recently, some new classes of heteroatomic PAH and PAH derivatives have also been investigated. Some sulfur-containing PAH analogs typical of combustion soots are moderately active [26], and amino-PAH isolated from coal liquefaction samples and nitro-PAH are generally stronger mutagens than the corresponding aza compounds [27]. Among the oxygen-containing PAH derivatives some phenols and quinones can be considered as moderate mutagens [28,29].

Thus mutagenicity testing has revealed the presence of classes of compounds in a variety of source emission and ambient air samples which are responsible for an excess mutagenicity not expected on the basis of their PAH content. The identification of the chemical structures of these species, both direct-acting and activatable mutagens, is currently a subject of intense research. Furthermore, the formation of direct-acting mutagens has been observed in laboratory exposure studies of single PAH to gaseous copollutants under simulated atmospheric conditions, and the compounds responsible for the biological effect have been isolated [30,31].

These results support the idea that contrary to some statements in the literature [32], PAH emitted as primary pollutants present on particulate matter can be subject to further chemical transformation through gas-particle interactions occurring either in stacks, emission plumes, exhaust systems, or during atmospheric transport. Therefore, this chapter will not be restricted to ambient gas-particle interactions alone but will also include some processes possibly occurring at or near the emission sources. Theoretically, some distinction between both types of transformations could be made on the basis of the chemical nature of the products, since primary gas copollutants (e.g., SO_x and NO_x) are more likely as reactions partners at the emission sources, whereas secondary gaseous pollutants [e.g., NO_2, O_3, peroxyacetylnitrate (PAN)] would be more effective during transport. Furthermore, besides thermal reactions of PAH with the gaseous copollutants, photochemical degradation should also be considered where molecular oxygen is an important reaction partner.

Chemical transformations of PAH will result in the introduction of polar functional groups into these molecules. The presence of certain functional groups is a prerequisite for the possible binding of the molecule to deoxyribonucleic acid (DNA). By analogy with the well-known binding of certain PAH metabolites, a similar interaction with atmospheric PAH transformation products is assumed to provide a proper molecular basis for the induction of frame shift mutations. Indeed, the biological activity of several potential BaP metabolites has been tested [33] and it was shown that several epoxides, phenols, and a dihydrodiolepoxide of BaP, which are formed enzymatically in mammalian cells, are direct mutagens [33,34]. By analogy, it can be expected that some PAH derivatives formed by electrophilic attack of chemical reagents would also meet the geometric and functional group requirements for possible direct mutagenic activity.

Finally, it is interesting to note that the direct-acting mutagenicity of ambient particulates is associated predominantly with the respirable particle size range ($< 2 \mu m$) [35], on which PAH preferentially accumulate through condensation processes at the emission source. This size fraction also provides the highest specific surface available for gas-particle interactions. Hence the particle size distribution of any secondary PAH-derived pollutant will essentially be similar to that of the parent PAH, and deposition of that compound upon inhalation in the pulmonary region of the respiratory tract will be comparable. However, the bioavailability will also depend on the resorption efficiency in the lung tissue. Increased water solubility of the polar PAH derivatives will probably promote resorption after inhalation, but will also facilitate wash-out removal from suspended particles in the air.

II. REACTIVITY OF PAH: THEORETICAL ASPECTS

In isolated molecule approximations the model of the PAH transition state is generally one in which the aromatic π system is perturbed to a relatively small degree by the attacking reagent. The tendency for reaction at different centers is determined by one or several indices defined in terms of the molecular orbitals (MO) of the original hydrocarbon [36].

For electrophilic aromatic substitution reactions the free-valence value F_r measures the extent by which the maximum valence of that carbon r is not satisfied by bonds or the amount of residual valency located on that carbon. Thus it depends on the bond orders of the adjacent carbon-carbon bonds. Carbon atoms with high F_r values have high residual affinity for the attacking reagent. Table 1 lists some free-valence values of PAH in descending order [36]. Other and more refined types of reactivity indices have been proposed in the literature. The nitration and sulfonation of aromatic systems are examples of electrophilic substitution reactions that follow this reactivity scale.

A different type of reaction involves double-bond reagents which initially form an adduct via a four-center mechanism (e.g., ozone). This reactivity can be understood in terms of localization energies. The para localization energy L_p is defined as the loss in bonding energy that results when two electrons are isolated from a network on atoms having mutual para orientation. Similarly, the bond localization energy L_b involves the isolation of two electrons on atoms having ortho orientation. Thus the ozonolysis of pyrene results in a fission of the 4,5 double bond (lowest L_b), and this is independent of the most reactive position in electrophilic substitution (carbon 1). Table 2 lists some L_b and L_p values for PAH [36]. Competition between 1,2 attack to yield ring cleavage products and 1,4 attack to yield quinones can be expected.

Table 1. Free Valence of Different Carbon Positions in PAH

Hydrocarbon	Position	F_r
Naphthalene	1	0.453
	2	0.404
Anthracene	9	0.520
	1	0.459
	2	0.409
Phenanthrene	9	0.452
	1	0.450
	4	0.441
Pyrene	1	0.468
	4	0.452
	2	0.393
Benz[a]anthracene	12	0.514
	7	0.503
Chrysene	6	0.457
Benzo[a]pyrene	6	0.530
Dibenz[a,h]anthracene	7	0.498

Source: Ref. 36.
Reprinted with permission.

Redox potentials of the quinone-hydroquinone system correlate well with both L_p and L_b energies [37].

The localized bond model provides a good description of the collective properties of conjugated molecules. It will give information about reactivities of compounds, but no information related to the actual mechanistic pathways. Transition states normally also contain systems other than π systems. The energies of these cannot be estimated by the localized bond model. Consequently, analysis of chemical reactivity based only on this parameter is not a completely satisfactory method.

III. THE REACTIVE GASEOUS SPECIES IN POLLUTED AIR

When emitted into polluted urban atmospheres, especially photochemical smog with its high oxidizing potential, particle-adsorbed PAH are exposed to a variety of gaseous copollutants. These include highly reactive intermediates, both free radicals and excited molecular species and stable molecules. The potential transformation reactions of PAH may show seasonal variations, as

Table 2. Localization Energies and Reactivity of PAH Toward Ozone (β units)

Hydrocarbon	Positions	L_b	L_p	Preferred ozoniza-tion positions
Naphthalene	1, 2	3.259		1, 2
	2, 3	3.729		
	1, 4		3.68	
Anthracene	1, 2	3.204		1, 2
	2, 3	3.786		
	1, 4		3.63	
	9, 10		3.31	9, 10
Phenanthrene	1, 2	3.321		
	2, 3	3.655		
	9, 10	3.065		9, 10
	1, 4		3.77	
Pyrene	4, 5	3.057		4, 5
Benz[a]anthracene	1, 2	3.358		
	5, 6	3.030		5, 6
	1, 4		3.78	
	7, 12		3.41	7, 12
Chrysene	1, 2	3.318		
	2, 3	3.712		
	5, 6	3.121		5, 6
	1, 4		3.74	
Dibenz[a,h]anthracene	1, 2	3.349		
	1, 4		3.79	
	5, 6	3.045		5, 6
	7, 14		3.51	

Source: Ref. 36. Reprinted with permission.

observed for the analytical data on PAH concentrations and the measured mutagenicity of POM [21]. A major pathway for PAH degradation in winter (conditions of low temperature and low irradiation) is probably the reaction with nitrogen oxides, sulfur oxides, and with the corresponding acids. Photochemical reactions with oxygen, and reactions with secondary air pollutants produced by photolysis such as ozone, peroxyacetylnitrate, and hydroxyl and hydroperoxyl radicals are expected to be important in summer (conditions of high temperature, intense irradiation). Oxides of nitrogen are crucial species in the formation of photochemical smog, since the photodecomposition of NO_2 is the only known source of anthropogenic ozone. The key steps are

$$NO_2 + h\nu \ (\lambda < 420 \text{ nm}) \longrightarrow NO + O \tag{1}$$

$$O + O_2 \xrightarrow{M} O_3 \tag{2}$$

The first reliable determination of the rate constant for the reaction of the hydroxyl radical with carbon monoxide [38] and a series of n-alkanes led to the conclusion that OH was a key intermediate in hydrocarbon oxidation to form HO_2 and RO_2 radicals, which are essential to the conversion of NO to NO_2. Thus chains such as

$$OH + CO \longrightarrow CO_2 + H \tag{3}$$

$$H + O_2 \xrightarrow{M} HO_2 \tag{4}$$

$$HO_2 + NO \longrightarrow OH + NO_2 \tag{5}$$

$$OH + RH \longrightarrow R + H_2O \tag{6}$$

$$R + O_2 \longrightarrow RO_2 \tag{7}$$

$$RO_2 + NO \longrightarrow RO + NO_2 \tag{8}$$

were formulated. The role of CO in the $NO \longrightarrow NO_2$ conversion in polluted air is likely to be minor compared to that of the organics.

The average concentration of hydroxyl radicals in the atmosphere at 60°N latitude is estimated to be about 1×10^6 radicals/cm^3 in summer. Different sources of OH have been suggested: the photolysis of nitrous acid [39],

$$NO + NO_2 + H_2O \rightleftharpoons 2HONO \tag{9}$$

$$HONO + h\nu \ (\lambda < 400 \text{ nm}) \longrightarrow HO + NO \tag{10}$$

the photolysis of hydrogen peroxide at $\lambda < 370$ nm, and the photolysis of ozone and subsequent reaction with water:

$$O_3 + h\nu \ (\lambda < 320 \text{ nm}) \longrightarrow O(^1D) + O_2(^1\Delta) \tag{11}$$

$$O(^1D) + H_2O \longrightarrow 2OH \tag{12}$$

Electronically excited singlet molecular oxygen $O_2(^1\Delta)$, which is produced simultaneously in this reaction can be formed by a number of other processes in polluted atmospheres. The major one uses a sensitizer molecule S_0 (eventually a PAH) to produce $O_2 \ (^1\Delta)$ be energy transfer [40]:

$$^1S_0 + h\nu \longrightarrow {}^1S^* \tag{13}$$

$$^1S^* + O_2(^3\Sigma) \longrightarrow {}^3S^* + O_2(^3\Sigma) \text{ or } O_2(^1\Delta) \tag{14}$$

$$^3S^* + O_2(^3\Sigma) \longrightarrow {}^1S_0 + O_2(^1\Delta) \tag{15}$$

Singlet molecular oxygen attack in the gas phase on simple olefins cannot significantly compete with OH attack. At the gas-particle interface however, reaction of $O_2(^1\Delta)$ can be of importance in the presence of an effective PAH sensitizer.

The hydroperoxyl radical species HO_2 may be produced by the three-body recombination of H atoms with O_2 or by H abstraction from certain free radicals, for example, alkoxy radicals:

$$RCH_2O + O_2 \longrightarrow RCHO + HO_2 \tag{16}$$

Hydrogen atoms are formed in the photolysis of formaldehyde:

$$HCHO + h\nu \; (\lambda < 370 \text{ nm}) \longrightarrow H + HCO \tag{17}$$

$$HCO + O_2 \longrightarrow HO_2 + CO \tag{18}$$

The reaction of OH with alkanes and with aldehydes proceeds by abstraction to produce alkyl or carbonyl radicals, which after further reaction with ground-state molecular oxygen form RO_2 and RCO_3. These may then again oxidize NO to NO_2. In addition, peroxyacylradicals may combine with NO_2 to form peroxylacyl nitrates (PAN):

$$CH_3CO + O_2 \longrightarrow CH_3C(O)OO \tag{19}$$

$$CH_3C(O)OO + NO_2 \longrightarrow CH_3C(O)OONO_2 \tag{20}$$

The OH radical reacts rapidly with olefins and simple aromatics, approaching the rate of diffusion controlled processes [$k \sim 10^9\text{-}10^{10}$ liters/(mol)(sec)]. In the gas phase, ozone reacts with olefins at a moderate rate [$k \sim 10^3\text{-}10^5$ liters/(mol)(sec)] and more slowly with gaseous aromatic hydrocarbons. However, ring opening occurs and products such as dicarbonyls can be important in the propagation of photochemistry [41].

Finally, trace-oxygenated sulfur- and nitrogen-containing pollutants are also of importance with respect to the reactivity of PAH adsorbed to particulate matter. Besides the well-known primary pollutants NO_x and SO_x, these include formaldehyde, formic acid, hydrogen peroxide, nitric acid, and sulfuric acid. Since soot particles and probably other types of airborne particulates are able to adsorb significant amounts of water, the latter acids may be involved in actual liquid-solid interactions, and electrophilic aromatic substitution in solution may be relevant to the atmospheric chemistry of POM.

IV. REACTIONS OF PAH WITH MOLECULAR OXYGEN

Gas-particle interactions between molecular oxygen and several PAH in the absence of irradiation appear to be very slow. Because ground-state molecular oxygen is in a triplet state and the PAH in the ground singlet state, such interaction is not expected. However, upon chemisorption of a PAH onto a carrier, it cannot be excluded that the populations of the electronic states or the electron density distribution in the molecule are perturbed sufficiently to induce some new species of different reactivity. Losses of benzo[a]pyrene up to 10%, observed over a period of 24 hr under high-volume sampling conditions in ambient air [42], can be explained by volatilization of BaP from the particles into the vapor phase, the so-called blow-off phenomenon. Long-range transport of PAH has been reported in the Nordic countries [43]. In the absence of or under irradiation with low-intensity light, little evidence for degradation of adsorbed PAH was obtained in those studies.

In the older literature, substantial evidence is found for the photochemical transformation of PAH, adsorbed on a variety of solid supports or particles. The National Academy of Sciences document [44] on particulate organic matter has stressed the importance of this reactivity in gas-particle interactions. Photochemical changes have been observed in thin-layer chromatograms of PAH. Thus Inscoe [45] studied the photomodification of 15 PAH, deposited on four different TLC adsorbents (silica gel, alumina, cellulose, and acetylated cellulose) upon exposure to actinic ultraviolet and room light. Only phenanthrene, chrysene, triphenylene, and picene did not react. However, on silica gel and alumina the other 11 PAH, including anthracene, benz- and dibenzanthracenes, pyrene, benzopyrenes, and perylene, underwent pronounced changes. On the less polar substrates cellulose and acetylated cellulose, conversion of PAH also occurred but changes in appearance of fluorescence were less extensive and developed more slowly.

The reactions of these PAH can be rationalized in terms of a free-radical formation in the presence of oxygen. Irradiation at 366 nm of BaP in benzene or carbon disulfide was shown to produce a 6-phenoxybenzo[a]pyrene radical by analysis of the hyperfine structure of its ESR signal [46]. Among others only these PAH, which contain the BaP skeleton, in which no reactive site to form the endoperoxide exists, showed this ESR signal. The mechanism of production of the phenoxyradical is not clear, but could involve an oxidation by singlet molecular oxygen, formed by energy transfer from the irradiated PAH as sensitizer [46,47]. Thus the reaction mechanism could be depicted as follows [$^1O_2^*$ is short for $O_2(^1\Delta)$]:

$$BaP \longrightarrow {}^1BaP^* \tag{21}$$

$$^1BaP^* \longrightarrow {}^3BaP^* \tag{22}$$

$$^3BaP^* + {}^3O_2 \longrightarrow BaP + {}^1O_2^* \tag{23}$$

$$^1BaP^* + {}^3O_2 \dashrightarrow {}^3BaP^* + {}^1O_2^* \tag{24}$$

$$BaP + {}^1O_2 \longrightarrow BaPOOH \tag{25}$$

$$2BaPOOH \longrightarrow 2BaPO + H_2O + (1/2){}^3O_2 \tag{26}$$

Most tetracyclic and pentacyclic PAH absorb the solar spectrum ($\lambda > 300$ nm) strongly and have high intersystem crossing quantum yields to the triplet state. Therefore, they are efficient sensitizers to produce singlet oxygen and can subsequently react with it [48]. The chemical reactions of singlet oxygen $^1O_2^*$ in solution are reasonably well understood in the case of olefins and conjugated dienes, yielding hydroperoxides and endoperoxides, respectively.

$$\text{(structure)} + {}^1O_2^* \longrightarrow \text{(structure)}OOH \tag{27}$$

$$\text{(structure)} + {}^1O_2^* \longrightarrow \text{(structure)} \tag{28}$$

Coomber and Pitts observed the production of singlet molecular oxygen in the gas phase, using benzaldehyde as the sensitizer under simulated atmospheric conditions. The singlet oxygen was detected by monitoring the hydroperoxide product of its reaction with 2,3-dimethylbutene-2 by long-path-length infrared (IR) spectroscopy [49].

Only limited information is available on the chemical reactions of singlet molecular oxygen with PAH. Geacintov [50] coated solid polystyrene fluffs with 20 PAH and irradiated them in the presence of oxygen and nitric oxide. Although no photodegradation products were isolated, efficient energy transfer from PAH to oxygen to form singlet molecular oxygen was observed. Grossman [51] performed the singlet oxygen oxidation of six PAH in three different systems:

1. A solution of the PAH to which rose bengal was added as the sensitizer was irradiated with filtered light ($\lambda \sim 550$ nm) in the presence of a pure oxygen flow.
2. An inert support material of high porosity and 100 mesh particle size was coated with single PAH and rose bengal in two different fractions. The coated particles were placed together in a fluidized-bed reactor and irradiated with filtered light in a flow of pure oxygen.
3. The latter experiment was repeated in the absence of the rose bengal sensitizer, but with irradiation by the total solar spectrum.

In all systems, transformation of PAH was observed. However, the kinetics of these reactions was not investigated and only structural information is available. Pyrene was the only PAH that did not react over a period of 2 hr. PAH that contained the anthracene skeleton reacted by cycloaddition to form endoperoxides and quinones. Both types of products could be isolated and identified by mass spectrometry. Other PAH, such as BaP, however, cannot form an endoperoxide and probably yield an unstable hydroperoxide [46], after rearrangement of the initial dioxetane adduct, which will further decompose to give quinones, the only products characterized by mass spectrometry.

Fox and Olive observed photodecomposition of anthracene finely dispersed into atmospheric particulate matter upon exposure to sunlight in ambient air [52]. Among the reaction products, the endoperoxide of anthracene was identified, thus providing evidence for the participation of singlet molecular oxygen, probably generated at the particle surface in the photooxidation.

The photooxidation of benz[a]anthracene and its 7,12-dimethyl homolog adsorbed onto different types of carbon black has been studied in aqueous suspension upon irradiation with long-wavelength ultraviolet (UV) light by De Wiest [53]. Combustion-derived soots contain oxygenated functional groups at their surface of acidic nature, which are efficient in adsorbing water molecules. Thus model studies of PAH degradation in aqueous suspensions may be relevant to the atmospheric fate of soot-associated PAH. Before adsorption, part of these carbon black particles were also chemically oxidized to modify their interfacial characteristics. From the experiments it followed that the active surface did not significantly affect the kinetic scheme. The half-life of benz[a]anthracene under realistic isolation conditions was estimated to vary between 6 and 18 hr depending on the weather conditions. Singlet molecular oxygen was again proposed at the reactive species in this process, but products were not characterized.

The earliest study of PAH photochemical degradation was conducted by Falk et al. [54]. A striking result of their measurements was the higher reactivity in air of the pure unadsorbed PAH compared to that for PAH ad-

Table 3. Percent Destruction of Polycyclic Aromatic Hydrocarbons Under Atmospheric Conditions

| | Pure unadsorbed compounds | | | | | Adsorbed on soot | |
| | Air in dark | | Air in light | | Smog in light 1 hr | Air in light 48 hr | Smog 1 hr |
	24 hr	48 hr	24 hr	48 hr			
Anthracene	44	49	42	44	–	5	55
Phenanthrene	39	61	34	60	–	–	–
Pyrene	24	43	20	42	83	1	58
Fluoranthene	17	20	16	24	59	4	59
Benzo[a]pyrene	0	0	21	22	50	10	18
Benzo[e]pyrene	–	–	–	–	–	7	51
Benzo[ghi]perylene	0	0	0	0	27	0	67
Coronene	0	0	0	0	5	–	–
Chrysene	0	0	11	0	15	–	–

Source: Ref. 54. Reprinted with permission from The Helen Dwight Reid Educational Foundation.

sorbed on soot particles. This effect was explained by the hypothesis that adsorption of PAH onto porous particles with high specific surface may result in protection from photooxidation. Similar differences in reactivity were not observed upon exposure of the PAH to photochemical smog; here significant conversion was noted both for the pure and the adsorbed compounds (Table 3). Obviously, other and different reaction mechanisms operate under conditions of photochemical smog (possibly the dark reaction with ozone).

Tebbens et al. [55] studied the chemical transformation of BaP and perylene in smoke, which was sent through a flow reactor. Upon irradiation, they found that 35-65% of the original PAH content had disappeared or had been transformed. Subsequently, Thomas, [56] employed the same procedure to measure the photochemical reactivity of BaP on soot particles. By analysis of samples from the entrance and the exit of the chamber he observed a 58% decrease of BaP upon irradiation. These results are obviously in quantitative disagreement with those of Falk [54], but the experimental conditions are not directly comparable.

In an effort to evaluate the photochemical reactivity of PAH associated with coal fly ash, Korfmacher et al. [57] performed the following experiments. Five PAH—anthracene, phenanthrene, fluoranthene, pyrene, and BaP—were adsorbed from the vapor phase onto the surface of fly ash collected from the electrostatic precipitators of coal-fired power plants. Upon irradiation in a rotating quartz flask, none of these PAH showed significant degradation over a period of 100 hr. However, BaP, pyrene, and anthracene all photolyze efficiently in liquid solution and BaP and anthracene are equally reactive adsorbed to activated alumina. Thus the photosensitivity of adsorbed PAH is strongly dependent on the nature of the surface on which the compound is adsorbed. According to the authors, this suppression of the photochemistry of PAH adsorbed onto coal fly ash may be related to a stabilization of their ground electronic state. Possibly the presence of trace transition metal ions at the fly ash surface can provide efficient quenching of the excited singlet and triplet states involved in photochemical degradation.

Other types of photolysis experiments have been described in the literature which may also be relevant to the discussion of PAH oxidation in air. Barovsky and Baum [58] exposed several PAH, adsorbed on carbon needle field desorption emitters, to ultraviolet radiation. BaP, pyrene, benz[a]anthracene, and anthracene all underwent photooxidation to carbonyl compounds. Indeed, polycyclic quinones can be isolated from ambient airborne particulate matter [59] and their presence is often postulated to result from photolysis of adsorbed PAH. Gibson and Smith [60] oxidized BaP adsorbed on silica gel in air in the presence of ^{60}Co γ radiation and identified several products. Besides the expected benzo[a]pyren-quinones, minor components of the reaction mixture were three dihydrodiols (4,5, 7,8, and 9,10 isomers), and the seco-BaP derivative 9-(2'-formylphenyl)phenalene-1-one. With exception of the latter compound, BaP oxidation products similar to those implicated in mammalian metabolism were obtained in these radiation-induced air-degradation experiments. Although these experiments were not designed to simulate environmental conditions, they may have some significance in this regard based on the observed mutagenic response in oxidized BaP preparations and similarities in the chromatography whether ionizing radiation, visible, or UV light was used to initiate oxidations.

Lane and Katz [61] examined the photodegradation of BaP and two benzofluoranthenes in the form of thin dispersions of pure solid compounds in glass dishes. Only solid BaP was highly photosensitive on irradiation with actinic

UV light (half-life of 5.2 hr), but a dramatic increase of degradation occurred for the three compounds in the presence of traces of ozone. However, in these experiments, the physical form of the PAH was not representative of atmospheric conditions.

V. REACTIONS OF PAH WITH OZONE

Before describing the actual experiments designed to evaluate the reactivity of PAH toward ozone under simulated atmospheric conditions, it is useful to review the general mechanism of ozonolysis of PAH in order to have some idea about the chemical nature of the expected products. It is generally accepted that ozone can degrade unsubstituted PAH according to two different mechanisms:

1. A one step nearly simultaneous electrophilic-nucleophilic attack on olefinic bonds with high electron density (usually the bond with lowest ortho localization energy) will originally yield a primary ozonide.

This intermediate is unstable at room temperature and will further decompose to result in ring opening with formation of aldehyde and/or carboxylic acid functional groups [62]. However, reclosure of the biradial or zwitterion intermediate may also result in the production of peroxides or α-keto-hydroperoxides [63,64]. These compounds are most likely to be isolated as methoxy derivatives when the ozonolysis is carried out in methanolic solution. Examples are shown below.

In the older literature on ozonolysis of PAH in solution, it was common practice to perform either on oxidative or reductive work-up of the reaction mixtures (e.g., using an alkaline hydrogen peroxide or potassium iodine solution). Thus peroxidic intermediates were destroyed and ring fission resulted in final products such as dicarboxylic acids or dialdehydes, respectively. Recent work in our laboratory [65] has shown this as an oversimplification of the ozonolysis scheme, and several labile intermediates with possible biological activity are present in the original ozonization mixtures. Their identification on the basis of mass spectral data alone is not totally unambiguous and their thermolability represents a problem for their separation and isolation as pure compounds.

2. A two-step electrophilic attack on the carbon atoms with lowest para-localization energy. Originally, a σ complex is formed here, and this is followed by a second nucleophilic 1,4 addition by the same

or by a second molecule of ozone to yield, finally, a quinone. Other intermediates may also be involved here (e.g., epoxides, phenols). Thus ozone attack on PAH will yield in part compounds similar to those observed in photooxidation.

The theoretical reactivity indices derived from molecular orbital calculations provide a good basis for the prediction of ring fission and quinone products. Thus benz[a]anthracene yields both benz[a]anthracen-7,12-dione and 3-(2'-carboxylphenyl)naphtalene-2-carboxylic acid [66,67]; pyrene yields 1,8-phenanthrene dicarboxylic acid [68].

BaP has been shown to yield three isomeric quinones (6,12-, 1,6-, and 3,6-), in accordance with the carbon positions of highest free valence as well as 7H-benz[de]anthracene-7-one-3,4-dicarboxylic acid and 1,2-anthraquinone dicarboxylic acid, depending on the molar ratio of ozone/PAH after oxidative workup [69].

Sawicki [70] has identified several carbonyl compounds derived from PAH (e.g., 7H-benz[de]anthracene-7-one and phenalene-9-one) in ambient particulate matter using a combination of TLC separation and fluorescence detection. Some high molecular weight quinones were discovered later in ambient air by Pierce and Katz [59].

Degradation studies of PAH by ozone in solution may not be relevant to the determination of their half-life on atmospheric particles. Based on kinetic measurements in solution and extrapolation to the gas phase, Radding et al. [71] estimated the half-life of BaP at ambient ozone levels to be 870 hr. However, Lane and Katz [61] showed the kinetics of the dark reaction of ozone toward several PAH to be very fast at nearly ambient ozone concentrations. They reported a half-life of 0.62 hr for pure BaP exposed as a quasi-monolayer in petri dishes to an ozone level of 200 ppb. Their major results are summarized in Table 4. In their experiments, irradiation did not seem to affect significantly the reactivity of the BaP-ozone system. They also observed that the two PAH-containing five-membered rings (i.e., benzo[k]-

Table 4. Half-Lives of Benzo[a]pyrene, Benzo[b]fluoranthene, and
Benzo[k]fluoranthene Under Various Reaction Conditions

Reaction conditions		Half-life (hr)		
Irradiation	Ozone concentration (ppm)	BaP	BbF	BkF
None	0.19	0.62	52.7	34.9
	0.70	0.4	10.8	13.8
	2.28	0.3	2.9	3.3
Quartzline	0.0	5.3	8.7	14.1
Q500T/CL	0.19	0.58	4.2	3.9
Lamp	0.70	0.2	3.6	3.1
	2.28	0.08	1.9	0.9

Source: Ref. 61. Reprinted with permission.

and benzo[b]fluoranthene) were far more resistant to oxidation than BaP, but irradiation accelerated their degradation. Experiments by Pitts et al. [31] have confirmed and extended the work of Lane and Katz. In these experiments, 8 x 10 in. Hi-Vol glass fiber filters were used as the substrate for BaP to study the kinetics as well as the products of its reaction with ozone in the dark. Fiber filters can provide a reasonable simulation of small particles with high surface area, but may induce some catalytic effects of the support in the transformations under study. BaP-coated filters (washed and fired) were exposed to concentrations of 0.1-0.2 ppm ozone in air for periods varying from 5 min up to 4 hr. For 0.2 ppm O_3 conversion yields of 50% after 1 hr and 80% after 4 hr were observed [31]. These values are in good agreement with the data of Lane and Katz [61]. Other almost equally reactive PAH toward ozone on glass fiber filters are benz[a]anthracene and pyrene.

By high-pressure liquid chromatography (HPLC) the major stable reaction products of the BaP-ozone exposure experiments were separated, isolated, and tentative chemical structures assigned to the peaks, based on the interpretation of low- and high-resolution mass spectral data, as well as on the comparison with reference spectra (Fig. 1). The reaction mixture consisted mainly of ring-opening compounds such as dialdehydes and dicarboxylic and ketocarboxylic acids (as methyl esters), but also contained disubstitution products such as diphenols and quinones. The benz[de]anthracene-7-one-3,4-dicarboxylic acid found in the reaction mixture was synthetized independently for comparison. It is believed to be the precursor to the benz-[de]anthracene-7-one, detected by Sawicki in the extract of ambient aerosols [70]. The reaction mixture from these experiments showed direct-acting mutagenic activity in the Ames test. The major stable contributor to this biological activity has been identified later as the benzo[a]pyren-4,5-epoxide, present in small yield only [31]. This compound, a photosensitive intermediate of the ozonolysis, is known as a DNA-binding metabolite of BaP in bio-

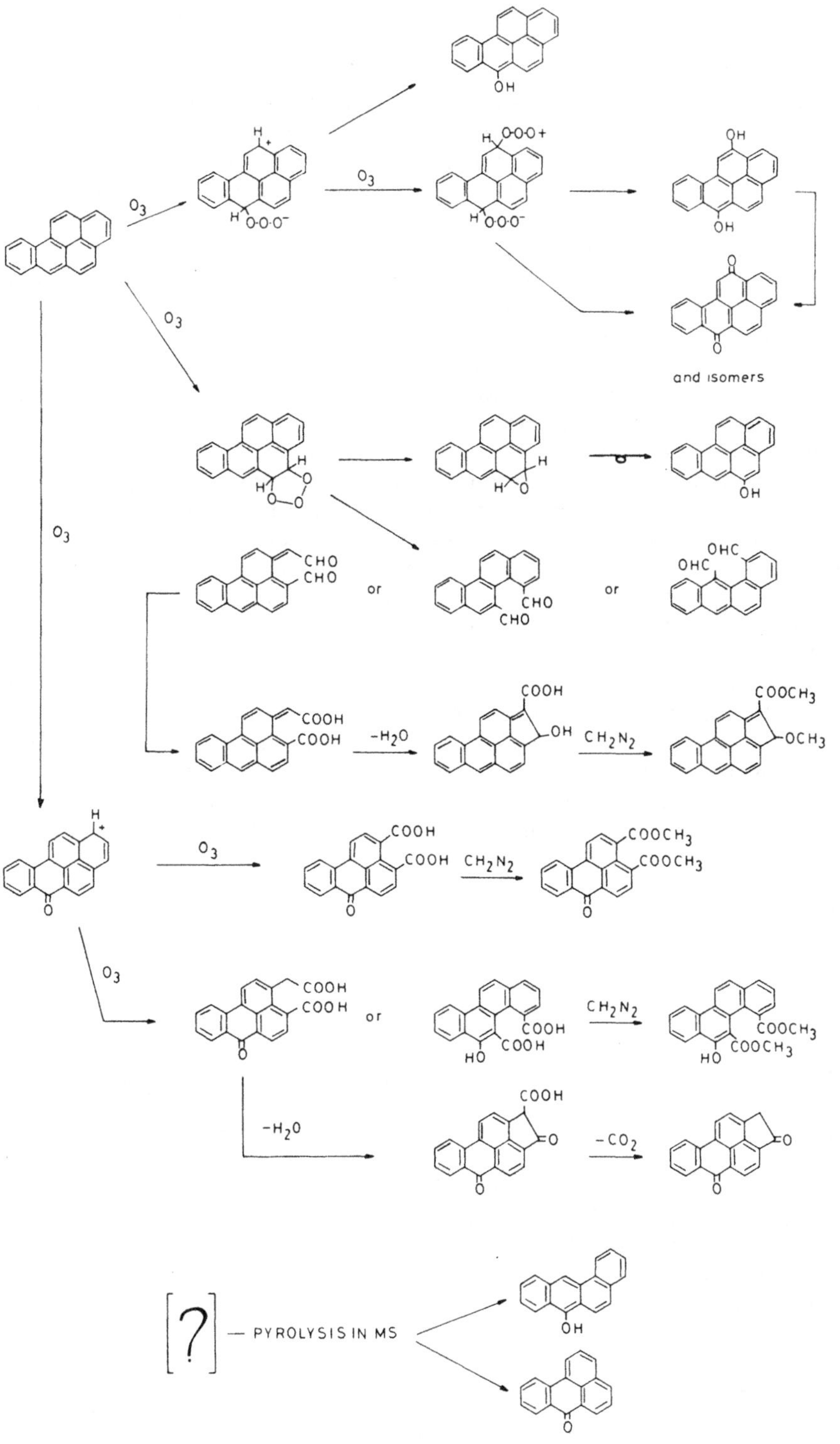

Figure 1. Major reaction products from the exposure of benzo[a]pyrene to 1 ppm ozone in air on glass fiber filters. (From Ref. 72.)

logical systems [73], a strong direct mutagen, and a weak carcinogen on mouse skin.

Recently, Peters and Seifert [42] used ^{14}C-labeled BaP-impregnated glass fiber filters to evaluate the BaP loss under high-volume sampling conditions. While losses up to only 10% were observed for ^{14}C in 24 hr exposures to ambient air (essentially due to the volatilization artifact), thin-layer chromatography determinations of BaP indicated that chemical reactions could account for losses up to 90% over that period. An inverse correlation was observed between the half-life of BaP and the measured ambient ozone concentrations (about 30 ppb). However, photodecomposition of BaP was also shown to be important. For irradiated BaP-impregnated filters decay curves were obtained similar to those of Lane and Katz [61] and Pitts [31] (estimated half-life of 2 hr). Interestingly, however, the authors noted a more pronounced difference in kinetics between the dark and irradiated exposures. Also, they observed little difference between the decay curves obtained for BaP-spiked dust-free and BaP-spiked dust-coated filters, suggesting that the kinetics of photodecomposition of BaP is not affected significantly by the presence of airborne particulate matter on the glass fiber filter.

Diesel exhaust particulate matter samples of aerodynamic diameter less than 0.5 μm, collected by the dilution tube technique on glass fiber filters, have been exposed to particle-free ozonized air (1.5 ppm of ozone) for periods varying between 0.5 and 4 hr. The conversion yields of 10 major nonvolatile PAH from benzo[ghi]fluoranthene up to benzo[ghi]perylene were determined by GC/MS single-ion monitoring [74]. Approximate half-lives are of the order of 0.5-1 hr (Fig. 2) for most PAH measured. This high reactivity of PAH toward ozone on a natural carbonaceous matrix is probably related to the large specific surface of diesel soot particles as well as to its high adsorptive capacity for several gaseous compounds [74]. Experiments under way also indicate significant conversion at lower, nearly ambient ozone levels

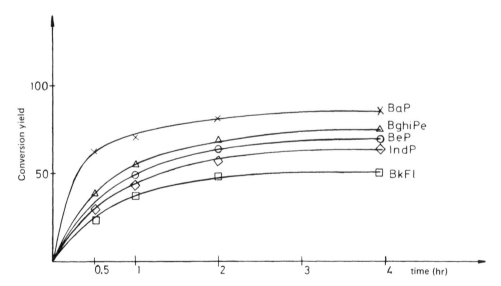

Figure 2. Kinetics of PAH degradation of diesel exhaust particulate matter upon exposure to 1.5 ppm of ozone under high-volume sampling conditions. BaP, benzo[a]pyrene; BghiP, benzo[ghi]perylene; BeP, benzo[e]pyrene; IndP, indeno[1,2,3-cd]pyrene; BkFl, benzo[k]fluoranthene.

(0.1 ppm). Differences in reactivity, especially between isomeric pairs of PAH, are in agreement with available reactivity parameters from MO calculations. In order to isolate and identify the reaction products of such systems, it is necessary to use small particles coated with single PAH in the exposure studies. Furthermore, the effects of different types of matrices (e.g., of mineral or carbonaceous nature) on the kinetics and product composition can be easily compared. In our laboratory, a fluidized-bed reactor is used to obtain efficient gas-particle interaction with model systems based on silica gel, alumina, fly ash, and carbonaceous particles to study the effects of ambient ozone levels and actinic UV light on the degradation of adsorbed PAH [65].

VI. REACTIONS OF PAH WITH FREE RADICALS

In the current literature, the information on the kinetics as well as the products of the reaction of condensed aromatic systems with the hydroxyl radical is scarce. Therefore, in the absence of more detailed studies, we can only expect the interactions of PAH with the hydroxyl radical to be analogous to these observed for simple aromatic hydrocarbons in the gas phase (e.g., benzene, toluene, xylenes, trimethylbenzenes). This extrapolation seems reasonable for the PAH of lower molecular weight, up to chrysene, where an actual gas-particle equilibrium is established. According to Atkinson and coworkers [75,76], the electrophilic character of the OH radical suggests that the major reaction path is addition of OH to the aromatic ring (II $\gg$ I, Fig. 3). Thus in the case of toluene (Fig. 3) the radicals formed by initial attack of OH will react further with oxygen and the formation of cresols and ring-opening products is observed (II). In addition, the cresols are more reactive toward OH than toluene itself and will ultimately form nitro- and hydroxy-nitrocresols (III). These compounds have indeed been identified in the aerosol fraction of a large outdoor smog chamber experiment with toluene [77]. As to PAH, the production of hydroxy and eventually nitrohydroxy derivatives as well as of ring-opening oxidation products can be anticipated. Some of the quinones identified in particulate matter may be the result of further oxidation of labile hydroxyl-containing species. However, the thermolabile character of the latter compounds is probably the reason only a few identification efforts have succeeded [78]. Conversion into stable trimethylsilyl derivatives is required to adequately separate and spectrometrically characterize phenolic or dihydrodiol derivatives of PAH [79].

The rate constant of hydroxyl radical attack on anthracenes, pyrene, and BaP has been measured recently in aqueous solutions [80] and was shown to exceed the diffusion rate constant for BaP. This was tentatively explained by the possible formation of hydroperoxyl radicals in the system. Naphthalene has been shown to produce a hydroxyl radical adduct by pulse radiolysis in aqueous solution [81]. Unfortunately, neither of these studies involved any product identification. The reactivity of alternant PAH toward hydroxyl radicals should be expected to be higher than that of benzene, but not by more than two orders of magnitude [80]. Based on an average atmospheric hydroxyl concentration at 60°N latitude of 1×10^6 molecules/cm^3 in summer [82] and a rate constant for OH attack of benzene of 1.2×10^{-12} cm^3/(molecule)(sec) [83], one can calculate a pseudo-first-order half-life of benzene of about 6 days as an upper limit in the absence of competitive processes. However, the reaction between the hydroxyl radical and phenolic derivatives seems to be faster than those of the parent aromatic compounds. Thus the half-life of the isomeric cresols would be reduced to less than 6 hr under similar atmospheric conditions [84].

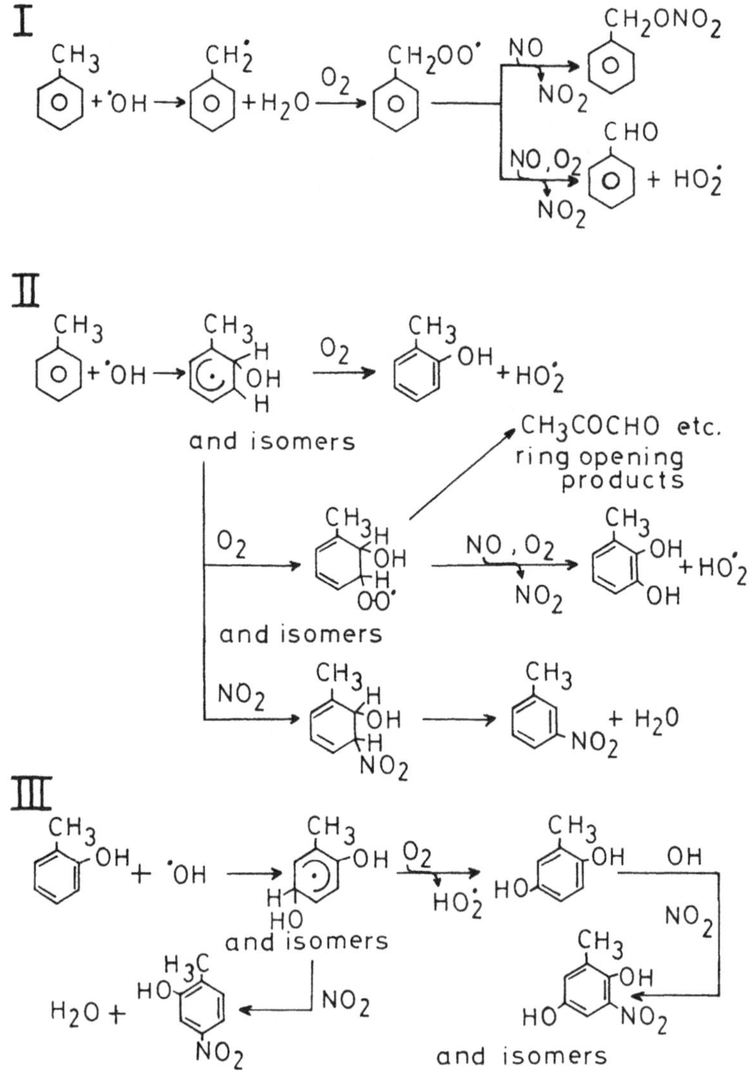

Figure 3. Reaction pathways of toluene with the hydroxyl radical in the presence of nitrogen dioxide.

Two types of free-radical processes may be important for particulate organic matter: the gas-particle interactions between OH radicals from the gas phase and particle-associated PAH, or a direct interaction of organic free radicals present at the particle surface (e.g., from soot with the adsorbed PAH) [85]. With respect to the former hypothesis, it is worthwhile noting the interrelation between the hydroxylation and nitration mechanisms proposed in Fig. 3. However, in the experiments of Jäger and Hanus [86], pyrene adsorbed onto silica gel, alumina, fly ash, and soot particles, exposed to ppm levels of nitrogen dioxide in air, did not show any significant difference in reactivity upon irradiation. This suggests that a photochemical nitration mechanism, as proposed for toluene, was not efficient under these experimental conditions.

VII. REACTION OF PAH WITH SULFUR OXIDES

The oxidation of sulfur dioxide to sulfate is an important atmospheric phenomenon. It is now recognized that both homogeneous and heterogeneous processes contribute to SO_2 oxidation in the atmosphere. In order of decreasing
importance, the homogeneous gas-phase reactions are primarily photochemical
in nature [87]:

$$SO_2 + OH \xrightarrow{M} HOSO_2 \qquad\qquad (29)$$

$$SO_2 + HO_2 \longrightarrow SO_3 + OH \qquad\qquad (30)$$

$$SO_2 + RO_2 \longrightarrow SO_3 + RO \qquad\qquad (31)$$

$$SO_2 + O(^3P) \xrightarrow{M} SO_3 \qquad\qquad (32)$$

$$SO_3 + H_2O \longrightarrow H_2SO_4 \qquad\qquad (33)$$

$$NH_3 + H_2SO_4 \longrightarrow NH_4HSO_4 + (NH_4)_2SO_4 \qquad\qquad (34)$$

Furthermore, the heterogeneous oxidation of SO_2 occurs primarily in aerosols,
to which SO_2 has been adsorbed [88], according to various mechanisms:

Liquid-phase oxidation of SO_2 by dissolved O_2
Liquid-phase oxidation of SO_2 by dissolved O_3
Metal-ion-catalyzed, liquid-phase oxidation
Catalytic oxidation of SO_2 on particle surfaces (e.g., soot) [89]

Relative humidity is a significant factor in the heterogeneous SO_2 oxidation process since it generally takes place in water droplets. The mechanisms
of oxidation in solution are not well understood even though empirical rate
equations have been proposed. The former two processes are probably important in clouds or fog, whereas the latter two are possibly substantial in
plumes. From the literature on the reactivity of sulfur-containing species
with PAH, it follows that ground-state SO_2 is not a reactive species, but in
view of the several heterogeneous oxidation mechanisms available, some conversion can be observed as the result of prior SO_2 transformation.

Benzene has been shown to react with SO_3 in the gas phase [90]. Pyrene
is transformed at room temperature by the action of concentrated sulfuric
acid [91] into a mixture of disulfonic acids. Other sulfur-containing functional groups can be introduced in a PAH molecule: In the presence of irradiation, excited SO_2^* may be involved in the formation of sulfinic acids, sul-
foxides, and sulfones. Catalytic oxidation by the particle surface will possibly
convert these intermediates into sulfonic acids.

Tebbens et al. [55] have studied the degradation of BaP adsorbed onto
soot particles in their flow reactor by SO_2 at various concentrations. At high
SO_2 concentration (50-80 ppm) approximately 50% conversion was observed,
whereas 8-10 ppm of SO_2 did not yield any degradation in the dark. However,
upon irradiation about 50% conversion was observed for both concentrations.
However, the nature of the reaction products was not investigated further.

In the experiments of Jäger and Rakovic [92], pyrene and BaP adsorbed
onto fly ash and alumina were exposed to concentrations of 10% sulfur dioxide
in air. These authors isolated many sulfur-containing PAH derivatives (e.g.,
the pyrene mono- and disulfonic acids) and a BaP-sulfonic acid.

Hughes et al. [93] used fly ash collected by electrostatic precipitation from coal-fired power plants and enriched with single PAH through vapor-phase adsorption [94] of about 100 ppm of SO_2 and SO_3. Although no reaction occurred in the presence of SO_2, many products were observed upon exposure to SO_3, as indicated by the HPLC separation of particulate matter extracts. Further identification attempts were not performed.

The levels of sulfur oxides used in all these experiments are unrealistically high: the latter exposure studies may be of relevance only to the conditions prevailing in a power plant emission plume, where typical SO_2 concentrations are about 10 ppm. Under the experimental conditions described, similar reaction patterns were observed on other substrates, such as silica gel, alumina, and activated charcoal.

Butler and Crossley [95] studied the reactivity of the PAH naturally present on soot particles generated in a flame from an ethylene-air burner in a chamber with ambient air containing only 5-10 ppm SO_2 and did not observe any significant loss of PAH on exposure over a period up to 3 months in the dark. In addition, the exposure of pure BaP, finely divided over glass fiber filters in air containing 1 ppm SO_2, in the absence of irradiation did not yield any significant conversion over 24 hr in Hi-Vol sampling equipment [96]. Nagai et al. [97] observed the conversion of anthracene into anthracen-9-sulfonic acid by the action of high SO_2 concentrations in the presence of irradiation in various solvents.

From the data available, it follows that SO_2 can be considered as a reactive gaseous pollutant toward PAH only when a prior catalytic conversion is possible at the particle surface either in liquid or solid phase, or when an excited electronic species of SO_2 can be effectively produced by irradiation. Heterogeneous conversion of SO_2 is generally recognized as an important process in ambient atmospheres.

Therefore, it seems surprising that no actual identification studies of PAH-derived sulfonic acids isolated from source-related or ambient particulate matter have been reported in the open literature. The high water solubility of sulfonic acids will probably result in poor extraction efficiencies from airborne particles using the conventional solvents. Furthermore, in most separation schemes of the organic extracts, water-soluble material is mostly discarded as being mainly inorganic in nature. Thus sulfur-containing PAH derivatives may have been overlooked in previous studies of airborne particulate matter.

VIII. REACTIONS OF PAH WITH NITROGEN OXIDES

Similarly to the atmospheric transformation of sulfur dioxide into sulfur trioxide, sulfuric acid, and sulfate, the nitrogen oxides (NO and NO_2) can also be converted into nitrous, nitric acid, nitrate, and peroxyacylnitrates. Nitrate formation can occur both by homogeneous and heterogeneous processes. The homogeneous nitrate formation involves reactions of NO and NO_2 with free radicals present in the atmosphere:

$$NO + OH \underset{h\nu}{\rightleftharpoons} HONO \tag{35}$$

$$NO_2 + OH \longrightarrow HNO_3 \tag{36}$$

$$NO + HO_2 \longrightarrow NO_2 + OH \tag{37}$$

$$RO + NO \overset{h\nu}{\rightleftharpoons} RONO \tag{38}$$

$$RO_2 + NO \longrightarrow NO_2 + RO$$

$$\longrightarrow RONO_2 \tag{39}$$

$$RCO + O_2 \longrightarrow RCO_3 \tag{40}$$

$$RCO_3 + NO \longrightarrow RCO_2 + NO_2 \tag{41}$$

$$RCO_3 + NO_2 \longrightarrow RCO_3NO_2 \ (PAN) \tag{42}$$

$$NH_3 + HNO_3 \longrightarrow NH_4NO_3 \tag{43}$$

As to the heterogeneous conversion processes, liquid-phase oxidation and catalytic oxidation of NO_x on particle surfaces are probable, but few hard data on their importance in the atmosphere are available. With respect to the atmospheric transformations of PAH by nitrogen-containing species, the interactions with NO_2, HONO, and HNO_3 should be considered. NO alone does not seem to react with polycyclic aromatic hydrocarbons. Peroxyacetylnitrate (PAN) levels of 1 ppm in air are known to convert BaP on a glass fiber filter and result in the formation of several products, but so far identification has not been attempted [96].

Electrophilic substitution of polycyclic aromatic systems in solution by the action of nitric acid in acetic acid or anhydride, yielding nitroarene derivatives, is a classical synthetic reaction with high yield [98]. In exposure experiments of BaP and perylene coated on glass fiber filters to 1 ppm of nitrogen dioxide in the dark, Pitts et al. [30] observed a significant conversion of these PAH into mononitro derivatives (three isomers for BaP) of 60% in 8 hr. Traces of nitric acid, present at the ppb level in the exposure gas mixture, were shown to be necessary for the reaction to proceed, but could not account for the observed conversion alone. The nitration products were direct-acting mutagens in the Ames test. These results, though obtained in a simplified model system, initiated several research efforts, both chemical and microbiological in nature, to assess the importance of nitroarenes in source emission and ambient aerosols. Previously, nitro derivatives of PAH had been detected in ambient aerosols only by Jäger [99], using fluorescence quenching after thin-layer chromatographic separation. Very soon, however, the presence of nitroarenes (mono and dinitro derivatives) was demonstrated in carbon blacks and toners [100,101], in diesel exhaust particulate matter [102], and in ambient aerosol samples [103,104]. Indirect evidence that the direct-acting mutagenicity of airborne particulate matter in polluted air might originate from nitro derivatives in substantial amounts was suggested by the decreased biological activity of the extracts of ambient aerosols in a nitroreductase-deficient bacterial strain derived from TA 98 [105,106]. However, the chemical structure of some of the compounds responsible is probably more complicated than that of the simple nitroarenes identified so far, which presently will account for only a part of the observed biological effect.

The presence, formation, and fate of nitro derivatives of PAH has been recently reviewed in detail by Nielsen [107]. A first process of formation involves radical reactions that would be typical of an atmosphere, containing high levels of photochemical oxidants and nitrogen dioxide, at high ambient temperatures. PAH which can be present in the vapor phase to a significant extent (i.e., up to tetracyclic compounds) could react according to a mechanism similar to that proposed for toluene [75,76]. At ambient temperature, addition of the OH radical is the predominant process for unsubstituted PAH leading to a cyclohexadienyl radical. Subsequent NO_2 attack leads to an adduct that decomposes to the nitro derivative by the loss of water (see Fig. 3, II). However, less than 5% of the toluene reacting with the OH radical will be converted into nitrotoluenes under conditions of heavy nitrogen dioxide pollution [107], and the significance of this mechanism for nitroarene formation can be questioned.

At temperatures above 100-200°C the dominant reaction of benzene with OH radicals appears to be hydrogen abstraction from the ring [108]. The corresponding phenyl radical could then possibly directly combine with NO_2 to form a nitroarene. Since in stacks and exhaust gases the nitrogen dioxide/ oxygen ratio will be much higher than in ambient atmospheres, nitration could become more important compared to oxidation under these circumstances.

Further transformation of hydroxy derivatives of PAH with NO_2 can also lead to hydroxynitro derivatives. Provided that they are truly present as a phenolic structure and not as a ketone, hydroxy-PAH can lead either to dihydroxycyclohexadienyl radical adducts or to the phenoxy radical by reaction with the OH radical, or to the phenoxy radical by reaction with the nitrate radical. Subsequent nitration of both species will ultimately produce hydroxynitro derivatives (Fig. 4). The nitrate radical has been identified spectroscopically at levels exceeding 100 ppt under conditions of low humidity and photochemical air pollution persisting during the night [109]. At a concentration of 10 ppt NO_3 the half-life of phenol is estimated at about 20 min [84]. The phenoxy radicals can further react with oxygen to form quinones or with NO_2 to yield nitrohydroxy derivatives.

A second process by which nitro-arenes can be formed is electrophilic substitution. Since soot particles [110] and probably also other types of airborne particles may adsorb significant amounts of water and humidity appears to affect the adsorption of NO_2 [111] onto particles, the nitration mechanism of particle-associated PAH may involve liquid-solid reaction steps [107]. Thus solution chemistry may become relevant to gas-particle interactions. The mechanism of electrophilic nitration of aromatic hydrocarbons is presently a subject of intense discussion [112]. The one-electron oxidation of PAH, resulting in a radical cation intermediate, is being reconsidered as an alternative pathway to nitro derivatives, in contrast to the classical nitronium cation attack on the parent PAH.

It has been observed that the decomposition rates of PAH in nitration reactions correlate with their oxidation and ionization potentials [113]. Thus the facile formation of cation radicals of reactive PAH has been utilized for preparing nitro derivatives [114]. In the presence of NO_2 of HONO, PAH are likely to be converted into this cation radical and subsequently to combine with the electron-acceptor species [113,115]. So far, no conclusive evidence for this cation radical mechanism has been put forward. Furthermore, in complex systems such as airborne particulate matter, these same radical cations may also undergo a number of other competitive reactions, and the final outcome with respect to nitrated PAH may not be as significant as simple model experiments would predict.

Based on measurements of the decomposition rates of PAH in weakly acidic solutions containing small concentrations of nitrate, nitrous acid, and dinitrogen tetroxide, Nielsen [116] has proposed a reactivity scale for electrophilic

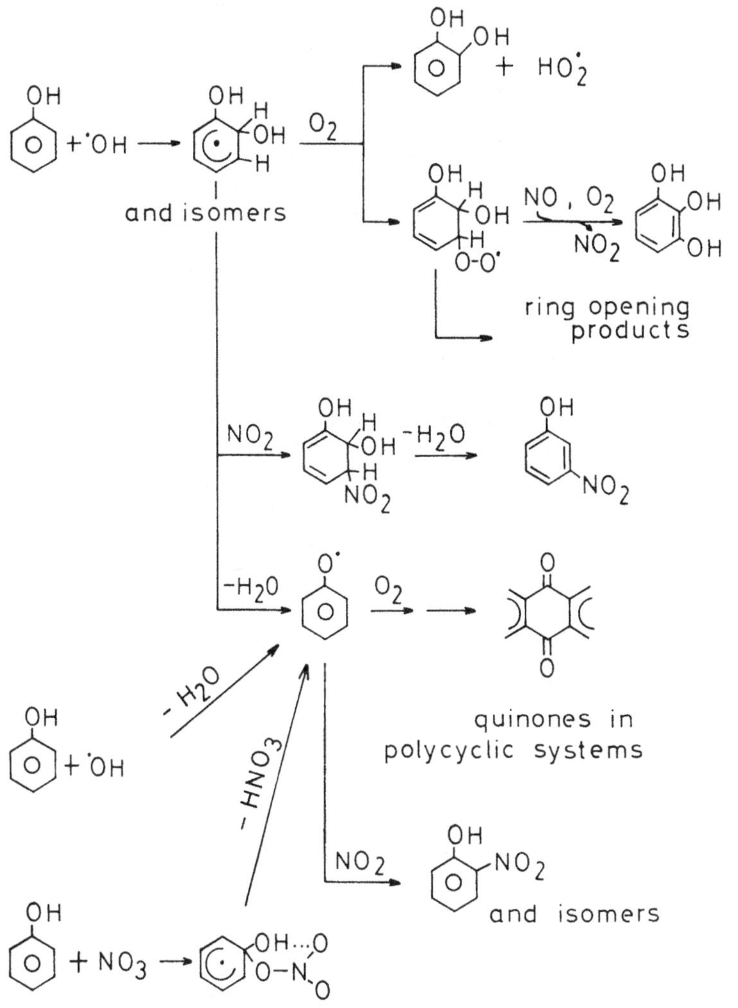

Figure 4. Reaction pathways of phenol with the hydroxyl and nitrate radicals and with nitrogen dioxide. (From Ref. 76.)

substitution of PAH. Whenever available, experimental results of other authors on the relative reactivity of adsorbed PAH seem to be in agreement with these measurements. Thus, referring only to the most common PAH found in particulate matter, this classification is as follows: BaP, perylene > benz[a]anthracene, benzo[ghi]perylene, cyclopenteno[cd]pyrene, pyrene > chrysene, benzo[e]pyrene, dibenzoanthracenes > fluoranthene, indeno[1, 2,3-cd]pyrene, benzofluoranthenes. Furthermore, as expected, mesomeric electron-donating substituents (e.g., hydroxy) will enhance the electrophilic reactivity of a PAH derivative, while mesomeric electron-withdrawing groups (e.g., keto) will desactivate the polycyclic aromatic system.

With respect to the exposure of PAH to NO_x directly on particle surfaces, some interesting experiments from the literature must be discussed. Jäger and Hanus [86] studied the interaction of gaseous nitrogen dioxide at a concentration of 1.3 ppm in air with four PAH adsorbed on silica gel, alumina,

coal fly ash, and carbon deposits in a glass reactor. The rate of nitration
was shown to decrease in the order silica gel, fly ash, alumina, and carbon
and was irrespective of the adsorbed PAH, thus illustrating the effects of the
carrier on the kinetics of the conversion. Hughes et al. [93] exposed coal
fly ash enriched with PAH through vapor-phase adsorption [94] to gaseous
concentrations of 100 ppm of NO_2 and demonstrated the formation of both
nitration and oxidation products. Similar conversion was also observed on
other types of particles (e.g., silica gel, alumina, and charcoal). Butler and
Crossley [95] used flame-generated soot as the carrier to study the conversion of the naturally present PAH in a reaction chamber with ambient air containing 5-10 ppm nitrogen oxides. The degradation of PAH by NO_x occurred
readily and resulted in half-lives varying between 7 and 30 days for BaP and
phenanthrene, respectively. In general, our knowledge of the chemistry and
fate of nitro derivatives of PAH is very limited. The most likely transformation of nitro-arenes is probably photodegradation [117]. However, surprising
differences in photochemical stability have been noted between isomers; the
1- and 3-nitrobenzo[a]pyrenes appear to be photostable, while the 6-nitro
isomer will degrade readily on irradiation [30,117]. The photodegradation of
nitro-PAH will involve a rearrangement of the nitro group to the corresponding aryl nitrate and subsequent elimination of nitrogen monoxide to form a
phenoxy radical [118,119]. These phenoxy radicals will then be available for
further oxidation to the quinones or for nitration to form nitrohydroxy derivatives. The latter class of compounds, accessible by several mechanisms
discussed before, has been identified very recently in diesel particulate matter [120].

IX. SAMPLING ARTIFACTS IN HIGH-VOLUME FILTRATION

A major problem in the assessment of atmospheric degradation reactions of
PAH is the possibility of transformation of parent PAH during the act of sampling of airborne particulate matter. The sensitivity of current analytical techniques (GC, LC, GC/MS) for the quantitative determination of the major PAH
normally requires the collection of airborne particulate matter during periods,
varying between 2 and 24 hr under high-volume flow conditions (20-60 m^3/
hr). Furthermore, the collection of similar amounts of PAH derivatives with
the aerosol is estimated to require sampling volumes increased by one additional order of magnitude. This means that a prolonged gas-particle interaction will occur during the act of sampling, during which particulate matter is
exposed to gaseous concentrations of several pollutants.

Since diurnal profiles of these gaseous pollutants are sometimes very
pronounced (e.g., with peaks in NO_2 and O_3 production in photochemical
smog), the chemical equilibrium existing between the gas and particulate
phase suspended in the air can be seriously displaced in a matter of hours,
and artifactual transformation of PAH on the sampling substrate may occur
when the rate constants are sufficiently high. The importance of this effect
is presently difficult to assess and will depend on the age of the collected
particles and their whole previous history of gas-particle interactions during
atmospheric transport. An interesting experiment particularly relevant to
this problem has been performed by Brorström and coworkers [121,122].
During 24 hr of high-volume sampling of urban particulate matter on glass
fiber filters in winter, the effect of adding traces of nitrogen dioxide (about
1 ppm) and ozone (about 0.2 ppm) was investigated in parallel sampling experiments. Upon exposure to nitrogen dioxide, degradation of pyrene, benz-

[a]anthracene, and BaP in the range 20%, 40%, and 60%, respectively, was observed over that sampling period. Furthermore, a three- to fourfold enhancement of direct-acting mutagenicity was detected both in nitroreductase proficient and deficient strains [123]. Thus both the chemical composition and the biological activity of the NO_2-exposed samples were seriously affected by artifactual reactions on the filter. Surprisingly, significant degradation of PAH upon addition of ozone seemed to occur in only one of the three experiments.

Little is known about the extent of possible reactions of PAH on the glass fiber filters widely employed for decades to collect ambient particulates versus that of actual atmospheric transformations. However, the results described suggest that they may indeed be significant. Elucidation of the role of possible filter artifacts in the collection of POM from various sources is of major importance, since historically most evaluations of the carcinogenic and mutagenic activity of organic particulate extracts have been based upon samples obtained by the Hi-Vol filtration method.

On one hand, when sampling on filters, the effective contact area between the particle-adsorbed PAH and the airflow will be rapidly reduced upon loading because of the penetration of particles between the fibers. On the other hand, gaseous pollutants have a tendency to adsorb effectively to particle surfaces or to the filter material, thus creating locally high levels of reactants.

Furthermore, a clear distinction should be made between the exposure experiments, in which PAH are finely distributed directly onto the filter fiber material and are thus in direct contact with it and those in which the filter serves only as a convenient support, with a high surface area for genuine aerosol particles and their associated PAH. In the first case filter-catalyzed reactions can be significant (e.g., the well-known artifactual nitrate formation on glass fiber) [124] and better deactivated filter materials have been recommended (quartz, Teflon-coated glass fiber, Teflon membrane filter) [125]. In contrast, in the case of diesel particulate matter sampling, no effect of filter type and increased filter loading was observed on the chemical composition and mutagenicity of the exhaust extracts [126]. The PAH associated with the diesel particles are adsorbed on an elemental carbon nucleus and do not seem to interact chemically with the filter substrate. Whether similar conclusions may be advanced for ambient particulate matter remains to be investigated.

Reactive gases can be efficiently eliminated from the airflow prior to the collection of the aerosol by the use of denuders. During laminar flow through a cylindrical tube both gas and particles diffuse to the walls, where they can be adsorbed into a film of liquid or solid coating. Because of the much higher diffusion coefficient of gases, their depletion is orders of magnitude higher than for particles, and thus further interaction between gases and particles is prevented [127]. However, the generalized application of this method with respect to sampling of POM has not found a breakthrough yet, in view of the design problems of versatile denuder systems, which will consecutively strip the airflow from all reactive gases at the high rates typical of high-volume sampling.

X. PARAMETERS INVOLVED IN REACTIVITY OF PAH

All the experiments described above give strong indications that transformations of PAH, adsorbed onto suspended particulate matter in the atmosphere, can occur by gas-particle interactions with reactive species, but the quantitative aspects of these degradations are highly uncertain or have not been determined at all. In many experiments designed to simulate atmospheric reactivity, both the particle material and the exposure conditions used are often chosen as a function of the available experimental facilities. Thus simplifications are introduced in the experimental techniques, which are certainly appropriate for fundamental studies (e.g., the choice of model supports, alumina, charcoal, coated with single PAH to study reaction products) but which make it difficult to extrapolate kinetic data obtained from laboratory exposures to the actual gas-aerosol chemistry of the atmosphere. Both in ambient and in source-emitted aerosols, the PAH released in combustion process will be incorporated into the particles by adsorption and condensation mechanisms. As a result, PAH are found almost exclusively in the accumulation mode of the aerosol (i.e., the respirable particles < 2 μm and largely in the submicron fraction) [2]. In fact, the particle size distribution of the mass concentration of PAH measured by cascade impaction corresponds very well with the surface distribution of the aerosol particles ($\Delta M/\Delta \log D_p$ and $\Delta S/\Delta \log D_p$).

Model experiments, in which a convenient support is spiked with a single PAH, are normally performed using coarse particles (e.g., alumina or fly ash, 40-60 μm) easy to manipulate both with respect to obtaining a uniform coating with PAH and as to further exposure conditions (filter, expanded bed). However, the specific surface of such particles is usually one order of magnitude less than that of ambient particulate matter. Therefore, this type of experiment should largely be reserved for product characterization studies, where only chromatographic separation of the reaction mixture from one PAH at a time must be achieved. When relevant kinetic data on PAH transformations must be obtained, genuine particulate matter of submicron size should be preferred.

It is useful to recognize some of the problems encountered in gas-particle exposure facilities. Whenever the aerosol under investigation cannot be produced directly on site and introduced into a reaction chamber (e.g., diesel exhaust, combustion soot), a collection step is required, typically yielding aerosol material finely divided over a filter surface. This usually means that filter samples must be used in further exposure studies. Fibers have some advantages: They can be held in a relatively rigid array, which is not subjected to the agglomeration and channeling problems encountered in contacting fine particles with gases. Furthermore, their surface-to-volume ratios are still realistic for the smallest practically attainable fiber diameters (in the micron range), compared to that of a suspended aerosol. However, the artifact considerations mentioned before can affect the kinetic results. Otherwise, the exposure of aerosols to gaseous pollutants can be performed in a continuous flow reactor or a static chamber. The use of a flow reactor [55] is restricted to relatively fast degradation processes and requires a stable aerosol input. The static chamber present two types of problems: Sufficient aerosol must be kept suspended in the chamber to yield a useful analytical sample, yet the concentration of aerosol must be low enough to keep coagulation rates within reasonable limits. Ultimately, deposition of particles will occur through different mechanisms: diffusion-controlled sedimentation and thermoforetic and electroforetic deposition. Calculations of the stability of a log-normally distributed aerosol with mediam diameter $\mu = 0.5$ μm ($\sigma = 2$) and initial concentration 10 mg/m^3 resulted in acceptable losses of 15% over an exposure period of 4 hr at 0.5 μm(μ) and up to 58% at 1.9 μm ($\mu \times \sigma^2$) [128].

The use of higher particle concentrations is an evident advantage of the flow reactor. Designed as a fluidized bed, this reactor has excellent capacity for semipreparative work, but does not produce well-expanded beds of submicron-size particles. Moreover, coagulation rapidly tends to enlarge the initial particle size distribution, thus making fluidization less effective. This phenomenon can be partly neutralized by producing a face velocity gradient in the reactor by giving it the shape of an inverted cone. Thus the requirement for a narrow particle size distribution is somewhat attenuated.

Each aerosol particle constitutes a matrix of a particular size, shape, and chemical composition. Thus fly ash consists mainly of spherical particles built from aluminosilicates and metal oxides; diesel exhaust consists of chain aggregates of carbon particles. To this matrix one or more surface layers of PAH are being adsorbed upon cooling of the emission. The chemical reactivity of PAH will therefore be affected by two factors, one of which is physical, the other chemical in nature. When a reactive compound is finaly divided over a surface, its accessibility for gaseous molecules will depend on the particle size and specific surface area. Although this parameter is probably more or less constant for fiber filters, coated with PAH from solutions it can vary significantly in the exposure experiments with particles. Better physical characterization of the matrix used in studying transformations of adsorbed PAH is required in terms of particle size, size range, and surface characteristics such as specific area and adsorbed water. Dissolution of reactive gases into a liquid film present on particles may be an additional factor in determining the further reactions of adsorbed PAH.

Most experiments performed so far have been conducted at low or unspecified humidity [86,95]. Yet the adsorption of nitrogen dioxide from the gas phase may be affected by the presence of water on carbonaceous particles [111]. Soot particles are able to adsorb relatively large amounts of water [110]. The nature of the adsorption sites of particle-associated PAH is not well defined [129,110], but some observations suggest that PAH are preferentially adsorbed on carbonaceous material [130]. This idea is supported by the high temperatures and the long extraction times required to obtain quantitative recovery of PAH deposited on diesel [131] and carbon black [100] particles and the observed high affinity of PAH for graphitized soot [132]. Furthermore, the chemical nature of freshly emitted soot particles is also changed in atmospheric oxidation processes during transport, producing a surface with more hydrophilic sites and free acids [129,133]. The effect of these transformations might be a higher susceptibility of the aged aerosol toward transformations of the adsorbed PAH than of fresh carbonaceous particles. Thus, in the experiments of Brorström [122], particulate sampling coincided with episodes of long-range transport of aged continental aerosols, and appreciable conversion of PAH by adding 1 ppm NO_2 during sampling was observed, especially for acidic particles. In contrast, De Wiest [53] did not observe any significant difference between the rate of degradation of benz[a]-anthracene adsorbed on the original and on a chemically oxidized commercial carbon black in photochemical oxidation experiments in aqueous suspension.

Catalytic effects of the particle matrix leading to a stabilization of the parent PAH toward photodegradation have also been observed (e.g., with fly ash from coal-fired power plants). This suppression of photochemical oxidation of PAH, adsorbed onto a fly ash matrix from the vapor phase [57], has been explained tentatively by the role of transition metal ions at the particle surface of fly ash in quenching excited electronic states of PAH formed upon irradiation, since photodegradation did occur readily on silica

and alumina supports. From these examples it follows that further investigations will be required to unravel the mechanisms by which the matrix can influence the reactivity of the adsorbed PAH.

Reactions that involve irradiation are even more complex. PAH which are
deposited inside porous particles or inner layers of PAH can easily be shielded from the incident light. Also, the light-absorbing or light-reflecting
characteristics of the carrier should be taken into account. Additionally,
these parameters may vary with the actual procedure of exposure (suspended
aerosol, reflectance of chamber walls, revolving flask, fluidized bed, filter
irradiation, etc.), and actinometric measurements will normally be required
to extrapolate the data obtained in irradiated chambers to atmospheric conditions [53]. At a given concentration, the distribution of PAH over the particle surface can correspond to less than one monolayer or to multilayers. For
most types of aerosols total PAH concentrations will be below the 1000 ppm
range by weight. Considering that the specific surface of submicron particles
is about 100 m^2/g, it follows that a homogeneous distribution would normally
not result in multilayers. However, other compounds classes are present as
well (e.g., n-alkanes in diesel particulate matter in 10 times higher concentrations). Since adsorption and condensation from the gas phase onto the
particles will normally occur in the order of increasing vapor pressure (PAH
< n-alkanes), this can result in multilayering. Obviously, this penomenon
will affect the heterogeneous reactivity in general, and only outer layer(s)
directly exposed to the gas phase may react, while inner layer(s) are being
protected from further attack by this initially formed surface layer of products. In that respect, the coating of aerosol particles with single PAH
("spiking") is often difficult to control either from solution or from the vapor
phase and may result in an inhomogeneous distribution of PAH. Data on the
reactivity of these spiked particles should therefore be considered with care.

XI. CONCLUSIONS

From the discussion above it follows that the characterization of the physical
and chemical environment of the reactive sites of a particle on a molecular
level will be required to better understand its fate and that of the adsorbed
PAH in the atmosphere. Exposure experiments should be continued, on two
different levels:

1. The exposure of PAH-spiked particles provides a simplified model system
   to study mechanisms and products of PAH transformation. Extrapolations
   of kinetic data from these experiments should be regarded with caution.
   However, isolation and spectroscopic identification of PAH derivatives
   from these samples provide baseline information for the search and eventual detection of these compounds in genuine aerosol samples, whose organic composition is often too complex for direct identification of polar
   PAH derivatives.
2. The exposure of natural combustion-related aerosols of various particle
   size and matrix composition to gaseous pollutants in reaction chambers
   should yield realistic kinetic information on PAH degradation in the atmosphere. Filter exposures of particulate matter will also give additional
   information on the extent of sampling artifacts.

In the literature, several experiments provide evidence for the high reactivity of various PAH both in photochemical degradation and in dark reactions with ozone or nitrogen dioxide under simulated atmospheric conditions.

In these laboratory experiments direct mutagens were produced in the transformation of PAH, which are likely to contribute also to the direct mutagenicity of ambient particulate matter. The important role of the physical and chemical properties of the particle matrix in PAH transformation has been recognized but is not well understood. Current experiments indicate at least surprising differences in rates of conversion. Since often full product identification has not been attempted, the effects of the matrix on mechanisms and chemical structures of the compounds are not clear. This problem is partially related to the urgent need for developing selective analytical methods for the qualitative and quantitative determination of different classes of PAH derivatives. The thermo- or photolabile character of several PAH derivatives will usually result in the application of sophisticated analytical methodology (e.g., LC/MS, MS/MS) to produce unambiguous structural information on these species [102]. Even then, the transformations occurring in exhaust systems or stacks cannot be clearly differentiated from those occurring during sampling and subsequent analytical protocol.

The nitroarenes are a first class of PAH derivatives for which the combined approach of model and aerosol exposure studies has resulted in their identification in polluted air. However, biological tests as well as theoretical considerations of the mechanisms of formation of nitro-arenes suggest that more complex nitro-PAH derivatives (e.g., nitrohydroxyarenes) are likely to be present in substantial amounts in winter aerosol samples.

PAH sulfonic acids and other SO_x-derived products have been detected only in laboratory exposure studies. Reactions are likely to occur under plume conditions but particle-catalyzed transformations, especially upon irradiation, have not been thoroughly investigated. Furthermore, fractionation schemes for the separation of organic extracts from aerosols do not provide a reliable scheme for characterization of water soluble derivatives.

Under conditions of photochemical air pollution, several mechanisms of oxidation can be advanced to account for fast degradation of PAH: OH radical attack, and reaction of singlet molecular oxygen and ozone. Major reaction products in the three cases seem to be hydroxy derivatives and quinones, but other reaction products may be typical of specific oxidation processes: endoperoxides, epoxides, and ring fission products. PAH derivatives containing keto and quinone functional groups have been identified in ambient aerosols and in a variety of emission samples. The reactions of PAH with peroxyacyl nitrates have not been studied. Finally, it should be stressed that it will remain a difficult task to make reliable predictions on the atmospheric transformations of particle-associated PAH with gaseous pollutants, even when the single gas-particle interaction processes have been characterized in chamber exposure studies, since the occurrence of synergetic effects of different pollutants on the transformation rates, and the possibility that chemical processes may change the chemical nature of the particles themselves during atmospheric transport, will also complicate the picture [134].

ACKNOWLEDGMENT

I am indebted to T. Nielsen, Risø National Laboratory, Denmark, for valuable reference material, especially on the formation and presence of nitro derivatives of PAH [107].

REFERENCES

1. L. Van Vaeck and K. Van Cauwenberghe, Atmos. Environ. *12*:2229 (1978).
2. L. Van Vaeck, G. Broddin, and K. Van Cauwenberghe, Environ. Sci. Technol. *13*:1494 (1979).
3. J. Leiter, M. B. Shimkin, and M. J. Shear, J. Natl. Cancer Inst. *3*: 155 (1942).
4. J. Leiter and M. J. Shear, J. Natl. Cancer Inst. *3*:455 (1943).
5. P. Kotin, H. L. Falk, P. Mader, and M. Thomas, Arch. Ind. Hyg. *9*:153 (1954).
6. W. C. Hueper, P. Kotin, E. C. Tabor, W. W. Payne, H. L. Falk, and E. Sawicki, Arch. Pathol. *74*:89 (1962).
7. E. Sawicki, Talanta *16*:1231 (1969).
8. K. C. Lao, R. S. Thomas, H. Oja, and L. Dubois, Anal. Chem. *45*: 908 (1973).
9. G. Grimmer, H. Böhnke, and A. Glaser, Zentralbl. Bakteriol. Hyg., I Abt. Orig. B *164*:218 (1977).
10. S. S. Epstein, S. Joshi, J. Andrea, N. Mantel, E. Sawicki, T. Stanley, and E. C. Tabor, Nature *212*:1305 (1966).
11. R. H. Kigdon and J. Neal, Tex. Rep. Biol. Med. *29*:110 (1971).
12. A. E. Freeman, P. J. Price, R. J. Bryan, R. J. Gordon, R. V. Gilden, G. J. Kelloff, and R. J. Huebner, Proc. Natl. Acad. Sci. USA *68*:445 (1971).
13. U. Mohr, H. Reznik-Schuller, G. Reznik, G. Grimmer, and J. Misfeld, Zentralbl. Bakteriol. Hyg., I Abt. Orig. B *163*:425 (1976).
14. R. J. Gordon, R. J. Bryan, J. S. Rhim, C. Demoise, R. G. Wolford, A. E. Freeman, and R. J. Huebner, Int. J. Cancer *12*:223 (1973).
15. B. N. Ames, J. McCann, and E. Yamasaki, Mutat. Res. *31*:347 (1975).
16. B. N. Ames and J. McCann, in *Screening Tests in Chemical Carcinogens*, IARC Scientific Publ. 12, Lyon, France, 1976.
17. J. N. Pitts, Jr., D. Grosjean, T. M. Mischke, V. F. Simmon, and D. Poole, Toxicol. Lett. *1*:65 (1977).
18. H. Tokiwa, H. Tokeyoshi, K. Morita, K. Takahashi, N. Soruta, and Y. Ohnishi, Mutat. Res. *38*:351 (1976).
19. R. Talcott and E. Wei, J. Natl. Cancer Inst. *58*:449 (1977).
20. K. Teranishi, K. Hamada, and H. Watanabe, Mutat. Res. *56*:273 (1978).
21. M. Møller and I. Alfheim, Atmos. Environ. *14*:83 (1980).
22. Y. Y. Wang, R. E. Talcott, R. F. Sawyer, S. M. Rappaport, and B. T. Wei, paper presented at the Symposium on Application of Short Term Bioassays in the Fractionation and Analysis of Complex Environmental Mixtures, U.S. EPA, Williamsburg, Va., 1978.
23. D. S. Barth and S. M. Blacker, paper presented at the 71st Air Pollution Control Association Annual Meeting, Houston, 1978.
24. C. E. Chrisp, G. L. Fisher, and J. E. Lammert, Science *199*:73 (1978).
25. D. A. Kaden and W. G. Thilly, in *Proceedings of the Workshop on Unregulated Diesel Emissions and Their Potential Health Effects*, U.S. Department of Transportation, Washington, D.C., 1978, p. 612.
26. R. A. Pelroy and D. L. Stewart, Mutat. Res. *90*:297 (1981).
27. C. H. Ho, B. Clark, M. Guerin, B. Barkenbus, T. Rau, and J. Epler, Mutat. Res. *85*:335 (1981).
28. B. J. Dean, Mutat. Res. *47*:75 (1978).

29. J. Brown and R. Brown, Mutat. Res. *40*:203 (1976).
30. J. N. Pitts, Jr., K. A. Van Cauwenberghe, D. Grosjean, J. P. Schmid, D. R. Fitz, W. L. Belser, Jr., G. B. Knudson, and P. M. Hynds, Science *202*:515 (1978).
31. J. N. Pitts, Jr., D. M. Lokensgard, P. S. Ripley, K. A. Van Cauwenberghe, L. Van Vaeck, S. D. Schaffer, A. J. Thill, and W. L. Belser, Jr., Science *210*:1347 (1980).
32. R. S. Berry and P. A. Lehman, Annu. Rev. Phys. Chem. *22*:47 (1971).
33. D. M. Jerina, H. Yagi, O. Hernandez, P. M. Dansette, A. W. Wood, W. Levin, R. L. Chang, P. G. Wislocki, and A. H. Conney, in *Polynuclear Aromatic Hydrocarbons: Chemistry, Metabolism, and Carcinogenesis,* R. I. Freudenthal and P. W. Jones (Eds.), Raven Press, New York, 1976.
34. H. R. Glatt and F. Oesch, Mutat. Res. *36*:379 (1976).
35. R. Talcott and W. Harger, Mutat. Res. *79*:177 (1980).
36. A. Streitwieser, Jr., *Molecular Orbital Theory for Organic Chemists,* Wiley, New York, 1961.
37. E. J. Moriconi, W. F. O'Connor, and F. T. Wallenberger, J. Am. Chem. Soc. *81*:6466 (1959).
38. J. T. Herron, J. Chem. Phys. *45*:1854 (1966).
39. R. A. Cox and R. G. Derwent, J. Photochem. *4*:139 (1975).
40. J. N. Pitts, Jr., A. W. Khan, E. B. Smith, and R. P. Wayne, Environ. Sci. Technol. *3*:241 (1969).
41. B. J. Finlayson and J. N. Pitts, Jr., Science *192*:111 (1976).
42. J. Peters and B. Seifert, Atmos. Environ. *14*:117 (1980).
43. A. Bjørseth, G. Lunde, and A. Lindskog, Atmos. Environ. *13*:45 (1979).
44. National Academy of Sciences, Committee on Biological Effects of Atmospheric Pollutants, *Particulate Polycyclic Organic Matter,* NAS, Washington, D.C., 1972.
45. M. N. Inscoe, Anal. Chem. *36*:2505 (1964).
46. M. Inomata and C. Nagata, Gann *63*:119 (1972).
47. P. A. Schaap (Ed.), *Singlet Molecular Oxygen,* Dowden, Hutchinson & Ross, Stroudsburg, Pa., 1976.
48. A. W. Khan, J. N. Pitts, Jr., and E. B. Smith, Environ. Sci. Technol. *1*:656 (1967).
49. J. W. Coomber and J. N. Pitts, Jr., Environ. Sci. Technol. *4*:506 (1970).
50. N. E. Geacintov, *Reactivity of Polynuclear Aromatic Hydrocarbons with O_2 and NO in the Presence of Light,* Publ. EPA-650/1-74-010, U.S. Environmental Protection Agency, Washington, D.C., 1973.
51. B. Grossman Parrondo, Photooxidation of polyaromatic hydrocarbons by singlet molecular oxygen. Ph.D. thesis, University of Antwerp, Wilrijk, Belgium, 1976.
52. M. A. Fox and S. Olive, Science *205*:582 (1979).
53. F. De Wiest, in *Physico-Chemical Behaviour of Atmospheric Pollutants,* Proceedings of the First European Symposium, B. Versino and H. Ott (Eds.), Commission of the European Communities, 1980, p. 185.
54. H. L. Falk, I. Markul, and P. Kotin, Arch. Environ. Health *13*:13 (1956).
55. B. D. Tebbens, J. F. Thomas, and M. Mukai, Am. Ind. Hyg. Assoc. J. *27*:415 (1966).

56. J. F. Thomas, M. Mukai, and B. D. Tebbens, Environ. Sci. Technol. 2:33 (1968).

57. W. A. Korfmacher, E. L. Wehry, G. Mamantov, and D. F. S. Natusch, Environ. Sci. Technol. 14:1094 (1980).

58. D. F. Barofsky and E. J. Baum, J. Am. Chem. Soc. 98:8286 (1976).

59. R. C. Pierce and M. Katz, Environ. Sci. Technol. 10:45 (1976).

60. T. L. Gibson and L. L. Smith, J. Org. Chem. 44:1842 (1979).

61. D. A. Lane and M. Katz, in Fate of Pollutants in the Air and Water Environments, Part 2, I. A. Suffet (Ed.), Wiley-Interscience, New York, 1977, p. 137.

62. R. Criegee, Rec. Chem. Prog. 18:111 (1957).

63. P. S. Bailey and F. Dobinson, Chem. Ind. 632 (1961).

64. H. E. O'Neal and C. Blumstein, Int. J. Chem. Kinet. 5:397 (1973).

65. M. Kemps, Ozonolysis by polycyclic aromatic hydrocarbons on model carriers under simulated atmospheric conditions. Ph.D. thesis, University of Antwerp, Wilrijk, Belgium, in progress.

66. P. S. Bailey, J. E. Batterbee, and A. G. Lane, J. Am. Chem. Soc. 90:1027 (1968).

67. E. J. Moriconi, W. F. O'Connor, and F. T. Wallenberger, Chem. Ind. 22 (1959).

68. C. Danheux, L. Hanoteau, R. H. Martin, and G. Van Binst, Bull. Soc. Chim. Belg. 72:289 (1963).

69. E. Moriconi, B. Rackoczy, and W. O'Connor, J. Am. Chem. Soc. 83: 4618 (1961).

70. E. Sawicki, Arch. Environ. Health 14:46 (1967).

71. S. B. Radding, T. Mill, C. W. Gould, D. H. Liu, H. L. Johnson, D. C. Bomberger, and C. J. Tojo, The Environmental Fate of Selected PAH, Stanford Research Institute, EPA Publ. 68-01-2681, 1976.

72. K. Van Cauwenberghe and L. Van Vaeck, in Advances in Mass Spectrometry, Vol. 8, A. Quayle (Ed.), Heyden, London, 1980, p. 1499.

73. P. L. Grover and P. Sims, Biochem. Pharmacol. 19:2251 (1970).

74. L. Van Vaeck and K. Van Cauwenberghe, Atmos. Environ., in press (1983).

75. R. Atkinson, K. R. Darnall, A. C. Lloyd, A. M. Winer, and J. N. Pitts, Jr., in Advances in Photochemistry, Vol. 11, J. N. Pitts, Jr., G. S. Hammond, and K. Gollnick (Eds.), Wiley, New York, 1979, p. 375.

76. R. Atkinson, W. P. Carter, K. R. Darnall, A. M. Winer, and J. N. Pitts, Jr., Int. J. Chem. Kinet. 12:779 (1980).

77. D. Grosjean, K. Van Cauwenberghe, D. R. Fitz, and J. N. Pitts, Jr., paper presented at American Chemical Society Meeting, Anaheim, Calif., 1978.

78. D. Schuetzle, F. S. C. Lee, T. J. Prater, and S. B. Tejada, Int. J. Environ. Anal. Chem. 9:93 (1981).

79. J. Jacob, A. Schmoldt, and G. Grimmer, Carcinogenesis 2:395 (1981).

80. V. Chekulaev and I. Shevchuk, Eesti NSV Tead. Akad. Toim. Keem. 30:138 (1981).

81. N. Zevos and K. Sehested, J. Phys. Chem. 82:138 (1978).

82. A. Volz, D. H. Erhalt, and R. G. Derwent, J. Geophys. Res. C86: 5163 (1981).

83. R. A. Perry, R. Atkinson, and J. N. Pitts, Jr., J. Phys. Chem. 81:296 (1977).

84. W. P. Carter, A. M. Winer, and J. N. Pitts, Jr., Environ. Sci. Technol. *15*:829 (1981).

85. J. E. Bennett, D. J. Ingram, and J. G. Tapley, J. Chem. Phys. *23*:215 (1955).

86. J. Jäger and V. Hanus, J. Hyg. Epidemiol. Microbiol. Immunol. *24*:1 (1980).

87. J. G. Calvert, F. Su, J. W. Bottenheim, and O. P. Strausz, Atmos. Environ. *12*:197 (1978).

88. R. J. Charlston, D. S. Covert, T. V. Larson, and A. P. Waggoner, Atmos. Environ. *12*:39 (1978).

89. T. Novakov, S. G. Chang, and A. B. Harker, Science *259*:186 (1974).

90. P. M. Heertjes, H. C. Van Beck, and G. I. Grimmon, Recl. Trav. Chim. *80*:82 (1981).

91. H. Vollmann, H. Becker, M. Corell, H. Streeck, and G. Langbein, Ann. Chem. *531*:1 (1937).

92. J. Jäger and M. Rakovic, J. Hyg. Epidemiol. Microbiol. Immunol. *18*:137 (1974).

93. M. M. Hughes, D. F. S. Natusch, D. R. Taylor, and M. V. Zeller, in *Polynuclear Aromatic Hydrocarbons: Chemistry and Biological Effects*, A. Bjørseth and A. J. Dennis (Eds.), Battelle Press, Columbus, Ohio, 1980, p. 1.

94. A. M. Miguel, W. A. Korfmacher, E. L. Wehry, G. Mamantov, and D. F. S. Natusch, Environ. Sci. Technol. *13*:1229 (1979).

95. J. D. Butler and P. Crossley, Atmos. Environ. *15*:91 (1981).

96. J. N. Pitts, Jr., unpublished results, 1978.

97. T. Nagai, K. Terauchi, and N. Takura, Bull. Chem. Soc. Japan *39*:868 (1966).

98. M. J. Dewar, T. Mole, D. S. Urch, and E. W. Warford, J. Chem. Soc. 3572 (1956).

99. J. Jäger, J. Chromatogr. *152*:575 (1978).

100. W. L. Fitch, E. T. Everhart, and D. H. Smith, Anal. Chem. *50*: 2122 (1978).

101. H. S. Rosenkranz, E. C. McCoy, D. R. Sanders, M. Butler, D. K. Kiriazides, and R. Mermelstein, Science *209*:1039 (1980).

102. D. Schuetzle, T. L. Riley, T. J. Prater, T. M. Harvey, and O. F. Hunt, Anal. Chem. *54*:265 (1982).

103. T. L. Gibson, Atmos. Environ. *16*:2037 (1982).

104. T. Nielsen, Anal. Chem. *55*:286 (1983).

105. C. Y. Wang, M. S. Lee, C. M. King, and P. O. Warner, Chemosphere *9*:83 (1980).

106. R. E. Talcott and W. Harger, Mutat. Res. *91*:433 (1981).

107. T. Nielsen, *Nitroderivatives of PAH: Formation, Presence and Transformation in Stacks, in Exhaust Gases and in the Atmosphere*, Risø-R-455, Risø National Laboratory, Roskilde, Denmark, 1981.

108. F. P. Tully, A. R. Ravishankara, R. L. Thompson, J. M. Nicovich, R. C. Shah, N. M. Kreutter, and P. H. Wine, J. Phys. Chem. *85*: 2262 (1981).

109. U. Platt, D. Perner, A. M. Winer, G. W. Harris, and J. N. Pitts, Jr., Geophys. Res. Lett. *7*:89 (1980).

110. F. De Wiest and P. M. Brull, in *Atmospheric Pollution*, Studies in Environmental Science, Vol. 8, M. M. Benarie (Ed.), Elsevier, Amsterdam, 1980, p. 227.

111. M. J. Matteson, in *Carbonaceous Particles in the Atmosphere*, T. Novakov (Ed.), LBL-9037, National Technical Information Service, Springfield, Va., 1979, p. 150.

112. G. A. Olah, S. C. Narang, and J. A. Olah, Proc. Natl. Acad. Sci. USA *78*:3298 (1981).

113. E. S. Pysh and N. C. Yang, J. Am. Chem. Soc. *85*:2124 (1963).

114. C. V. Ristagno and H. J. Shine, J. Am. Chem. Soc. *93*:1811 (1971).

115. J. F. Coetzee, G. H. Kazi, and J. C. Spurgeon, Anal. Chem. *48*: 2170 (1976).

116. T. Nielsen, *A Study of the Reactivity of PAH*, Nordic PAH-project, Report 10, Central Institute for Industrial Research, Oslo, 1981.

117. J. N. Pitts, Jr., and Y. A. Katzenstein, Atmos. Environ. *15*:1782 (1981).

118. Y. Ioki, J. Chem. Soc. Perkin Trans. II, 1240 (1977).

119. O. L. Chapman, D. C. Heckert, J. W. Keasoner, and S. P. Thackaberry, J. Am. Chem. Soc. *88*:5550 (1966).

120. T. E. Jensen, D. Schuetzle, T. L. Kiley, and I. Salmeen, *Proceedings of the 8th International Symposium on Polynuclear Aromatic Hydrocarbons*, Battelle, Columbus, Ohio, 1983.

121. E. Brorström, P. Grennfelt, A. Lindskog, A. Sjödin, and T. Nielsen, in *Proceedings of the 7th International Symposium on Polynuclear Aromatic Hydrocarbons*, Battelle, Columbus, Ohio, 1982.

122. E. Brorström, P. Grennfelt, and A. Lindskog, Atmos. Environ. *17*: 601 (1983).

123. G. Löfroth, R. Toftgard, J. Carlstedt-Duke, J. A. Gustafsson, E. Brorström, P. Grennfelt, and A. Lindskog, paper presented at EPA's Diesel Emissions Symposium, Raleigh, N.C., 1981.

124. W. R. Pierson, R. H. Hammerle, and W. W. Brachaczek, Anal. Chem. *48*:1808 (1976).

125. F. S. C. Lee, W. R. Pierson, and J. Ezike, in *Polynuclear Aromatic Hydrocarbons: Chemistry and Biological Effects*, A. Bjørseth and A. J. Dennis (Eds.), Battelle Press, Columbus, Ohio, 1980, p. 543.

126. R. A. Gorse, Jr., I. T. Salmeen, and C. R. Clark, Atmos. Environ. *16*:1523 (1982).

127. J. L. Durham, W. E. Wilson, and E. B. Bailey, Atmos. Environ. *12*: 883 (1978).

128. L. Van Vaeck, unpublished calculations, 1981.

129. B. D. Tebbens, M. Mukai, and J. T. Thomas, Am. Ind. Hyg. Assoc. J. *32*:365 (1971).

130. J. D. Butler, P. Crossley, and D. M. Colwill, Sci. Total Environ. *19*:179 (1981).

131. G. Grimmer, in *Symposium on Polycyclic Aromatic Hydrocarbons*, Conference Papers, Verein Deutscher Ingenieure, Düsseldorf, West Germany, 1979.

132. E. V. Kalashnikova, A. V. Kiselev, R. S. Petrova, K. D. Shcherbakova, and D. P. Poshkus, Chromatographia *12*:799 (1979).

133. R. W. Coughlin and F. S. Ezra, Environ. Sci. Technol. *2*:291 (1968).

134. T. Nielsen, T. Ramdahl, and A. Bjørseth, Environ. Health Perspect. *47*:103 (1983).

11

Reference Materials for the Analysis of Polycyclic Aromatic Compounds

WALTER KARCHER / Materials Department, Joint Research Centre of the Commission of the European Communities, Petten, The Netherlands

I.   Introduction   385

II. Preparation   386

    A. Selection and purity criteria   386
    B. Synthesis   387
    C. Purification   390
    D. Solutions and matrix materials   394

III. Characterization and Certification   396

    A. Homogeneity testing   396
    B. Purity analysis   396
    C. Evaluation of data   398

IV. Applications   403

    A. Analysis   403
    B. Microbiological testing   404

    References   405

I. INTRODUCTION

The accurate and reproducible analysis and control of polycyclic aromatic compounds in the environment (air, water, sediments, soils), in industrial exposures and in foods presents a very difficult task because of the great number of individual components that have to be separated, identified, and determined quantitatively at trace levels.

Owing to the large variability in isomer formation, polycyclic aromatic hydrocarbons can appear in combustion processes in great numbers. For instance, for the series possessing six aromatic rings, 82 isomeric configurations are possible, increasing to 333 isomeric species for PAH with seven aromatic rings [1].

In addition, a CH group may be replaced by -N-, -S-, or >NH, which in turn leads to a still higher figure for the possible isomeric configurations, since the N or S atom can appear in many different positions in the aromatic ring. The number of compounds observed in environmental matrices is multiplied further by alkyl substitutions of an H atom on the aromatic ring. Recently, NO_2- and NH_2-substituted polycyclic aromatic compounds have become of importance since pollutants of these types, exhibiting appreciable mutagenic or carcinogenic activity, were found in car exhaust (especially from diesel engines) [2] and coal liquefaction processes, respectively [3,4].

It is clear that the identification and quantitative determination of this vast number of closely related and often isomeric species at trace levels presents an almost impossible task to the analyst if most of the individual compounds to be identified are unavailable and if other relevant analytical data, such as retention indices, ultraviolet (UV) and fluorescence data, and mass spectra, are absent.

To bridge this apparent gap and provide essential elements for the reliable assessment and control of the health effects of human exposure to PAH and related compounds, a long-term project was undertaken in the frame of the R&D programs of the Commission of the European Communities (Directorate-General for Science, Research and Development: Community Bureau of References and Joint Research Centre) to develop and certify high-purity compounds for the calibration of analytical methods and apparatus [5].

The availability of a comprehensive range of well-characterized calibration and reference materials and related analytical data should considerably facilitate the reliable analysis and control of these hazardous and ubiquitous pollutants of the human environment. In the longer term, this work may also have an impact on the development and application of harmonized and standardized procedures of analysis, and on the reporting of health assessment for this important class of pollutants.

II. PREPARATION

A. Selection and Purity Criteria

The choice of reference materials to be prepared was guided primarily by the following considerations:

Environmental and occupational pollution aspects (occurrence)
Carcinogenic and toxicological aspects
Nonavailability from commercial suppliers
National and international regulations or recommendations [European Economic
    Community (EEC), World Health Organization (WHO), and U.S. Environ-
    mental Protection Agency (EPA)]
Analytical requirements

Before embarking on such a long term and costly exercise, some exploratory research was carried out to ensure that the compounds selected could be prepared homogeneously to the purity standards required; that sufficiently reliable and reproducible methods of analysis were available; and that the materials were sufficiently stable for a minimum storage time of 5 years. Some of these questions were addressed in a pilot study, which selected the most suitable techniques for homogeneity determinations, purity analysis, and quality control. The minimum purity level was set at a mass fraction of 0.99 g/g with the additional requirement to identify major impurities (>0.001 g/g) by means of mass spectrometry and/or nuclear magnetic resonance spectrometry. The batch size of the reference material varied between 10 and 50 g, according to factors such as difficulty of preparation and expected demand, to establish a bank of materials (at a sample size of 100 mg) sufficient for an extended period of use.

Reference materials for calibration and quality control in polycyclic aromatic hydrocarbon (PAH) analyses can be prepared in various forms:

1. As solid compounds of certified purity
2. As solutions containing certified or specified PAH concentrations
3. As environmental matrices containing certified PAH levels

For the preparation of solid reference materials two routes can be followed:

1. Multistage synthesis
2. Purification of commercial samples

The ultimate choice depends largely on availability and cost, and in most cases synthesis was preferred because it proved to be less expensive and tedious than the purification of commercial materials of doubtful origin and history. A survey of the 40 compounds available at present at a certified purity, including some S and N heterocycles, is given in Table 1, and Table 2 summarizes the candidate reference materials in preparation.

B. Synthesis

In most cases, scaling up of the synthetic procedures described in the literature resulted in acceptable yields and purity levels for the required batch size. However, in a few instances, alternative synthetic routes were developed, either because the yield of conventional routes was too low or because of the consistent occurrence of side products that could not easily be separated from the main component [6].

A relevant example is the synthesis of cyclopenta[cd]pyrene in the gram range. Many of the different synthetic routes found in the literature for this compound start with pyrene and involve between 8 and 10 individual steps. In a first attempt, the procedure described by Gold et al. [7] was tried (see Fig. 1). However, it was soon apparent that the overall yield decreased considerably during scale-up (possibly due to an autocatalytic side reaction), resulting in an overall yield of less than 1%.

Instead, an alternative approach was developed by Tintel et al. [8]. This yields the key intermediate product, pyrene-4-yl acetic acid, in a single step with a yield of more than 90% by adding sodium to a solution of pyrene in diethyl ether/liquid ammonia (1:1) at -78°C and reacting it with sodium iodoacetate. Also, for the cyclization step, the difficult reaction with HF was replaced by one with polyphosphoric acid, followed by reduction with $NaBH_4$ and dehydration by toluene-p-sulfonic acid (see the reaction scheme in Fig. 2).

This new approach allowed the preparation of sufficient quantities of cyclopenta[cd]pyrene. Impurities present were identified as pyrene, dihydrocyclopenta[cd]pyrene, and probably, 3-oxo-3,4-dihydrocyclopenta[cd]-pyrene (intermediate product) (see also Table 5).

Examples of the syntheses of heterocyclic compounds are presented in Figs. 3 and 4, illustrating the preparation of benzo[b]naphtho[2,3-d]thiophene and dibenz[c,h]acridine, respectively. The preparation of benzo[b]-naphtho[2,3-d]thiophene required five individual steps: an addition of maleic acid anhydride to dibenzothiophene in a mixture of tetrachloroethane and nitrobenzene in the presence of aluminum chloride, followed by reduction with hydrazine at 160°C. Ring closure was achieved in either via reaction with thionyl chloride and tin tetrachloride, after which a second reduction with hydrazine was followed by the final dehydrogenation with amorphous selenium [9-12]. The main impurity (present at about 0.2 wt %) was found to be another benzonaphthothiophene isomer.

Table 1. Pure Polycyclic Aromatic Compounds Available for Calibration and Reference Purposes[a]

Ref. matl. no.	Compound	Certified purity
46	Benzo[b]chrysene	0.995 ± 0.003
47	Benzo[b]fluoranthene	0.995 ± 0.003
48	Benzo[k]fluoranthene	0.995 ± 0.003
49	Benzo[j]fluoranthene	0.995 ± 0.003
50	Benzo[e]pyrene	0.993 ± 0.004
51	Benzo[a]pyrene	0.990 ± 0.005
52	Benzo[ghi]perylene	0.990 ± 0.004
53	Indeno[1,2,3-cd]pyrene	0.990 ± 0.005
77	1-Methylchrysene	0.990 ± 0.003
78	2-Methylchrysene	0.992 ± 0.003
79	3-Methylchrysene	0.992 ± 0.003
80	4-Methylchrysene	0.992 ± 0.003
81	5-Methylchrysene	0.995 ± 0.003
82	6-Methylchrysene	0.998 ± 0.002
91	Anthanthrene	0.995 ± 0.003
92	10-Azabenzo[a]pyrene	0.995 ± 0.004
93	1-Methylbenz[a]anthracene	0.993 ± 0.004
94	Dibenz[a,c]anthracene	0.995 ± 0.003
95	Dibenz[a,j]anthracene	0.997 ± 0.003
96	Dibenzo[a,l]pyrene	0.996 ± 0.003
97	Benzo[a]fluoranthene	0.995 ± 0.003
133	Dibenzo[a,e]pyrene	0.996 ± 0.004
134	Benzo[c]phenanthrene	0.997 ± 0.002
135	Benzo[b]naphtho[2,1-d]thiophene	0.995 ± 0.003
136	Benzo[b]naphtho[2,3-d]thiophene	0.994 ± 0.005
137	Benzo[b]naphtho[1,2-d]thiophene	0.997 ± 0.003
138	Dibenz[a,h]anthracene	0.990 ± 0.005
139	Benzo[ghi]fluoranthene	0.994 ± 0.003
140	Benzo[c]chrysene	0.995 ± 0.005
153	Dibenz[a,h]acridine	0.998 ± 0.002
154	Dibenz[a,j]acridine	0.997 ± 0.002
155	Dibenz[a,c]acridine	0.997 ± 0.002
156	Dibenz[c,h]acridine	0.994 ± 0.001

Table 1. Continued

Ref. matl. no.	Compound	Certified purity
157	Benz[a]acridine	0.997 ± 0.002
158	Benz[c]acridine	0.998 ± 0.002
159	Dibenzo[a,h]pyrene	0.992 ± 0.004
160	Fluoranthene	0.995 ± 0.004
161	Cyclopenta[cd]pyrene	0.992 ± 0.003
168	1-Nitropyrene	0.990 ± 0.006
177	Pyrene	0.997 ± 0.002

[a]These materials can be obtained through the Community Bureau of References (BCR), Directorate-General XII, Commission of the European Communities, 200 rue de la Loi, B-1049 Brussels.

The synthesis of dibenz[c,h]acridine followed the route reported by André et al. [13]: the fusion of α-naphthisatin (benzo[g]isatin) with 1-tetralone in boiling KOH/ethanol, followed by decarboxylation at 300°C with copper chromite and reduction at 300-320°C under nitrogen with a Pd/C mixture. The older, simpler method described by Senier and Austin [14], involving the fusion of α-naphthylamine with methylene chloride at 225°C, was discarded because of the consistent appearance of methyldibenz[c,h]acridine as a side product (5-10%), which is very difficult to separate from the main component.

A more general scheme which can be used for the synthesis of benz- and dibenzanthracenes and aza derivatives was described recently by Doadt et al. [15] utilizing the preparation of the corresponding anthraquinones or acridones, via ortholithiated tertiary benzamides. Anthraquinones are subsequently reduced to the corresponding PAH with aluminum tricyclohexoxide

Table 2. PAH and Related Reference Materials in Preparation

Benz[a]anthracene	9-Nitroanthracene
Chrysene	
Triphenylene	3-Nitrofluoranthene
Dibenzo[a,i]pyrene	6-Nitrochrysene
Indeno[1,2,3-cd]fluoranthene	6-Nitrobenzo[a]pyrene
Dibenzo[a,e]fluoranthene	
Coronene	1,8-Dinitropyrene
Dibenzo[c,g]carbazole	
Dibenz[a,i]acridine	
Picene	

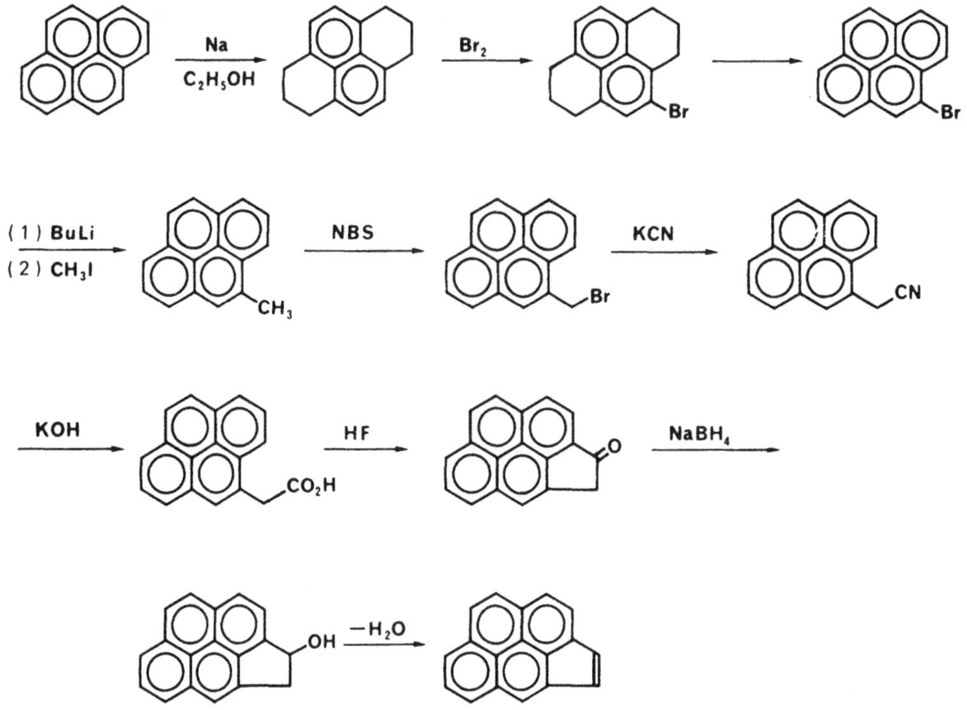

Figure 1. Synthesis scheme for cyclopenta[cd]pyrene. (From Ref. 7.)

in cyclohexanol. Acridones can be converted to the corresponding acridines following conventional reduction methods [16].

C. Purification

Various techniques are available for the purification of these materials:

Recrystallization
Sublimation

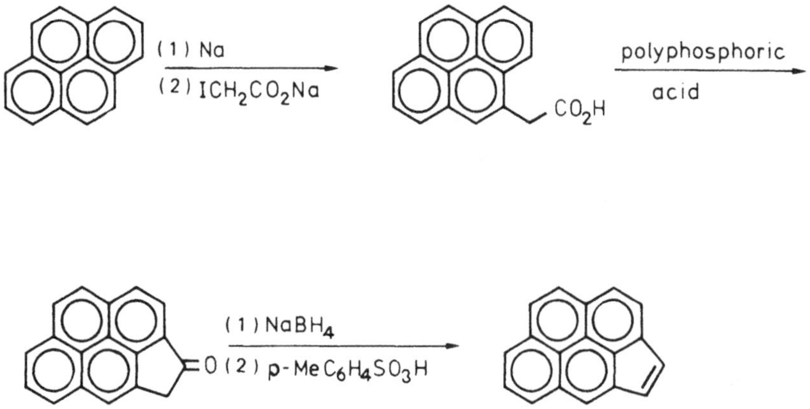

Figure 2. Short synthetic route for cyclopenta[cd]pyrene. (From Ref. 8.)

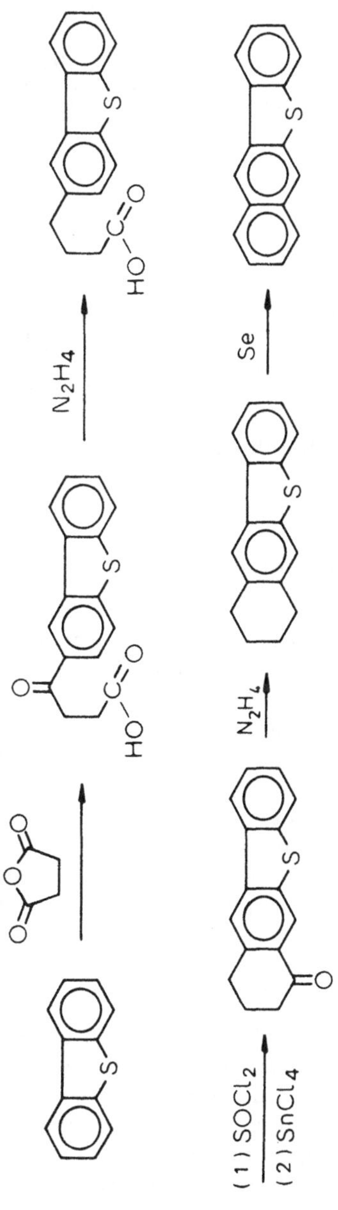

Figure 3. Synthesis scheme for benzo[b]naphtho[2,3-d]thiophene. (From Refs. 9 to 12.)

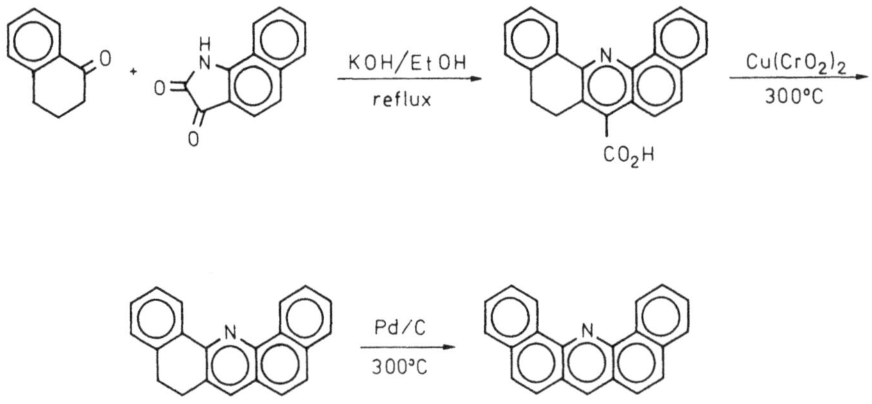

Figure 4. Synthesis scheme for dibenz[c,h]acridine. (From Ref. 13.)

Column chromatography
Preparative high-performance liquid chromatography (HPLC)
Zone melting

Generally, a combination of several of these techniques presents the most suitable approach, ensuring optimum results in terms of yield, attainable purity, and expenditure of time. Normally, the experimental approach to be selected depends critically on the batch size and purity of the impure bulk material. Thus procedures for the purification of a custom-synthesized product containing minor impurities differ markedly from the approach for a commercial batch, which often includes one or more major impurities.

In the first case, a combination of recrystallization and sublimation is often sufficient, whereas the presence of major impurities usually requires a chromatographic purification step prior to recrystallization and/or sublimation.

1. Purification of Custom-Synthesized Materials

In this case, yield is paramount since only a limited batch is available. Therefore, a combination of recrystallization and sublimation offers a suitable means of obtaining the purity standard desired in those cases where no major impurities are present, which would otherwise require chromatographic preseparation. Good results are normally obtained by a presublimation of the synthesis product, followed by one or more recrystallization steps and then final sublimation of the end product to remove impurities that cannot be vaporized (e.g., inorganics). Optimum sublimation conditions are best tested for each substance in advance. A suitable sublimation rate is on the order of 1 g/min.

Recrystallizations were carried out in benzene, xylene, toluene, ethanol, cyclohexane, acetone, or binary mixtures of these solvents. For impurities that were difficult to remove from the main component, column chromatography, usually on alumina, was used.

2. Purification of Commercial Materials

When suitable commercial PAH compounds were available as feedstock, column chromatography and/or preparative HPLC were usually applied first, followed by recrystallization and/or sublimation. In many instances, this procedure

Table 3. Heterocyclic Compounds Identified in PAH

Compound	Formula	Mass
Phenanthro[4,5-bcd]thiophene	$C_{14}H_8S$	208
Benzo[a]carbazole	$C_{16}H_{11}N$	217
Benzo[c]carbazole	$C_{16}H_{11}N$	217
Benzo[b]naphtho[2,1-d]thiophene	$C_{16}H_{10}S$	234
Benzo[b]naphtho[2,3-d]thiophene	$C_{16}H_{10}S$	234
Benzo[2,3]phenanthro[4,5-bcd]thiophene	$C_{18}H_{10}S$	258
Chryseno[4,5-bcd]thiophene	$C_{18}H_{10}S$	258
Benzo[1,11]tripheyleno[4,5-bcd]thiophene	$C_{20}H_{12}S$	282
Dinaphtho[2,1-b:1',2'-d]thiophene	$C_{20}H_{12}S$	284
Dinaphtho[2,3-b:2',3'-d]thiophene	$C_{20}H_{12}S$	284

facilitated the isolation and identification of the main impurities. A number of heterocyclic compounds having the same number of aromatic rings as the main component were consistently found in commercial samples. For example, benzocarbazoles and benzonaphthothiophenes were found in chrysene; phenanthro[4,5-bcd]thiophene in pyrene; benzo[12,3]phenanthro[4,5-bcd]thiophene, chryseno[4,5-bcd]thiophene, and dinaphthothiophenes in benzo[a]pyrene; and benzo[1,11]triphenyleno[4,5-bcd]thiophene in benzo[ghi]perylene (see Table 3). In most cases one six-membered aromatic ring of the "parent" compound is replaced by a thiophene or pyrrole ring residue [17].

To isolate impurities by column chromatography, the equipment shown in Fig. 5 proved to be of considerable value [18]. About 5 g of the compound to be purified was intimately mixed with the alumina layer on top of the column and the eluent (hexane containing 3–5% benzene) recycled through the column by means of reflux distillation. In this way, the main component can be purified to a purity level exceeding 0.995 g/g in a period of about 10 hr, isolating, at the same time, impurities in sufficient quantity to permit structural identification by NMR. This procedure not only limits the exposure of operators to these hazardous materials but also considerably reduces solvent consumption. (A comparison of gas-liquid chromatograms of a commercial product before and after purification is given in Fig. 6.)

In some cases, when isolation of separated impurities was not necessary, full-scale preparative recycle HPLC was preferred. In a typical experiment, 30 g of a material with a purity of 0.95 g/g was refined to a level between 0.99 and 0.995 g/g in 40 hr at a yield of 70 to 75%, using dried silica as column packing and hexane or a mixture of hexane and chloroform as solvent [17,18]. Preferably, direct sunlight should be excluded from all operations in order to minimize photodecomposition. Also, oxygenated solvents, such as

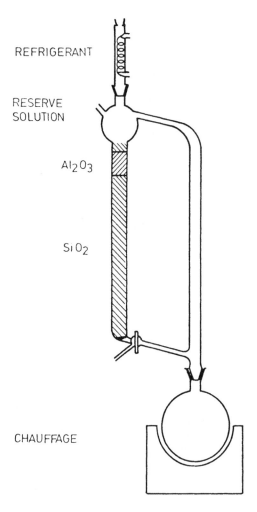

REFRIGERANT

RESERVE
SOLUTION

Al$_2$O$_3$

SiO$_2$

CHAUFFAGE

Figure 5. Purification column for continuous elution. (From Ref. 18.)

acetone, should be avoided since photosensitive PAH appear to be less stable
in these solutions than in pure hydrocarbons.

   Finally, zone melting techniques, which are generally of considerable
scope for the preparation of very pure substances, do not appear to have
been widely applied for the purification of polycyclic aromatic compounds.

D. Solutions and Matrix Materials

In response to the demand for calibration and reference materials with certified PAH concentrations which may be appropriate to environmental and occupational conditions, various solutions* or matrix materials with certified or in-

---

*Solutions in cyclohexane containing certified concentrations of the six PAH
which are the subject of regulations for drinking water (EEC Directive 1975
[19] and WHO Standard [20]) are at present being prepared for the Community Bureau of References of the European Communities.

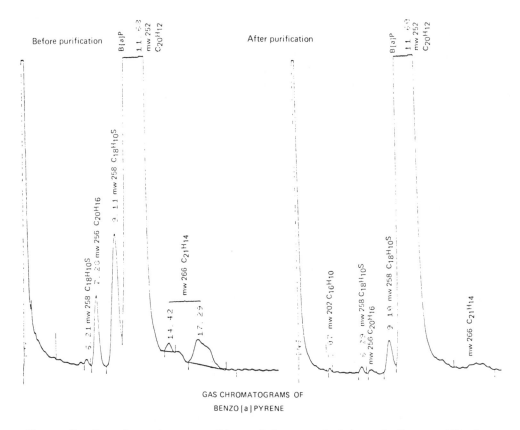

Figure 6. Gas chromatograms of benzo[a]pyrene before and after purification.

dicated PAH levels have been developed. (For a review of specimens available as Standard Reference Materials for PAH analysis through the National Bureau of Standards, see Chap. 5, Sec. VII.A.)

Solutions with certified PAH levels are sometimes preferred by users, for reasons of safety and convenience, since they eliminate the necessity of storing and handling larger quantities of carcinogenic or mutagenic substances. On the other hand, their useful shelf life may be considerably shorter than is the case for solid reference materials, due particularly to increased sensitivity to photodecomposition.

As is also the case with solid PAH reference materials, homogeneity and stability aspects are very important criteria in the preparation of these standards. This applies specifically to matrix materials, such as the urban dust that was collected over an extended period in St. Louis, where factors such as grain size, distribution, and surface area; chemical variations in composition of the matrix; and the adsorption, evaporation, and extraction behavior of the trace components to be certified may seriously affect the homogeneity and long-term stability of the candidate reference materials [21].

Therefore, homogenization or other pretreatments (e.g., aging) represent important steps in the preparation of these matrix materials. For solutions of PAH, solubility and stability data are essential requirements for the preparation of certifiable samples. Therefore, the preparation stage is normally preceded by solubility and stability studies, since the necessary data are often unavailable in the literature for the specific solvents selected.

The actual preparation of the reference solutions is carried out preferably in four phases:

1. Preparation of concentrated solutions by weighing and dissolving the selected PAH compounds in a suitable volume of solvent
2. Dilution to the required concentration level
3. Analytical determination of the individual PAH concentrations
4. Statistical evaluation and certification

III. CHARACTERIZATION AND CERTIFICATION

Before subjecting the solid candidate reference materials to the certification exercise each compound is prescreened to ensure identity, homogeneity, and approximate purity level. In most cases, the selected synthetic routes are sufficiently proven to allow the unambiguous preparation of individual compounds. In special situations, however, where similar reaction paths are employed for obtaining different but closely related isomers, confirmation of structural identity by an independent method, for instance NMR spectrometry, is desirable. As an example, it was recently found that the method described in the literature [22] for the synthesis of dibenz[a,i]acridine yields in fact another isomer, dibenz[a,j]acridine [23].

A. Homogeneity Testing

All candidate reference materials, including those purified by preparative HPLC or sublimation, were subjected to homogenization in solution by single-batch recrystallization or evaporation to dryness. Subsequently, all bulk materials were mechanically shaken for 30 min just prior to sampling for the homogeneity control. For these tests, six samples (sample size 1.5-2.5 mg) were taken at random from each material. Usually, differential scanning calorimetry (DSC) was used to estimate the distribution of purity, but in the case of materials where DSC was unsuitable because of unreproducible melting behavior and/or decomposition at the melting point, homogeneity was determined by gas chromatography (GC) analysis.

The results usually indicated no evidence of significant inhomogeneity likely to hinder the use of the materials as calibrants. DSC is also used to monitor homogeneity and long-term stability of the certified reference materials.

As a bonus, the homogeneity measurements by DSC also permitted estimation of the melting points. (The instrument was calibrated against the melting points of pure indium and lead [24].) Naturally, the melting points thus obtained are related to the purity level of the specific compounds and cannot be considered as definitive values.

B. Purity Analysis

The purity measurements for the certification exercise were carried out in the frame of a round-robin exercise, involving 12 laboratories of the European Community [25,26]. Each laboratory was supplied with a sample, selected at random, of 50-70 mg of the compound to be certified. The laboratories were requested to analyze these samples by means of at least three independent analytical methods chosen from the following:

Gas-liquid chromatography (GLC): capillary and/or packed columns
HPLC: adsorption and/or reversed phase
Mass spectrometry (MS): GC/MS or direct inlet MS, using different ionization
  potentials, etc.

It must not be forgotten in these methods that purity determinations are
essentially the estimation of total impurities detected and subtraction of this
total from unity. Efficiency of separation is therefore as important as detector
response characteristics.

1. Gas-Liquid Chromatography

This technique was used extensively for the certification analyses since it
was available in all participating laboratories. The advantages of GLC methods
are high resolution for impurities and similar response factors for most carbon
compounds if not too many heteroatoms are present and flame ionization detectors are being used.

In a typical GLC analysis, a solution of the PAH in a suitable organic solvent (dichloromethane, hexane, diethyl ether, chloroform, toluene, xylene,
CS_2) was chromatographed at temperatures between 180 and 305°C. Column
length varied between 1 and 10 m in packed columns (diameter 2-4 mm) and
20 to 30 m (diameter 0.2-0.5 mm) in capillary columns. Often, temperature
programming was employed (in the range 60-300°C) in place of isothermal
conditions.

Systematic errors in GLC analysis may be due either to retention of impurities on the column, insufficient resolution between impurities and the main
component, or partial decomposition of the main component during analysis or
during sample preparation. Whereas the first two phenomena would lead to an
overestimation of purity, the third effect would result in values of purity
which are too low. However, since various capillary and packed columns with
different column packings were used in any one exercise, it is unlikely that
organic impurities with sufficient volatility would systematically escape detection.

2. High-Performance Liquid Chromatography

Complementary to GLC, HPLC methods permit the purity determination of organic materials with high boiling points or of those which are subject to decomposition at elevated temperatures. Another advantage is the ease of separation and isolation of impurities for further analysis. However, quantitative
analysis by HPLC in combination with UV detectors necessitates the determination of the UV response factors. This in turn requires the availability of a
sizable quantity of each impurity to be determined (preferably at least 1 mg).
For the HPLC analyses, adsorption as well as reversed-phase methods were
used as far as possible.

In HPLC analysis column length was either 25 or 30 cm with internal diameters in the range 4-4.6 mm. Similar PAH concentrations were used as in
GLC analysis (around 1% by mass) and the injected sample size was between
0.2 and 20 μl. The UV absorbance was usually measured at 254 nm, and in
some cases, fluorescence detection was used in addition.

As in GLC, analytical errors in HPLC may be caused by retention of impurities in the columns or by their insufficient separation from the main component peak. A third source of error is found in the mode of quantification.
Notwithstanding the remarks above, only rarely could impurities be obtained

in sufficient quantity to permit the determination of detector response factors and consequently, a response ratio of 1 between the impurity and the main peak was assumed. This inevitably leads to errors in the calculation of the impurity content. For accurate quantitative determination of impurities, the UV spectra of individual impurities should be available. However, the close correspondance of purity determinations by the different methods is evidence that this is not a significant source of inaccuracy for very pure materials.

3. Mass Spectrometry

In addition to the combination with gas chromatography, mass spectrometry was used in the direct inlet mode for the purity analyses. Impurities that are strongly retained by GLC or HPLC columns, and therefore can be overlooked in these other techniques, can often be detected. In contrast, however, impurities with the same m/z as the major component (e.g., isomers) will not be detectable.

By using low ionization potentials (20 eV or less) and thermal programming from room temperature, a simple mass spectrum can be obtained in which the parent ions predominate [27]. In addition, some indication of chemical structure was given by the mass spectra and when the chemical nature of minor constituents was similar to the major compound, quantitative analysis was possible without determining individual response factors. When coupled to GLC, similar sources of errors may be expected as those which have been discussed for GLC. Additional errors might be caused by thermal decomposition or overfragmentation of the main component, especially if electron charge ionization is being used. Similar overfragmentation might occur in the same proportion with impurities of structure similar to that of major components, leading to an underestimation of the purity of the reference material.

4. Assessment of Organic and Inorganic Impurities

Organic impurities that were detected by more than one laboratory were reported individually for each of the certified reference materials (for an example, see Table 4). For an assessment of inorganic impurities that may be present, the ash content and some volatile elements were determined gravimetrically. In most cases, the ash mass fraction was found to be less than 0.0002 g/g, and the summed mass fractions of all inorganics were less than 0.003 g/g.

C. Evaluation of Data

In the early series, the certified purity levels were calculated from the average of individual results and the uncertainties were based on the standard deviations of individual measurements.

In line with general Community Bureau of Reference (BCR) practice, in the latest series the certified purity values were derived from the intermediate mean results for individual methods in each laboratory and the uncertainty is based on the 95% confidence limits of the mean purity. Besides overall purity, the certificate, which is issued together with the certification report with each certified compound, also states the upper limit of organic and inorganic impurities.

Before calculating certifiable values, the distribution of the accepted intermediate mean results were tested for normality. Inspection of the intermediate mean data suggested, especially for some apparently high purity compounds, that the skewed distributions would be more consistent with a log-normal distribution (i.e., one in which the logarithms of the mass fractions of the impurities are distributed normally).

Accordingly, the intermediate mean results, expressed as the mass fractions of the impurities, and in a separate calculation, their logarithms, were tested for conformity to a normal distribution using the Kolmogorov-Lilliefors test. In general, it was indeed found that results for the higher purity values could be fitted more satisfactorily to a log-normal distribution. This is not surprising in view of the high purity of these materials, in which impurities (when detected) were consistently at or near the detection limit. The materials of lower purity usually conformed more satisfactorily to a normal distribution.

Where a normal distribution for impurities was indicated, the confidence limits of the overall mean were calculated according to normal practice for small sample statistics using the relation

$$\bar{X} \pm \frac{ts}{p} \tag{1}$$

in which $\bar{X}$ is the overall arithmetic mean purity, given by

$$\bar{X} = 1 - \frac{1}{p} \sum_{i=1}^{p} \bar{Y}_i \tag{2}$$

where
$\bar{Y}_i$ = mean value of the impurity content found by one laboratory using one technique and p is the number of accepted intermediate mean results
   t = Student's t factor (at 5% significance level) taken from tables for p - 1 degrees of freedom
   s = calculated standard deviation of the overall mean $\bar{X}$

Where a log-normal distribution for intermediate means was indicated, corresponding formulas employing the logarithms of the intermediate means and leading to asymmetric confidence limits were used.

Table 5 gives an example for the evaluation of the analytical results of dibenz[c,h]acridine. For a pictorial appreciation of the results, a bar chart of the intermediate mean values together with their extreme intervals is given in Fig. 7.

As a final test of the overall homogeneity of the results for a whole series of compounds, the nonparametric Friedman test was applied to the rankings of the laboratories' means ($\bar{X}_l$) and also to the method means ($\bar{X}_m$). This showed that all laboratories were equally likely to report a high or a low result for any material analyzed (i.e., no laboratory consistently produced high or low results). Comparing the methods of analysis, it appeared that packed column GC, GC/MS, and direct inlet MS were more likely to give high results for purity than were HPLC methods and capillary column GC.

The total estimated purity of the material was calculated by a linear combination of inorganic and organic impurity results. Although this treatment yields the highest purity that can be supported technically, it inevitably leads to a somewhat pessimistic estimate of total purity.

Table 4. Summary of Organic Impurities Detected

Ref. mat'l no.	Name	Impurity m/z	mg/g	±s(n)	Identification	
153	Dibenz[a,h]acridine	293	1.7	±0.9 (9)	$C_{22}H_{15}N$	Me-
154	Dibenz[a,j]acridine	295	0.3[a]		$C_{22}H_{17}N$	H_2, Me-
155	Dibenz[a,c]acridine	295	3[a]		$C_{22}H_{17}N$	H_2, Me-
156	Dibenz[c,h]acridine	293	6.0	±1.6 (24)	$C_{22}H_{15}N$	Me-
		296	0.4[a]		$C_{21}H_{12}O_2$	?
157	Benz[a]acridine	231	0.66	±0.21(29)	$C_{17}H_{13}N$	H_2-
		243	0.21	±0.10(18)	$C_{18}H_{13}N$	Me-
		263	0.58	±0.33 (6)		?
159	Dibenzo[a,h]pyrene	306	0.70	±0.54(10)	$C_{24}H_{18}$	H_4-
		316	1.06	±0.32(16)	$C_{25}H_{16}$	Me-
		336	3.6	±1.3 (17)	$C_{24}H_{13}Cl$	Cl-
		392	0.9[a]			?

160	Fluoranthene	166	0.55	±0.07(10)	$C_{13}H_{10}$	Fluorene?
		178	0.54	±0.14(10)		?
		192	0.28	±0.04(10)		?
		204	2.8	±0.96(21)	$C_{16}H_{12}$	H_2-
		216	0.35	±0.25(11)		?
161	Cyclopenta[cd]pyrene	202	3.8	±1.4 (24)	$C_{16}H_{10}$	Pyrene
		216	1.1	±0.21(15)	$C_{17}H_{12}$	Me-pyrene?
		226	0.60	±0.27(18)	$C_{18}H_{12}$	H_2-
		240	0.25	±0.06 (5)		?
		242	0.5[a]			?
168	1-Nitropyrene	253	2.6	±1.2 (16)	$C_{16}H_{15}NO_2$	H_6-
		292	1.2	±0.8 (13)	$C_{16}H_8N_2O_4$	Dinitropyrene(s)
177	Pyrene	190	0.24	±0.11(25)	$C_{15}H_{10}$	4H-Cyclopenta[def]phenanthrene?
		204	0.65	±0.23(21)	$C_{16}H_{12}$	H_2- ?
		208	0.68	±0.28(20)	$C_{14}H_8S$	Phenanthro[4,5-bcd]thiophene

[a]Number of individual results too few for meaningful average and standard deviation.

Table 5. Mean Results for Dibenz[c,h]acridine (g/g) (BCR Ref. Mat'l 156)[a]

Lab.	GC packed	GC capillary	HPLC ads.	HPLC rev.	GC/MS	MS	$\bar{X}_1$	s_1
1	0.9940	—	0.9986	0.9900	—	—	0.9942	0.0043
2	0.9953	—	—	0.9950	—	—	0.9951_5	0.0002_1
3	—	0.9931	—	—	—	—	0.9931	—
4	0.9942	0.9936	0.9958	0.9943	0.9954	—	0.9946_6	0.0009_1
6	0.9916	—	—	—	0.9925	0.9950	0.9930	0.0018
8	0.9917	0.9983	—	—	—	0.9912	0.9937	0.0040
11	0.9952	0.9934	—	0.9916	—	0.9976	0.9944	0.0026
$\bar{X}_m$	0.9937	0.9946	0.9972	0.9927	0.9940	0.9946	—	—
s_m	0.0016	0.0025	0.0020	0.0023	0.0021	0.0032	—	—

[a] $\bar{X}$ = 0.9942, CL_+ = 0.9952, CL_- = 0.9932, and p = 21.

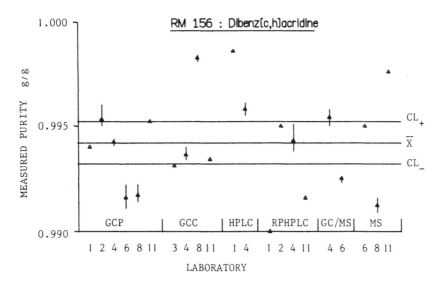

Figure 7. Distribution of analytical results for dibenz[c,h]acridine.

IV. APPLICATIONS

Certified PAH reference materials and matrices are intended primarily for the
calibration and quality control of analytical techniques. In addition, high-
purity PAH compounds may be used for various microbiological tests where
purity of the substances investigated is essential. An indication of some of
the fields of application is given in Table 6.

A. Analysis

In view of the presence of a great number of individual compounds in PAH
analysis, the availability of well-characterized reference compounds is of great
help for their unambiguous identification and reproducible quantification.
Thus retention times can be ascertained for both GC and HPLC techniques.
For quantitative analysis, the certified reference materials permit the determination of UV and fluorescence response factors at any chosen wavelength,
which is of special importance in HPLC analysis.

In some instances PAH reference materials have been used also to optimize techniques in order to separate PAH isomers which are not distinguished
by conventional techniques [28], and they can be applied for recovery studies
in PAH analysis. Similarly, the availability of an extended range of PAH
standards is very useful for testing and validating methods of analysis, for
quality control of routine laboratory work and in round-robin exercises, as
well as implementation of regulations where limits have been set for PAH
levels in various environmental and occupational matrices [18,19]. Some compounds that are not usually observed in the environment or in occupational
situations were included in the program in view of their use as internal standards.

In a special field of application, the PAH reference materials have been
used for determining and collecting the molecular spectra of a large number of
individual compounds encountered in environmental and occupational analyses
[29]. This collection includes the UV, infrared (IR), fluorescence, MS, and

Table 6. Fields of Application for PAH Certified Reference Materials

Subject	Application	Regulations
Ambient air	Control of PAH levels in air	Recommended level for BaP in West Germany [34]
Coal combustion/ conversion	Monitoring of PAH emissions to ambient air	Tentative limiting value [35]
Surface waters	Quality control of drinking water	EEC Directive 75/440 [18] WHO Standard [19] EPA Quality Criteria for water [36]
Occupational exposure	Control of exposure limits	U.S. limit for coke oven emissions [37] (recommendation)
Foods	Quality control	BaP limit for chewing gum in West Germany [38]
	Control of PAH levels in food chains, recovery studies	BaP limit for smoked meat in West Germany [39]
Biological activity	Testing for mutagenicity, carcinogenicity and cocarcinogenicity, and related toxicological tests	

nuclear magnetic resonance (NMR) spectra and should be of considerable assistance for the analyst since they were taken under standardized conditions, wherever possible, and are based on an identical set of well-characterized reference compounds.

B. Microbiological Testing

The availability of high-purity reference compounds can also be of considerable advantage for testing the microbiological activity. Discrepancies that are apparent in literature data may be attributable not only to differences in experimental conditions but also in part to the purity and composition of the material used, since impurities may significantly influence the outcome of the tests.

In the Ames test, for instance, the presence of an impurity that is much more active than the main component can lead to an overestimation of mutagenic activity. In other cases, impurities that are toxic to the bacterial strain being used may mask to a considerable extent the mutagenic activity of the substance under test. Since the PAH reference materials are well characterized for purity and known impurities are indicated, intereference can be minimized in these tests. A number of the certified materials, especially heterocyclic compounds, have been used to test mutagenic activities on various bacterial strains [30].

Another use for high-purity reference compounds is the study of potential cocarcinogenic effects. Since this effect depends on arylmonooxygenase activity, which converts PAH compounds into the ultimate carcinogenic transdihydrodiolepoxide, stimulation or induction of this enzyme system plays a vital role in carcinogenesis [31].

In this context, some of the high-purity PAH reference materials have been used to determine induction potentials on the cytochrome P_{448} monooxygenase system. For example, induction factors from 4 to 6.5 were found for benzo[a]pyrene and benzo[b]-, benzo[k]- and benzo[j]fluoranthenes on the substrate benz[a]anthracene [32]. In similar experiments, chrysene, benz[a]anthracene, and anthanthrene were identified as potent inductors or cocarcinogens in the P_{448} system [33].

Finally, high-purity PAH reference materials may be used for other biochemical experiments which do not require the use of large quantities of the test material.

ACKNOWLEDGMENT

The author wishes to thank R. J. Fordham, JRC Petten Establishment, for valuable discussions and suggestions in the preparation of this chapter and in particular for the section on the statistical evaluation of certification analyses.

REFERENCES

1. M. Zander, in *Polycyclische aromatische Kohlenwasserstoffe*, VDI-Bericht 358, VDI-Verlag, Düsseldorf, West Germany, p. 11.
2. D. Schüetzle, F. S. C. Lee, T. J. Prater, and S. B. Tejada, Int. J. Environ. Anal. Chem. 9:93 (1981).
3. B. W. Wilson, R. A. Pelroy, and J. T. Cresto, Mutat. Res. 79:193 (1980).
4. M. R. Guerin, C.-H. Ho, T. K. Rao, B. R. Clark, and J. L. Epler, Environ. Res. 23:42 (1980).
5. J. Jacob, W. Karcher, and P. Wagstaffe, Fresenius' Z. Anal. Chem., in press (1985).
6. P. Studt, Liebigs Ann. Chem. 1443 (1979).
7. A. Gold, J. Schultz, and E. Eisenstadt, in *Polynuclear Aromatic Hydrocarbons*, P. W. Jones and P. Leber (Eds.), Ann Arbor Science, Ann Arbor, Mich., 1979, pp. 695-704.
8. K. Tintel, J. Lugtenburg, and J. Cornelisse, J. Chem. Soc., Chem. Commun. 185 (1982).
9. H. Gilman and A. L. Jacoby, J. Org. Chem. 3:118 (1938).
10. E. Campaigne and S. W. Osborn, J. Heterocycl. Chem. 5:655 (1968).
11. N. P. Buu-Hoi and P. Cagniant, Chem. Ber. 76:1269 (1943).
12. E. G. G. Warner, Recl. Trav. Chim. 68:523 (1949).
13. J. André, P. Jaquignon, F. Périn, and N. P. Buu-Hoi, Bull. Soc. Chim. France 3908 (1970).
14. A. Senier and P. C. Austin, J. Chem. Soc. 89:1390 (1906).
15. E. G. Doadt, M. Iwao, J. N. Reed, and V. Snieckus, in *Proceedings of the 8th International Symposium on Polynuclear Aromatic Hydrocarbons*, Columbus, Ohio, 1984, p.413.
16. R. M. Acheson (Ed.), *Acridones*, 2nd ed., Wiley, New York, 1973.

17. W. Karcher, R. Depaus, J. van Eijk, and J. Jacob, in *Polynuclear Aromatic Hydrocarbons—and Biology—Carcinogenesis and Mutagenesis*, P. W. Jones and P. Leber (Eds.), Ann Arbor Science, Ann Arbor, Mich., 1979, p. 341.

18. R. Depaus, J. Chromatogr. *176*:337 (1979).

19. EEC Council Directive 74/440/EEC, O.J. L 194, July 25, 1975.

20. *WHO International Standard for Drinking Water*, 3rd ed., WHO, Geneva, 1971.

21. S. A. Wise, S. L. Bowie, S. N. Chesler, W. F. Cuthrell, W. E. May, and R. E. Rebbert, in *Polynuclear Aromatic Hydrocarbons: Physical and Biological Chemistry*, M. Cooke and A. Dennis (Eds.), Battelle Press, Columbus, Ohio, 1982, p. 919.

22. N. P. Buu-Hoi and P. Jaquignon, J. Chem. Soc. 2964 (1951).

23. W. Schmidt, J. Jacob, P. Glaude, and W. Karcher, unpublished results.

24. A. P. Gray, in Purity determination by DSC, Therm. Anal. News Lett., Nos. 5 and 6, Perkin-Elmer (1966).

25. W. Karcher, J. Jacob, and L. Haemers, EUR Report 6967, 1980.

26. W. Karcher, R. Fordham, and J. Jacob, EUR Reports, EUR 7175, 1981; EUR 7812, 1982; EUR 8497, 1983.

27. F. Belsito, L. Boniforti, R. Dommarco, and G. Laguzzi, in *Quantitative Mass Spectrometry in Life Sciences*, Vol. 2, A. P. Leenheer and R. R. Ronucci (Eds.), Elsevier, Amsterdam, 1977, p. 431.

28. W. Karcher, in *Mobile Source Emissions Including Polycyclic Organic Species*, D. Rondia, M. Cooke, and R. K. Haroz (Eds.), D. Reidel, Dordrecht, The Netherlands, 1983, p. 127.

29. W. Karcher, R. J. Fordham, J. J. Dubois, P. G. J. M. Glaude, and J. A. M. Ligthart, *Spectral Atlas of Polycyclic Aromatic Compounds*, D. Reidel, Dordrecht, The Netherlands, 1985.

30. W. Karcher, J. Dubois, R. Fordham, P. Glaude, R. Barale, and D. Zucconi, in *Proceedings of the 8th International Symposium on Polynuclear Aromatic Hydrocarbons*, Columbus, Ohio, Oct. 1983 (published in *Polynuclear Aromatic Hydrocarbons*, 1984, p.685).

31. F. J. Wiebel, in *Carcinogenesis*, Vol. 5: *Modifiers of Chemical Carcinogenesis*, T. J. Slaga (Ed.), Raven Press, New York, 1980.

32. J. Jacob, G. Grimmer, and A. Schmoldt, Hoppe-Seyler's Z. Physiol. Chem. *360*:1525 (1979).

33. J. Jacob, G. Grimmer, and A. Schmoldt, in *Polynuclear Aromatic Hydrocarbons: Chemistry and Biological Effects*, A. Bjørseth and A. J. Dennis (Eds.), Battelle Press, Columbus, Ohio, 1980, p. 807.

34. G. Grimmer (Ed.), *Berichte 1/79 Luftqualitätskriterien für ausgewählte polyzyklische aromatische Kohlenwasserstoffe-Umweltbundesamt*, Erich Schmidt Verlag, Berlin, 1979.

35. N. E. Bolton, C. L. Hunt, T. A. Lincoln, and W. E. Porter, *Workplace Carcinogens*, Health Protection and Planning, OSHA, Mar./Apr. 1977, p. 30.

36. EPA, Fed. Regist. 69464 (Dec. 3, 1979); rev. Apr. 1978.

37. U.S. Department of Labor, OSHA, Fed. Regist. *41*:46742 (Oct. 22, 1976).

38. Verordnung über Zulassung von Zusatzstoffen für die Herstellung von Kaugummi, Bundesgesetsblatt I.S. 1825, Sept. 20, 1972.

39. Fleischverordnung, Bundesgesetzblatt I.S. 1003, July 4, 1978.

Index

A

Abietic acid, 82
Acid-base extraction, 287
Aerosol particle, 377
Agricultural burning, 11, 77
Airplanes, 13
Air pollution, 4
    carcinogenicity, 351
    diurnal profiles, 374
    reactive gaseous species, 354—357
    regulatory strategies, 83
    from wood combustion, 82
Alcohol, as fuel, 13, 107
Alumina adsorption chromatography,
        292—299
    picric acid doping, 299
    recoveries, 294
Aluminum production, 6, 237
    plant workers, 245
Amberlyt resin, 291
Ames test, 255, 260, 352, 371, 404
Amine-220, 317, 319
2-Aminonaphthalene, 283
Amino-PAH, 210, 220, 222, 272,
        352
    analysis, 312
    biological activity, 283—286
    chemical properties, 272
    chromatogram, 334
    in coal liquids, 222, 265
    GC retention data, 338
    mass spectra, 210, 211
    physical properties, 272
    in solvent refined coal, 330, 338
    spectroscopic properties, 272
Aminophenanthrenes, 199, 222
Ammonia chemical ionization, 210,
        222, 313
Analysis
    alternate fuels, 72, 107
    automobile exhaust, 87—111
    biodirected, 283
    body fluids, 237—252

[Analysis]
    combustion units, 22, 28—54, 64—
        77
    gas chromatography, 5, 147, 193—
        236, 315—337
    high performance liquid chromatog-
        raphy, 113—191, 337—339
    mass spectrometry, 205, 208—212,
        309—315
    of metabolites, 245, 253
    neutron activation, 301
    oils, 103, 168
    particulate matter, 82, 152, 351
    source emissions, 22, 64
    urine, 243
Analytical techniques
    gas chromatography, 5, 147, 193,
        315
    high performance liquid chromatog-
        raphy, 113, 337
    open column chromatography, 5,
        113, 291
Aniline, 275
Anthanthrene, 150
Anthracene, 31, 70, 148, 246, 354,
        355
    methyl-, 31, 70, 148, 209
Anthracene oil
    chromatograms of, 325
    PANH in, 326—327
Anthra[2,1,9-def]quinoline, 271
Anthropogenic sources of PAH, 4
APAH (see Amino-PAH)
Apiezon L&M, 318
Asphalt production, 9
Atmospheric reactions, 351—384
Automobiles, 13, 87
    PAH emissions from, 13, 87—111
Azaarenes, 210, 272
    mass spectra, 210, 211
1-Azabenzo[a]pyrene, 271
4H-Azacyclopenta[def]phenanthrene,
        271
3-Azaphenanthrene, 271

B

BaP (*see* Benzo[a]pyrene)
Benz[e]acenaphthylene, 70
Benz[c]acridine, 351
Benz[a]anthracene, 32, 70, 120,
    137, 149, 246, 354, 355
  carcinogenicity, 351
  methyl-, 149
  photooxidation, 359
Benzene, 275, 369
Benz[h]isoquinoline, 269
Benzo[b]chrysene, 118, 150
Benzo[c]chrysene, 118
Benzo[a]fluoranthene, 118
Benzo[b]fluoranthene, 33, 71, 118,
    149
Benzo[ghi]fluoranthene, 74, 118,
    149
Benzo[j]fluoranthene, 118, 149
Benzo[k]fluoranthene, 33, 118, 149
Benzo[a]fluorene, 71, 148, 246
Benzo[b]fluorene, 71, 148
Benzo[b]naphtho[2,1-d]thiophene,
    77, 219
Benzo[b]naphtho[2,3-d]thiophene,
    387
  synthesis of, 391
Benzonitrile, 275
Benzo[ghi]perylene, 33, 71, 118,
    150
Benzo[c]phenanthrene, 149
Benzo[a]pyrene, 33, 71, 118, 149,
    206, 246, 354
  adducts with DNA, 248, 364
  in blood, 246
  carcinogenicity, 351
  emission factors, 6−13
  formation, 2
  in human urine, 243, 246
  as indicator substance for PAH,
    238
  intratracheal instillation, 242
  metabolism of, 240, 254
  methyl-, 123, 127
  ozonization, 365
  photooxidation, 358, 361
  radioactive, 238
    excretion studies with, 239
  sources, 6−13
  sulfonic acid, 369
Benzo[e]pyrene, 33, 71, 118, 149,
    206, 246

3,4-Benzpyrene (*see* Benzo[a]pyrene)
Bio-Beads, 291
Biomass combustion, 11, 61−85
  bark, 61
  charcoal, 61
  firewood, 61
  straw, 61
  wood chips, 61
Biomass combustion units, 62−77
Birch, 65
Blood lymphocytes, 246
BMBT liquid crystal, 198, 319
Body fluids, 237−252
Briquettes, 76

C

Capillary column gas chromatography
    (*see* Gas chromatography)
Carbazoles, 221, 266
Carbon-14, 82
Carbon black
  production, 9
  reactions on, 359
  as sampling medium, 88
Carbowax, 196, 220, 227, 317, 318
Carcinogenicity, 1, 253, 351, 352
  "excess," 352
  of N-PAC, 266, 283, 351
  of PAH, 1, 88, 237, 253, 283,
    351, 405
  of POM, 351
  structure-activity correlations,
    283
Catalytic combustion
  automobiles, 101
  charcoal, 65
  wood stoves, 65
Cellulose, 62
Central heating boilers, 65
Charcoal, 65
Charring, 63
Chemical ionization, 208, 212, 313
Chemiluminescence, 212, 214
Chloro-PAH, 207
Chromosorb-102, 93
Chrysene, 33, 70, 137, 246, 354,
    355
  methyl-, 203, 218
Cigarette smoke condensate, 288
Coal-derived materials, 265-349
  conversion, 266

[Coal-derived materials]
  gasification, 265
  solvent refining, 265
Coal gasification plants, 237
Coal liquefaction plants, 237
Coal liquids, 134, 265
Coke production, 9, 237
Collision-induced dissociation, 315
Column chromatography, 392 (see
    also Liquid chromatography)
Commercial PAH, 392
  impurities in, 393
Condenser, 24, 64, 89, 91, 95, 96
Confidence limits, 399
Conifers, 62
Conjugates, 240
Connectivity index, 215, 217
Constant volume sampler, 95
Contemporary carbon, 82
Coronene, 75
CPSil5, 224, 318
Cryogenic sampling, 96, 98
Cyano-PAH, 272, 331
  chromatogram, 334
Cyclone-fired power plant, 49
4H-Cyclopenta[def]phenanthrene,
    73, 148
Cyclopenta[cd]pyrene, 70, 77, 219,
    387
  synthesis of, 390

D

DB-5, 223, 224, 318
Denuder, 375
Detectors
  chemiluminescence, 212, 214
  electrochemical, 339
  electron capture, 208–212, 327
    oxygen doped, 328
  flame ionization, 205–214, 220
    alkali, 327
  fluorescence, 93, 243, 248, 279,
    337
  Fourier transform infrared spec-
    troscopy, 23, 212, 339
  glow discharge, 328
  Hall electrolytic conductivity, 327
  infrared, 212
  mass spectrometry, 23, 208–212,
    223, 241, 312, 315, 328, 351
  nitrogen-phosphorous, 220, 327

  photoionization, 208–210
  thermoionic emission, 327
  thin-layer electrochemilumines-
    cence, 145
  ultraviolet, 145, 212
    scanning, 339
  videofluorometry, 145
Dexil-400, 224, 319
Dibenz[c,h]acridine, 387
  synthesis of, 392
Dibenz[a,c]anthracene, 123, 150
Dibenz[a,h]anthracene, 118, 150,
    246, 354, 355
Dibenz[a,j]anthracene, 123
4H-Dibenzo[b,def]carbazole, 271
Dibenzo[a,l]pyrene, 243
Diesel engines, 13, 87, 351, 366
Differential scanning calorimetry,
    396
Dilution tunnel, 94, 95
7,12-Dimethylbenz[a]anthracene,
    239
District heating plant, 76
DNA, 241, 253
Drinking water, 1
Driving cycles, 101

E

Electric Power Research Institute,
    22
Electrochemical detectors, 339
Electron attachment, 206
Electron capture chemical ionization,
    212
Electron capture coefficient, 205
Electron capture detector, 205–208,
    327
  oxygen doping, 328
  response enhancements, 207
Electron impact ionization, 205, 313
  mass spectra of
    amino-PAH, 210, 313
    azaarenes, 210, 313
    nitro-PAH, 223
Electrophilic aromatic substitution,
    357, 371, 372
Electrostatic precipitator, 28
Emission factors, 1–20, 27–54, 65,
    105, 107
Emission of PAH, 1–20
  from automobiles, 87–111

[Emission of PAH]
    affecting factors, 100
    from biomass combustion, 64—85
    from coal-fired plants, 21—59
      affecting factors, 22
    total annual, estimation, 17
Endoperoxides, 358, 359
Environmental Protection Agency
    (EPA), 23
Epidemiological studies, 238
Epoxidases, 240, 254
1,2-Epoxy-3,3,3-trichloropropane,
    256
Eukaryotic systems, 254
Exposure to PAH
    effects, 241
    occupational, 237
Extraction, 23, 64
    emission samples, 23
    Soxhlet, 23
    vacuum sublimation, 23

F

Fast atom bombardment, 313
Feces, 239
Ferroalloy industry, 9, 237
FFAP, 319
Field desorption, 313
Field ionization, 313
Filter temperature, 25, 64, 89
Fireplace, 64
Fixed-grate combustion, 40
Flame ionization detector, 205—214,
    220
Flames, 2
Florisil, 291
Flow reactor, 376
Fluidized bed combustion, 11, 21,
    40
Fluoranthene, 32, 70, 148, 246
Fluorene, 148, 246
Fluorescence detection, 137—145,
    279, 338, 351
    wavelength selection, 140, 280—
    282
Forest fires, 11, 77
Formation of PAH
    by carbonization, 4
    in flames, 2
    model, 2
    from organic material, 2—4, 76

[Formation of PAH]
    by pyrolysis, 2
Fossil pine, 82
Fourier transform infrared spectros-
    copy, 23, 212, 339
Free-valence value, 353
Front-walled-fired power plant,
    49
Fused silica columns, 194

G

Gas chromatography, 5, 147, 193—
    236, 315-337
    applications for PAC determination,
      26, 219, 224, 315
    columns, 194—196
    detectors, 205—214, 327—328
    injection methods, 204
      variability, 205
    mass spectrometry, 23, 208—212,
      223, 312, 315, 328, 351
      of air particulate extract,
      147
      of amino-PAH, 312, 328
      of diesel exhaust, 351
      of gasoline exhaust, 351
      of metabolites, 241
    metabolism studies with, 241
    purity analysis with, 397
    stationary phases, 196—204, 224—
      227, 318—319
      bonded, 196
      free-radical crosslinking,
      197
      gums, 196, 216
      liquid-crystals, 123, 198—204,
      217, 218
      mesogenic, 198, 204, 217
      retention mechanism, 215
      selective, 198, 329
Gas combustion, 11
Gasoline engines, 13, 87, 351
Gas phase sampling, 96
Gas trap, 88, 91, 93, 97
GC (see Gas chromatography)
Gel permeation chromatography,
    291, 292
Glucuronic acid, 240, 259
β-Glucuronidase, 259
Glutathione, 240, 259
Gum phases, 196, 216

H

Hamster embryonic fibroblasts, 256
Hantzsch-Widman nomenclature system, 266, 269
Heat generation, 9
  bark-fired boilers, 76
  industrial boilers, 9, 28
  peat-fired boilers, 72
  wood-fired boilers, 65
Hemicellulose, 62
Hemoglobin, 249
Hepatobiliary system, 239
High performance liquid chromatography, 113–191, 291, 300–301, 337–339
  applications for PAC determination, 181, 245, 253, 300, 337, 364
  capacity factors, 301, 302
  columns
    amine, 128
    diamine, 128
    monomeric C_{18}, 120
    narrow-bore microparticulate, 136
    polymeric C_{18}, 120
  detectors, 137–176, 338–339, 397
  high resolution, 136
  mass spectrometry, 339
  multidimensional, 137, 147–176, 301–309
  normal-phase, 128–129, 301, 337
  of N-PAC, 337–339
  of PAH, 113–191
  preparative, 392
  purity analysis with, 397
  reversed-phase, 114–128, 337
  very high speed, 137
High pressure liquid chromatography (see High performance liquid chromatography)
High-resolution liquid chromatography, 136
High-resolution mass spectrometry, 273, 312
High-volume sampling (see Sampling)
Homogeneity testing, 396
HPLC (see High performance liquid chromatography)
Hydroperoxides, 358
Hydroperoxyl radical, 357

Hydroxylamines, 255
Hydroxy-6-nitrobenzo[a]pyrene, 257
Hydroxyl-PAC, 286
Hydroxyl radical, 356, 367
1-Hydroxypyrene, 245

I

Immunoassays for PAH, 247
Impingers, 64, 88
Incineration, 11
  automobile tires, 11
  coal refuse, 11
  municipal, 11
Indeno[1,2,3-cd]pyrene, 33, 150
Indoles, 266
Industrial PAH emissions, 6–9
Injection techniques, 204
  cold on-column, 205
  split, 204
  splitless, 205
  variability, 205
International Union of Pure and Applied Chemistry, 266
Ion exchange chromatography, 291, 300
Ionization potential, 205, 372
Ionization techniques, 205, 208–212, 223, 313
Iron works, 9, 237
Isolation methods, 286–309
Isoquinoline, 269

K

Kovats retention index, 214, 324

L

LC (see Liquid chromatography)
Length-to-breadth ratio (L/B), 118, 123, 218
Lignin, 62
Liquefied petroleum gas, 13, 107
Liquid chromatography, 113, 291–301 (see also High performance liquid chromatography, Column chromatography)
  advantages, 291
  alumina, 291

[Liquid chromatography]
  Amberlyt, 291
  Bio-Beads, 291
  cellulose acetate, 291
  Florisil, 291
  picric acid doping, 299
  Sephadex LH-20, 291, 304–305
  silica gel, 291
  silicic acid, 291, 307
Liquid crystals, 123, 198–204, 217,
    218
Localization energy, 353
Lung cancer, 242

                    M

Mammalian cells, 253
Mass-analyzed ion kinetic energy
    spectrometry (MIKES), 315
Mass spectrometry, 205, 208–212,
    223, 309–315
  chemical ionization, 208–212, 313
  collision-induced ionization, 315
  direct inlet probe, 309
  electron impact ionization, 205,
    210, 223, 313
  fast atom bombardment, 313
  field desorption, 313
  field ionization, 313
  high resolution, 223, 273, 309,
    312, 315
  high voltage, 315
  laser ionization, 209
  low resolution, 223, 309
  low voltage, 313, 315
  MIKES, 315
  negative ion detection, 212
  purity analysis with, 398
  tandem, 273
Matrix materials, 394
Melting point, 396
Metastable ions, 315
Methylbenzoquinolines, 222
3-Methylcholanthrene, 239
4,5-Methylenephenanthrene (see
    4H-Cyclopenta[def]phenan-
    threne)
1-Methyl-7-isopropylphenanthrene
    (see Retene)
Microsomes, 253, 258
Mobile sources of PAH, 13, 87
  alcohol, 13, 107

[Mobile sources of PAH]
  diesel, 13, 87
  gasoline, 13, 87
  liquefied petroleum gas, 13, 107
Molecular orbitals, 353
Molecular polarizability, 215, 217
Molecular size separation, 174
Monooxygenases, 239, 253, 405
Motorcycles, 13, 87
Moving-grate combustion, 21, 35
Multidimensional LC techniques, 147–
    176, 301–309
Mutagenicity
  of amino-PAH, 222, 283, 352
  direct acting, 352
  of nitro-PAH, 223, 352, 371
  of N-PAC, 266, 283
  of oxygenated PAH, 352
  of PASH, 352
  structure-activity correlations,
    283

                    N

Naphthalene, 148, 354, 355
Natural sources of PAH, 4
Nematic liquid crystals, 198
Neutron activation analysis, 301
Nitrate radical, 372
Nitric acid, 370
9-Nitroanthracene, 223
7-Nitrobenz[a]anthracene, 223
6-Nitrobenzo[a]pyrene, 253
  metabolism of, 253–264
3-Nitrofluoranthene, 223
Nitrogen dioxide, reactions of, 24,
    64, 370–374
Nitrogen oxides, 370
Nitro-PAH, 206, 223, 352, 370
  in ambient air, 212, 371
  in carbon black, 212, 371
  in diesel exhaust, 212, 371
  fate, 371, 374
  formation, 370–374
  gas chromatography of, 212
    chromatogram, 213
  metabolism of, 253–264
  mutagenicity of, 223, 352, 371
1-Nitropyrene, 223
2-Nitropyrene, 223
Nitroreductases, 256
Nitrous acid, 370

Nomenclature, 266—272
Normal-phase LC, 128
Norway, total annual PAH emission
        in, 17
N.P., 318
N-PAC (see PAC, nitrogen contain-
        ing)
NPAH (see Nitro-PAH)
Nuclear magnetic resonance, 279

                O

Occupational exposure
        to N-PAC, 266
        to PAH, 237
On-line measurements, 93
Open burning, 77
Open tubular column, 193
OV-1, 196, 318
OV-7, 318
OV-17, 196, 224, 318
OV-61, 318
OV-73, 196, 224, 318
OV-101, 196, 318
OV-225, 318
OV-1701, 196
Oxidation potential, 372
Oxygen, 357-362
        singlet molecular, 356—357
Ozone, 362—367
        formation, 356
Ozonolysis, 355
        mechanism, 362

                P

PAC, 194, 385
        bioavailability, 238, 353
        nitrogen-containing, 220, 265—
                349 (see also Amino-PAH,
                PANH)
        chemical properties, 272
        in coal derived materials, 265—
                349
        infrared spectra, 278
        isobaric, 273
        IUPAC structures, 270
        nomenclature, 269—272
        physical properties, 272
        spectroscopic properties,
                272

[PAC]
        oxygen-containing, 228, 359, 361,
                363
        reference materials for analysis,
                385—405
                applications, 403
                preparation, 386
        retention index system, 214, 324
        sulfur-containing (see PASH)
        synthesis, 387
PAH
        adsorption of, in animal systems,
                238
        in ambient air, 162, 351
        atmospheric reactions, 351—384
        in blood, 246
        boiling points, 278
        cation radicals, 372
        chlorinated, 207
        complexes with serum albumin, 247
        determination of, in body fluids,
                237
        distribution of, in animal systems,
                238
        electrophilic substitution, 353, 371
        emission factors, 1—20, 27—54, 65,
                105, 107
        excretion of, in animal systems,
                238
        formation of, 1, 76, 237
        free-valence value, 353, 354
        high molecular weight, 196, 197
        in human urine, 243
        inhalation exposure to, 238
        IUPAC structures, 268
        localization energy, 353
        metabolism of, 239
        methyl derivatives, 3
        nomenclature, 266—268
        number of isomers, 385
        occupational exposure, 237
                biological monitoring of, 243
        in oil, 103
        photodecomposition, 359
        profiles, 2
        reactions with
                free radicals, 367—368
                molecular oxygen, 357—362
                nitrogen oxides, 370—374
                ozone, 362—367
                sulfur oxides, 369—370
        reactivity of, 353—354
                effects of carrier, 373—374

[PAH]
parameters affecting, 375–378
photochemical, 361
reduction of metabolites of, 245
reference materials, 385–405
applications, 403
in shale oil, 168
sources, 1–20
Standard Reference Materials, 176–180
PAH GC chromatograms
ambient air, 81, 153–161
carbon black, 197
coal combustion, 201
coal gasification condensate, 195
coal tar, 204
impurities in benzo[a]pyrene, 395
peat combustion, 76
wood combustion, 69, 80
PAH HPLC chromatograms
ambient air, 138, 141, 146, 151
human urine, 247
shale oil, 169, 170
PANH, 215, 217, 220–223, 265–349
acid-base qualities, 275
biological activity, 283–286
boiling points, 278
chemical properties, 272
chromatograms, 332, 333
in coal-derived materials, 265
in coal gasification tar, 335, 336
in coal liquids, 220
dipole moments, 275
infrared spectra, 278
number of isomers, 273
in petroleum, 220
physical properties, 272
in shale oils, 220
spectroscopic properties, 272
Particle sampling (see also Sampling)
cyclones, 25, 99
electrostatic precipitators, 99
filters, 25, 99, 375
Particulate organic matter, see POM
Partition chromatography, 287–291
PASH, 130, 228
in coal liquids, 134, 202
GC separation, 203, 228
LC separation, 129
in synfuels, 228
Peat, 76
PEG-400, 227, 319
PEG-1500, 227, 319

Pellets, 76
Pentafluoropropyl amides, 222, 312, 328
Peroxyacetylnitrate, 352, 357, 370
Perylene, 33, 71, 118, 150
Petroleum catalytic cracking, 6, 237
Phenalen-1-one, 228
Phenanthrene, 31, 70, 148, 239, 246, 354, 355
methyl-, 31, 32, 70, 148, 209
Phenanthro[3,4-c]phenanthrene, 119
Photochemical smog, 351, 354
Photodecomposition, 359, 393
Photoionization detector, 208–210
Picene, 118, 150
derivatives, 77
Pine needles, 78
Pitch, 1
Pit coal, 22
Pluronic L64, 319
Pluronic L68, 319
Polycyclic aromatic compounds (see PAC)
Polycyclic aromatic hydrocarbons (see PAH)
Polycyclic aromatic nitrogen heterocycles (see PANH)
Polycyclic aromatic sulfur heterocycles (see PASH)
Poly S-179, 220, 319
POM, 351, 355, 357, 374, 375
Power generation, 9, 21–59
coal-fired plants, 9, 21, 77
oil-fired plants, 9
peat-fired plants, 77
Precombustion, 76
Priority pollutant PAH, 37
Proportional sampling, 91
Proteins, 241, 253
Pulverized-coal-fired plants, 28
Purification, 390
Purity analysis, 396
Purity of standards, 386
Pyrene, 32, 70, 136, 148, 246
methyl-, 32, 71
Pyridines, 266, 275
Pyrroles, 266, 275

Q

Quinolines, 266
Quinones, 359, 361, 363

R

Recrystallization, 392
Regulations, 386
Reoplex-400, 319
Residential heating, 6, 64—65
  coal, 6
  oil, 6
  wood, 6, 64
Respirable particles, 376
Retene, 82, 219
Retention data, 324
  gas chromatography
    amino-PAH, 338
    azaarenes, 322
    nitro-PAH, 223
    PAH, 216, 218
    PANH, 322, 334, 336
  liquid chromatography
    PAH, 148
    PASH, 130
Retention index system, 214, 324
  for N-PAC, 324
  for PAC, 214
  for PAH, 214
Reversed-phase LC, 114—118, 327
RNA, 240, 253
Road dust, 88

S

Salmonella typhimurium assay, 255,
      260, 352, 404
Sampling, 22—27, 64, 88—100, 374—
      375
  artifacts, 24, 64, 357, 374
  automobile exhaust, 88—100
    diluted exhausts, 93—95
    undiluted exhausts, 88—93
  blow-off, 358
  combustion units, 22, 64
  criteria, 22, 88
  cryogenic techniques, 96
  degradation during, 24, 64, 374
  denuders, 375
  equipment, 22
  filters, 375
  gas-phase sampling, 24, 64, 96—
      99
  high-volume, 357
    losses during, 357, 366
  Method 5, 23, 25

[Sampling]
  on-line measurements, 93
  particle trapping, 25, 64, 99
  proportional, 91—93
  SASS train, 23, 24, 25
    modification, 24
  strategy, 65
Scanning ultraviolet detector, 339
SE-30, 196, 225, 318
SE-52, 196, 223, 224, 318
SE-54, 196, 223, 224, 318
Selectivity, 114, 198
Sephadex LH-20, 291, 304—305
Shpol'skii spectroscopy, 176, 279
Sieving, 291
Silica adsorption chromatography,
      292—299
  recoveries, 294
Silylation, 194, 367
Singlet molecular oxygen, 359
Sister chromatid exchanges, 246
Size exclusion chromatography, 174
Skin painting, 242
Skin tumors, 283
Smectic liquid crystals, 198
Smokers, 243, 245
Solid sorbents, 23, 64, 88, 93
  long-term stability, 23
Solutions of PAH, 394
Solvamin-20, 319
Solvent extraction, 286
Soot, 1, 357, 369, 370
Source Assessment Sampling System
      (see Sampling, SASS train)
Sources of PAH, 1—20
  anthropogenic, 1
    mobile, 4
    stationary, 4
  natural, 1
Source-specific PAH, 77, 219
SP-2100, 318
SP-2250, 224, 318
SP-2340, 318
Spectroscopic techniques,
  fluorescence, 93, 243, 248, 279,
      351
  infrared, 212, 278
    Fourier transform, 23, 212,
      339
  Shpol'skii, 176, 279
  ultraviolet, 145, 212
    scanning, 339
  videofluorometry, 145

Spreader-stoker-fired power plant, 49
Spruce, 65
Standard Reference Materials, 176–180
Stationary phases, 196–204, 318–319
    bonded, 196, 320
    free-radical crosslinking, 197
    gums, 196, 216, 320
    liquid crystals, 123, 198–204, 217, 218
    mesogenic, 198, 204, 217
    retention mechanism, 215
    selective, 198, 329
    thermal stability, 317
Steel industry, 237
Straw
    barley, 77
    rice, 77
Structure-activity correlations, 283
Sublimation, 392
Substitutive nomenclature, 271
Sulfate esters, 240
Sulfur dioxide, 369
Superox, 224, 317, 319
Sweden, total annual PAH emission in, 17
Swedish Coal-Health-Environment Project, 22, 24
Synfuels, 220
Synthesis of PAC, 387

T

TA-98, 352
    nitroreductase-deficient, 371
Tandem mass spectrometry, 273
Tar, 1, 83, 89, 208
Tangentially-fired power plant, 49
Tenax GC, 23, 64
    decomposition products, 23
Tetrabenzo[a,c,f,h]naphthalene, 119

Thermal degradation, 63
Toluene, 367
Trifluoroacetyl amide, 222, 312, 328
Trimethylsilyl derivatives, 367
Triphenylene, 70, 149
Triton X-305, 319
Tumorigenicity, 283

U

UCON, 220, 319, 324
United States, total annual PAH emission in, 17

V

Vapor phase, 24, 27, 64, 87, 96
Vehicular traffic, 17, 87
Versamid-900, 319

W

Washing flask, 88
Window diagram, 201–202
Wood chemistry, 62
Wood combustion, 62
Wood pyrolysis, 62
Wood stove, 64
Working atmospheres, 1, 237, 266

X

XAD-2, 23, 64, 96, 97, 291
    cleaning, 25
    decomposition products, 23

Z

Z-numbers, 309-312
Zone melting, 392

For Product Safety Concerns and Information please contact our EU
representative GPSR@taylorandfrancis.com Taylor & Francis Verlag GmbH,
Kaufingerstraße 24, 80331 München, Germany

Printed and bound by CPI Group (UK) Ltd, Croydon, CR0 4YY

01/05/2025

01858601-0001